Viscous
Hypersonic
Flow

Viscous Hypersonic Flow

Theory of Reacting and Hypersonic Boundary Layers

William H. Dorrance

Dover Publications, Inc.
Mineola, New York

Bibliographical Note

This Dover edition, first published in 2017, is an unabridged republication
of the work originally published by the McGraw-Hill Book Company, Inc.,
New York, in 1962.

Library of Congress Cataloging-in-Publication Data

Names: Dorrance, William H., 1921–
Title: Viscous hypersonic flow : theory of reacting and hypersonic boundary
    layers / William H. Dorrance.
Description: Mineola, New York : Dover Publications, Inc., [2017] | "This
    Dover edition, first published in 2017, is an unabridged republication of
    the work originally published by the McGraw-Hill Book Company, Inc.,
    New York, in 1962." | Includes bibliographical references and index.
Identifiers: LCCN 2017000037| ISBN 9780486812885 | ISBN 048681288X
Subjects: LCSH: Boundary layer. | Aerodynamics, Hypersonic. | Gas dynamics.
Classification: LCC TL574.B6 D64 2017 | DDC 629.132/306—dc23 LC record
available at https://lccn.loc.gov/2017000037

Printed in Canada
81288X02   2024
www.doverpublications.com

Preface

The primary objective of this book is to provide a description of theories for treating the laminar and turbulent boundary layers of reacting gas mixtures. The theories are developed from fundamentals, and all related chemical, thermodynamic, and physical concepts are described so as to make the book a self-contained entity.

A secondary objective is to provide for those who are concerned with engineering applications useful equations for calculating heat transfer between reacting gas boundary layers and reacting, melting, sublimating, and otherwise decomposing surfaces.

It is intended that this book serve as a suitable text for an upper-level or graduate course in modern boundary-layer theory. In addition, it is believed that the book can also serve as a suitable reference for those whose training came before the subject matter of this book became of significance. Furthermore, it is believed that the book will serve as a reference and point of departure for those engaged in research in this important area.

The book not only contains a thorough description of the theories for treating reacting laminar and turbulent boundary layers but also describes the underlying theories and methods of calculating transport and thermodynamic properties for dilute gas mixtures. Certain useful concepts from thermochemistry are also described in detail. The point of view taken by the author is that of a fundamentalist; that is, an attempt is made to develop all theories presented beginning from the fundamentals. As such, much is made of the basic physics of interactions between like and unlike particles and the roles these interactions play in determining the transport properties of reacting gas mixtures.

The book has the advantage for the reader of presenting the laminar-boundary-layer theory with its many ramifications within the framework of one set of differential equations resulting from a convenient transformation of the conventional statement of the boundary-layer equations. The theory for the reacting turbulent boundary layer is also presented within the framework of one set of boundary-layer equations. Tedious conversions back and forth between transformed equations and the conventional boundary-layer equations are thereby avoided. No loss in generality results, since the effects of chemical reactions, mass transfer, surface melting, shock-wave–boundary-layer interaction, body shape, and pressure gradient are all accounted for without resorting to different transformations for differing boundary conditions.

Aspects of kinetic theory, the molecular theory of gases, thermodynamics, physical chemistry, quantum theory, wave mechanics, and statistical mechanics are all briefly described as they relate to the main topic of this book. Indeed, Chap. 2 is devoted to the origin of the boundary-layer equations and their derivation, beginning with the molecular theory of gases. Chapter 9 is devoted to thermodynamics of gas mixtures and the methods of quantum theory, spectroscopic analysis, and statistical mechanics as they are applied to determining thermodynamic properties and equilibrium composition of gas mixtures. Chapter 10 is devoted to transport properties and the role intermolecular forces play in determining them.

The book was originally intended to be a portion of a joint venture by Professor Ting Y. Li and the author embracing inviscid-flow theory as well as boundary-layer theory. The study of the flow of reacting gas mixtures burgeoned during the writing of the book, and both authors found they had much more to say than could be included in one book. Further complications arose because of the 2000 miles separating them. Therefore, it was decided to write two books: the present one on viscous-flow problems and a later book on the inviscid flow of reacting gas mixtures in preparation by Professor Li. It is intended that the two books complement each other and together provide a comprehensive and up-to-date introduction to the theory of the flow of reacting gas mixtures.

The author wishes to acknowledge the assistance of several colleagues in preparing this book. Professor Ting Y. Li was a constant source of encouragement and constructive criticism during the preparation of this book. Mr. Robert Lowen was particularly helpful in reading and commenting on Chap. 9. Mr. Merwin Sibulkin undertook the burden of reading the entire first draft and constructively criticizing it. By no means the least contribution was made by Mrs. Maxine Matson, who typed the manuscript.

WILLIAM H. DORRANCE
AEROSPACE CORPORATION

Contents

vii

1

An Introduction to Hypersonic Heat-transfer Problems

1-1. General Considerations. This book is written in an era of greatly accelerated application of the fruits of scientific research to engineering problems. Not the least evident of these applications are those related to the development of space-exploration vehicles and long-range missiles. Early in the development of such devices, problems arose which could be solved only by applying the results of the most advanced research in gas dynamics or, in some cases, by furthering the research in this time-honored field of science. Before engineers can reliably design devices to survive flight through or into an atmosphere at hypersonic speeds, they must somehow provide for, avoid, or otherwise accommodate the enormous heat-transfer rates to the vehicle engendered by such flight speeds. This book deals with the details of some of the theories used to cope with this problem.

In particular, this book is concerned with viscous-gas-flow problems involving reacting gas mixtures. Starting from the statement of the boundary-layer equations for a mixture of gases, the theory is developed for both the laminar and turbulent boundary layers. Solutions for heat transfer to and from two- and three-dimensional shapes are developed taking into account the effects of dissociation of the boundary-layer gas, the effects of chemical reactions other than dissociation, the effects of mass transfer, the effects of a melting surface, and interactions between the boundary-layer gases and the surfaces over which they flow. An effort is made to provide fundamental ideas for the research scientist who wishes to push beyond the present state of this science while also providing some results which can be applied by those engaged in the design of hypersonic

1

vehicles. The book is intended to cover hypersonic and reacting boundary-layer theory only. Those interested in problems of hypersonic inviscid flow and experimental techniques will not find those subjects covered in this book. Rather, it is intended that the various problems associated with a hot, reacting, compressible boundary layer will be thoroughly treated in the present book.

1-2. The Hypersonic Heating Problem. The balance of this book beyond the present chapter is devoted to the details of hypersonic or reacting boundary-layer theory. It is the purpose of the present chapter to provide a less detailed, broader description of the "hypersonic heating problem" and its implications.

When an object penetrates or flies within an atmosphere at hypersonic[1] Mach numbers, considerable drag work is done by the object upon its surroundings. Such an object might be a long-range ballistic or boost-glide missile, a reentering satellite, a meteor, or any other object of extraterrestrial origin encountering an atmosphere. The drag work done ultimately shows up as heat. This heat, in turn, is found in the increased temperature of the air in the wake of the object or in heat which was transferred to the object itself. This drag work may be done over a long period of time, as by a boost-glide vehicle, or over a short period of time, as by a ballistic missile nose cone, reentering earth satellite, or meteor. Let us calculate, to a first approximation, the magnitude of the heat transferred to a hypersonic object owing to the viscous heating of the boundary-layer air.

Newton's first law applied to an object decelerating within an atmosphere at hypersonic speed in the absence of a thrust force yields

$$\frac{W}{g}\frac{dV}{dt} = -D \qquad (1\text{-}1)$$

where W = weight of the object, assumed constant to the approximation of this analysis

V = flight velocity

g = gravitational constant

D = total drag force or resistance due to atmosphere

and where we have neglected the contribution due to gravity as

---

[1] By "hypersonic Mach numbers" in this book we mean flight Mach numbers high enough that the gas molecules begin to dissociate in appreciable numbers in the high-temperature regions of the gas surrounding the body. For O_2, a major constituent of earth's atmosphere, that Mach number is about 6. For CO_2, a possible major component of Venus's atmosphere, that Mach number is about 10. For CH_4, a possible major component of Jupiter's atmosphere, that Mach number is about 5.5.

being much less than drag in the region of interest. Furthermore, define a heat-transfer coefficient $\bar{C}_H$ by the relation

$$\frac{dQ}{dt} = \bar{C}_H \rho V S \, \Delta I \tag{1-2}$$

where Q = heat transferred to the object
ρ = atmospheric density
S = area of object exposed to high-temperature gas
ΔI = enthalpy potential representing temperature difference which results in heat transfer from hot gas to cooler body
That is, define ΔI as

$$\Delta I = I_\infty - I_w \tag{1-3}$$

where I_∞ is the stagnation enthalpy defined as

$$I_\infty = C_p T_\infty + \frac{V^2}{2} \tag{1-4}$$

I_w is the enthalpy of the air mixture at the object's surface temperature as given by

$$I_w = \int_0^{T_w} C_p \, dT \tag{1-5}$$

where C_p is the specific heat at constant pressure of the boundary-layer gas mixture and T_∞ is the ambient gas temperature before being heated by the passage of the object.

Equation (1-2) indicates that the heat-transfer rate is directly dependent upon body geometry, density or flight altitude, and flight velocity. In this book we are concerned chiefly with problems related to high heat-transfer rates which, in turn, occur within the relatively dense atmosphere under most circumstances. Thus we deal with continuum-flow problems exclusively.

Now, because by our definition of the hypersonic speed range we are concerned with values of flight Mach number exceeding 6, then, for air

$$\frac{V^2}{2C_p T_\infty} = \frac{\gamma - 1}{2} M^2 \geq 7.2 \qquad \text{for } \gamma = 1.4$$

where

$$\gamma = \frac{C_p}{C_v}$$

$$C_p - C_v = \bar{R}$$

and M is flight Mach number, where

$$M^2 = \frac{V^2}{\gamma \bar{R} T_\infty}$$

Since T_w is of the order of T_∞, Eq. (1-3) becomes

$$\Delta I \simeq \frac{V^2}{2} \tag{1-6}$$

in view of the orders of magnitude of the various terms involved in ΔI. Calculations of the type to be described in later chapters show that the heat-transfer coefficient $\bar{C}_H$ is proportional to the skin-friction coefficient $\bar{C}_{D_f}$, where $\bar{C}_{D_f}$ is defined by the equation

$$D_f = \bar{C}_{D_f} \frac{\rho V^2}{2} S \tag{1-7}$$

where D_f is the drag force due to viscous shear stress at the exposed surface of the object and

$$\bar{C}_H \propto \bar{C}_{D_f} \tag{1-8}$$

For a smooth flat plate the constant of proportionality is $\frac{1}{2}$. For convenience we assume $\frac{1}{2}$ here. Thus, Eqs. (1-1), (1-2), and (1-6) can be combined to give

$$dQ = -\frac{1}{2} \frac{\bar{C}_{D_f}}{\bar{C}_D} d\left(\frac{W}{2g} V^2\right) \tag{1-9}$$

where D is the total drag, that is,

$$D = D_p + D_f = \bar{C}_D \rho \frac{V^2}{2} S$$

D_p is the drag force due to normal stresses over the exposed surface, and we assume to a first approximation that the ratio of $\bar{C}_{D_f}$ to $\bar{C}_D$ is independent of velocity. Integrating Eq. (1-9) we obtain

$$Q_f = \frac{1}{2} \frac{\bar{C}_{D_f}}{\bar{C}_D} \frac{1}{2} \frac{W}{g} V_i^2 \tag{1-10}$$

where the subscript f denotes the final value of heat transferred to the body after it has spent its kinetic energy and the subscript i denotes the initial value. We assume that $V_f = 0$ and $Q_i = 0$ to a first approximation.

Thus we see that the heat transferred to an object decelerating at hypersonic speeds within an atmosphere is directly proportional to the product of its initial kinetic energy and the ratio of the friction drag coefficient to the total drag coefficient. We note that only half of this product represents heat transferred to the object. The other half is transferred to the surrounding cooler gas. Equation (1-10) immediately suggests that, if Q_f is to be minimized, the ratio of

$\bar{C}_{D_f}/\bar{C}_D$ should be minimized; that is, the fraction of the total resistance to motion represented by friction drag should be minimized.

Equation (1-10) was used to calculate total heat transferred to decelerating bodies as a function of initial velocity. The results of

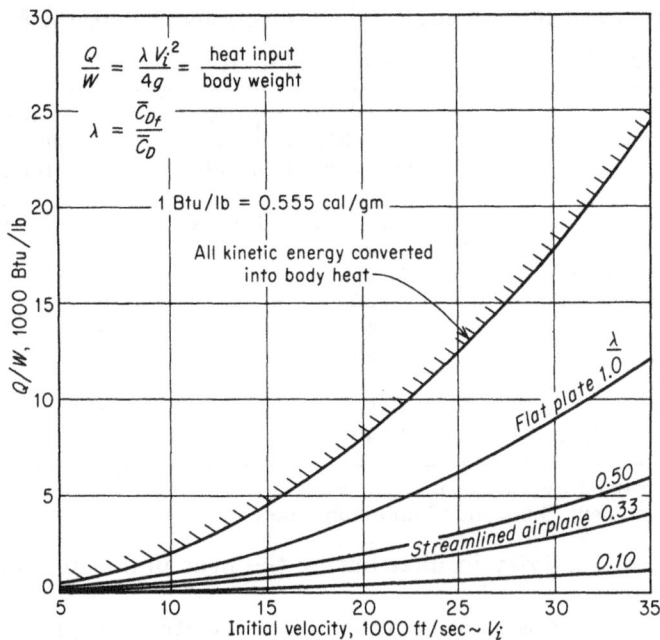

FIG. 1-1. Heat transferred to a decelerating body at hypersonic speeds.

this calculation are shown in Fig. 1-1. Some typical values of the ratio $\bar{C}_{D_f}/\bar{C}_D$ are given in Table 1-1.

TABLE 1-1. SOME TYPICAL VALUES
OF THE RATIO $\bar{C}_{D_f}/\bar{C}_D$

Object	$\bar{C}_{D_f}/\bar{C}_D$
Sphere	0.01
Streamlined airplane	0.33
Flat plate	1.0

Figure 1-1 shows that the total heat transferred to a decelerating body can exceed the heat capacity of most known refractory materials if the body is streamlined and the initial velocity is high enough. The heat capacities of several refractory materials, including the heats of fusion and evaporation, are given in Table 1-2. Without regard to the unfavorable manufacturing characteristics of some of these materials, it is apparent that few of them provide sufficient heat capacity for use as a structural material of a streamlined decelerating object without precautions being taken somehow to avoid or dissipate the heat. Furthermore, the theoretical maximum heat capacity of any material is never realized in applications of the type being discussed here owing to the limitations of materials with finite

TABLE 1-2. HEAT CAPACITIES OF SOME REFRACTORY MATERIALS

Material	Total heat capacity from 312°K through vaporization, in cal/g	Total heat capacity† from 560°R through vaporization, in Btu/lb
Graphite	15,950 (up to 3980°K)	28,700 (up to 7160°R)
Beryllium oxide	7,450	13,400
Magnesium oxide	5,600	10,090
Silicon carbide	3,920	7,050
Titanium	2,145	3,865
Molybdenum	1,990	3,580
Zirconium	1,522	2,740
Tantalum	1,232	2,220
Tungsten	1,040	1,870

† Heat capacities except for that of graphite provided by AVCO Research and Development Division, Wilmington, Mass.

thermal conductivity in distributing the heat uniformly through the material.

Of course, steps can be taken to alleviate this problem. This book will present the details of some of these steps as they relate to the reacting boundary layer surrounding the object.

1-3. Heat-transfer Effects upon Hypersonic-vehicle Design. It is too early in this book to describe the details of the interaction of the hot gas boundary layer with the surface material of the object over which it flows. This will be done in later chapters. However, we have seen enough to anticipate some of the steps which can be taken by a designer of a hypersonic vehicle to deal best with the hypersonic heating problem. Equation (1-10) and Fig. 1-1 immediately suggest that the object might be made rather blunt and unstreamlined if large amounts of heating of the object are to be avoided. Thus a sphere might successfully serve as the shape of such an object. However, the use of a blunt shape has its disadvantages depending upon the use to which the object is put. For example, a sphere has high-pressure drag accompanied by strong bow shock waves which result in a large volume of gas being heated as the gas passes through the shock wave. This heating can be accompanied by ionization and luminosity of the gas in the wake of the object.   Communication with the object may thereby be compromised.[1]   Such an object may well be difficult to conceal if that is also an objective.

Streamlining the object, on the other hand, may result in excessively high heat-transfer rates, since deceleration will thereby be delayed until denser regions of the atmosphere are penetrated.   The

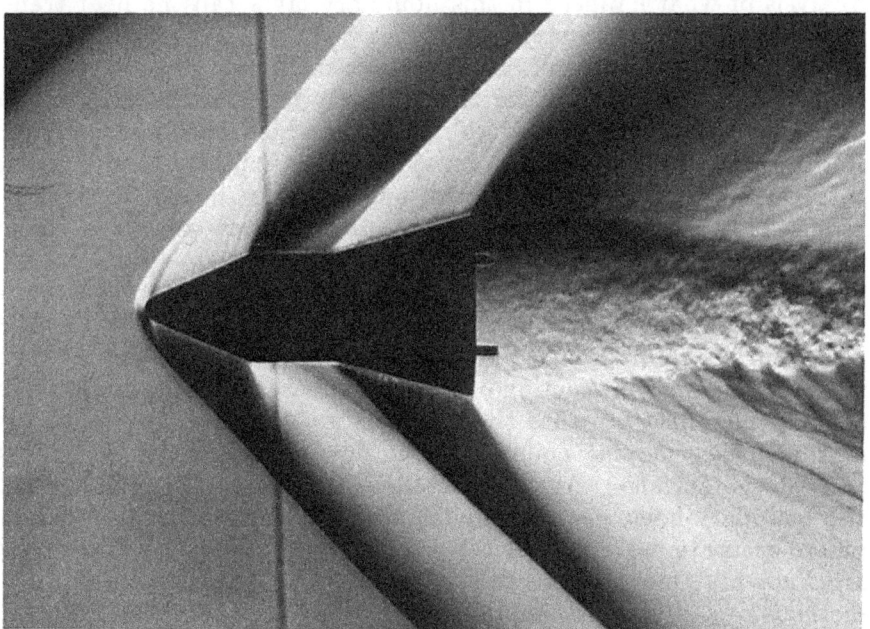

Fig. 1-2. Model reentry shape in ballistic range.   The model has been photographed a fraction of a second after being fired from a light-gas gun, using a shadowgraph technique.   (*Photograph courtesy of AVCO Corporation, Research and Advanced Development Division, Wilmington, Mass.*)

designer finds himself judiciously seeking a compromise between the heat-protection system and the shape of the object.   No hard and fast rules are available to a designer, and an optimum compromise is continually being sought as the designer exercises his skills.

Figure 1-2 is a shadow photograph of a model of a typical missile nose-cone shape in flight down a ballistic range.   The nose is blunted in order to reduce heat-transfer rates in that region.   The aft-cone frustrum   provides   aerodynamic   stability.[2]   Note   the   turbulent

[1] See, for example, W. H. Dorrance, *Aerospace Eng.*, vol. 17, no. 5, pp. 40–43, May, 1958.

[2] See, for example, W. H. Dorrance, *J. Aeronaut. Sci.*, vol. 18, no. 8, pp. 505–512, Aug., 1951.

nature of the wake. As the turbulent eddies are reduced in size with the passage of time through the effects of viscosity, the temperature of the gas in the wake rises, so that after the object has passed, the average temperature of the gas in and surrounding the wake has increased. Most of the viscous heating of the gas occurs in the relatively thin boundary layer next to the body. Only by detailed analysis of what happens in this boundary layer can the heat transferred to the object be predicted with accuracy.

We must not overlook the importance of heat-transfer rate to hypersonic-vehicle design, for it is not only the amount but the *rate* at which heat is transferred to the body which affects the design. For example, a chosen material may have the capacity to absorb the total amount of heat transferred but lack the heat conductivity required to distribute the heat to the interior fast enough to avoid overheating the material near the surface. Various schemes have been devised to deal with this problem, including the use of sublimating or reacting materials which are deliberately allowed to ablate[1] at the surface in an orderly and predictable manner. Such materials may soak up heat in reacting or changing phase and may also provide a "blocking effect" in that they provide a convective current of gas at the surface moving counter to the flow of heat being conducted toward the surface. Again a detailed analysis of the behavior of the boundary-layer gas flow under such circumstances is necessary before accurate heat-transfer rates to the solid undersurface material are known. One purpose of this book is to describe appropriate methods of analysis of this problem.

1-4. The Scope of This Book. Whenever a surface material is heated or cooled by a viscous gas stream, the methods of this book will apply. The heated surface may be that of a rocket nozzle throat, a wind-tunnel wall, or a surface undergoing decomposition owing to phase changes or surface combustion. Many surface-material–gas-layer interactions will be analyzed in detail for both laminar and turbulent boundary layers. The effects of surface combustion, gas-layer combustion, surface mass transfer, dissociation, and variation of gas properties with temperature and composition upon the characteristics of both laminar and turbulent boundary layers will all be treated in later chapters of this book. Because the thermodynamic and transport properties of gas mixtures are required to apply the results of the boundary-layer analyses, separate chapters are devoted to each of these subjects. This book is intended

---

[1] Ablation is the process of decomposition of a surface material due to the absorption of heat energy and includes the processes of melting, vaporization, and surface chemical reactions according to the usage of this book.

to be a reasonably complete reference for the equations and gas property values required to calculate heat-transfer rates to or from a compressible viscous boundary layer under various surface conditions, including those of a chemically reacting surface material.

The subject matter of this book touches on several scientific disciplines, including those of gas dynamics, thermodynamics, thermochemistry, molecular physics, quantum theory, and statistical mechanics.   Because of this, an annoying problem of duplication of symbols and nomenclature customarily used in these different scientific disciplines presented itself.   Wherever possible, we retained the convention used in the discipline related to the subject under discussion.   Some duplication occurs, but whenever it occurs, it is believed that the text makes it obvious which meaning is attached to the ambiguous symbol.   Most of the symbols are explained in the appendix.

A special effort was made to cite appropriate references for contributions not yet commonplace in the literature.   The open literature was relied on almost exclusively.   Occasionally reference is made to an unclassified reference privately published, which, however, should be available in most technical libraries.   In some cases these references can be obtained by writing to the agency which published the document.   It is regrettable that not all useful references have appeared in the open literature, and it is hoped that by citing them in this manner we may somehow stimulate their publication in the open literature.

2

The Boundary-layer Equations

2-1. Introduction. The concept of a thin, viscous layer next to a body over which a fluid is flowing is due to Prandtl.[1] According to Prandtl, the fluid velocity relative to the surface of the body increases from zero at the surface to its maximum value away from the surface in a thin region called the boundary layer. This concept is well established and has become a fundamental postulate of fluid dynamics. The concept has been directly confirmed by careful measurements of velocity distribution through the boundary-layer region. Indirect confirmation is provided by the excellent agreement with measurements of solutions to the boundary-layer equations—a set of differential equations derived from the more general equations of motion using the boundary-layer concept to drop certain terms from the more general equations.

In this chapter we shall postulate the existence of a boundary layer and develop the differential equations describing the flow of a reacting gas mixture within the boundary layer. First the origin of the parent equations of gas dynamics will be examined in order to bring out the contributions due to chemical reactions, such contributions being absent in the conventional statement of the boundary-layer equations. A method of seeking similar solutions to the laminary-boundary-layer equations will be presented and used to reduce the partial differential equations to ordinary differential equations. Some particular integrals of the boundary-layer equations will be presented. The appearance of certain dimensionless transport parameters will be discussed, and the significant parameters defined. The equations developed in this chapter will serve as the starting point for the theories presented in Chaps. 3 through 8.

---

[1] See, for example, L. Prandtl, *NACA TM* 452, 1928, a reprint of Prandtl's original paper delivered in 1904.

2-2. Gas-dynamics Equations. The fundamental equations of gas dynamics include equations for the conservation of species, conservation of mass, conservation of momentum, and conservation of energy along with the equation of state for the gas mixture. In their most general form they consist of a set of nonlinear partial differential equations with four independent variables: three in space and one in time. The equations follow from the application of the fundamental laws of classical mechanics and thermodynamics to the flow of fluids and gases. With very few exceptions these equations have not been integrated in closed form for boundary conditions appropriate to physically sensible problems. The exceptions are usually rather simple physical situations possessing a natural symmetry which permits the dropping of many terms in the full equations before a solution is sought.

Historically there have been two different approaches to developing the equations of gas dynamics. One method is the phenomenological approach wherein certain relations between shear and strain, heat flux and temperature gradient, and diffusion flux and concentration gradient are postulated and the equations then developed using the laws of classical mechanics and heat flow. The method is somewhat unsatisfactory in that it leaves the transport coefficients, that is, the constants of proportionality between shear and strain, heat flux and temperature gradient, and diffusion flux and concentration gradient, undefined with no method other than direct measurements available to determine their values. In this book we shall identify these transport coefficients by the symbols μ, k, and D_{ij}, where, for example,

$$\mu = \frac{\tau_{yx}}{\partial u/\partial y} \qquad \text{coefficient of viscosity}$$

$$k = \frac{-\dot{q}_y}{\partial T/\partial y} \qquad \text{coefficient of thermal conductivity} \qquad (2\text{-}1)$$

and $$D_{ij} = \frac{-\bar{V}_{iy}C_i}{\partial C_i/\partial y} \qquad \text{binary diffusion coefficient}$$

where τ_{yx} = shear stress on surface perpendicular to y axis acting in x direction in cartesian coordinate system

$-\dot{q}_y$ = heat flux in y direction of cartesian coordinate system

$\bar{V}_{iy}$ = component of diffusion velocity due to concentration gradients in y direction of cartesian coordinate system for a species i diffusing into a mixture of species i and j

$C_i = \rho_i/\rho$, mass fraction of species i

T and ρ are the temperature and density of the gas mixture, respectively.

The phenomenological approach is thoroughly treated in a book by Schlichting,[1] which is recommended to the interested reader.

The alternative approach is that of the mathematical theory for nonuniform gases. This method yields the fluid-dynamic equations with the transport coefficients defined in terms of certain integral relations which involve the dynamics of colliding particles. Some empiricism is involved in specifying the interparticle forces needed to evaluate the collision integrals but even here, in principle, recourse to theory will narrow the margin of uncertainty involved. Chapter 10 treats this subject.

The mathematical theory of nonuniform gases, essentially a kinetic theory approach, is described in comprehensive detail in books by Chapman and Cowling[2] and Hirschfelder, Curtiss, and Bird.[3] The transport coefficients μ, k, and D_{12} are found to be functions of the gas-mixture temperature, gas-species molecular weights, and certain parameters of the interparticle force field as is described in Chap. 10. Because this theory serves as a basis for the transport coefficient equations presented in Chap. 10 and because it results in the equations of gas dynamics for reacting gas mixtures, the elements of this theory will be outlined in brief detail here.

The Molecular Theory of Gas Flow. The mathematical theory for the molecular flow of gases is concerned with describing the macroscopic features of a gas mixture which is not in thermodynamic or chemical equilibrium based upon a postulated microscopic behavior of the constituent gas particles. We begin by postulating a molecular velocity distribution function $f_i(x,y,z,v_x,v_y,v_z,t)$ which gives at time t the probable number of particles of kind i in volume $dx\,dy\,dz$ at point x, y, z with velocity components between v_x and $v_x + dv_x$, v_y and $v_y + dv_y$, and v_z and $v_z + dv_z$. The space (x,y,z,v_x,v_y,v_z) is called phase space, and in that which follows, the position coordinates and velocity coordinates are to be treated as independent variables. Obviously, the total number of particles in a volume $dx\,dy\,dz$ at time t must be

$$n_i(\mathbf{r},t) = \int_{-\infty}^{\infty} \int_{-\infty}^{\infty} \int_{-\infty}^{\infty} f_i\,dv_x\,dv_y\,dv_z \qquad (2\text{-}2)$$

[1] Hermann Schlichting, "Boundary Layer Theory," 4th ed., pp. 42–54 and 288–292, McGraw-Hill Book Company, Inc., New York, 1960.

[2] Sidney Chapman and T. G. Cowling, "The Mathematical Theory of Nonuniform Gases," 2d ed., Cambridge University Press, New York, 1958.

[3] Joseph O. Hirschfelder, Charles F. Curtiss, and R. Byron Bird, "Molecular Theory of Gases and Liquids," chap. 7, pp. 441–513, John Wiley & Sons, Inc., New York, 1954.

and the density of particles of kind i is

$$\rho_i = n_i m_i = m_i \int f_i \, d\mathbf{v}_i \tag{2-3}$$

where m_i is the mass of one particle of kind i, n_i is the number density of particles of kind i, and where we shall use the following definitions in that which follows:

$$\mathbf{r} = x\mathbf{i} + y\mathbf{j} + z\mathbf{k} \tag{2-4}[1]$$

$$\mathbf{v} = v_x\mathbf{i} + v_y\mathbf{j} + v_z\mathbf{k} \tag{2-5}$$

and $$d\mathbf{v} = dv_x \, dv_y \, dv_z \tag{2-6}$$

All integrations of integrals of the type given in Eq. (2-3) will be between the limits $-\infty$ to $+\infty$ unless otherwise noted. The other thermodynamic and kinematic variables of interest to us can be expressed in terms of integrals of the type given in Eq. (2-3). That is, for example, the average value of any dependent variable associated with particles of kind i will be

$$\bar{\theta}_i(\mathbf{r},t) = \frac{1}{n_i} \int f_i \theta_i \, d\mathbf{v}_i \tag{2-7}$$

Then it follows that

$$\bar{\mathbf{v}}_i(\mathbf{r},t) = \frac{1}{n_i} \int f_i \mathbf{v}_i \, d\mathbf{v}_i \tag{2-8}$$

is the average velocity of particles of kind i. The gas-dynamic or mass average velocity will be

$$\mathbf{v}_0(\mathbf{r},t) = \frac{\sum_i n_i m_i \bar{\mathbf{v}}_i}{\sum_i n_i m_i} \tag{2-9}$$

or, since $\sum n_i m_i = \rho$, the mass average velocity is

$$\mathbf{v}_0(\mathbf{r},t) = \frac{1}{\rho} \sum_i m_i \int f_i \mathbf{v}_i \, d\mathbf{v}_i \tag{2-10}$$

when only one kind of gas particle is present: $\mathbf{v}_0 = \bar{\mathbf{v}}_i$. The peculiar velocity of particles of kind i is defined as

$$\mathbf{V}_i = \mathbf{v}_i - \mathbf{v}_0 \tag{2-11}$$

and is the velocity of a particle of kind i with respect to a coordinate

---

[1] Vectors will appear in boldface type in this book. Cartesian coordinates only will be used. In Eq. (2-4), for example, $\mathbf{i}$, $\mathbf{j}$, $\mathbf{k}$ are unit vectors parallel to the x, y, z axes, respectively.

system moving with the mass average velocity $\mathbf{v}_0$ of the fluid. The diffusion velocity of particles of kind i is defined as

$$\overline{\mathbf{V}}_i(\mathbf{r},t) = \frac{1}{n_i} \int f_i \mathbf{V}_i \, d\mathbf{v}_i = \frac{1}{n_i} \int f_i(\mathbf{v}_i - \mathbf{v}_0) \, d\mathbf{v}_i \qquad (2\text{-}12)$$

This velocity is zero when only one kind of particle is present.

From kinetic theory the temperature is related to the peculiar velocity by

$$\frac{3}{2} nkT = \sum_i \frac{n_i m_i}{2} \overline{V_i^2} \qquad (2\text{-}13)$$

where k is Boltzmann's constant. Thus, making use of Eq. (2-7),

$$T(\mathbf{r},t) = \frac{1}{3nk} \sum_i m_i \int f_i V_i^2 \, d\mathbf{v}_i \qquad (2\text{-}14)$$

where $n = \Sigma n_i$. Now we have equations relating all the macroscopic variables to the microscopic distribution function. That is, from Eqs. (2-2), (2-10), (2-14), and the equation of state for a perfect gas

$$\rho(\mathbf{r},t) = \sum n_i m_i = \sum_i m_i \int f_i \, d\mathbf{v}_i \qquad (2\text{-}15)$$

$$T(\mathbf{r},t) = \frac{1}{3nk} \sum_i m_i \int f_i V_i^2 \, d\mathbf{v}_i \qquad (2\text{-}14)$$

$$\mathbf{v}_0(\mathbf{r},t) = \frac{1}{\rho} \sum_i m_i \int f_i \mathbf{v}_i \, d\mathbf{v}_i \qquad (2\text{-}10)$$

$$p_i = \rho_i \frac{k}{m_i} T \qquad \rho_i = n_i m_i \qquad (2\text{-}16)$$

and
$$p = \sum p_i \qquad (2\text{-}17)$$

and it is apparent that, if the distribution function f_i could be determined, the problem of determining the flow-field variables would be completely specified. The distribution function is determined by Boltzmann's equation which describes the variation of f_i in phase space. This equation is,[1] assuming no external forces on the particles,

$$\frac{\partial f_i}{\partial t} + v_{ix} \frac{\partial f_i}{\partial x} + v_{iy} \frac{\partial f_i}{\partial y} + v_{iz} \frac{\partial f_i}{\partial z} = \sum_j J_{ij} \qquad (2\text{-}18)$$

[1] See, for example, Hirschfelder, Curtiss, and Bird, op. cit., pp. 444–449.

where f_i is the single-particle or singlet distribution function[1]

$$J_{ij} = 2\pi \int \int (f_i'f_j' - f_if_j)g_{ij}b \, db \, d\mathbf{v}_j \qquad (2\text{-}19)$$

where $f_i'f_j'$ = distribution functions of colliding particles of kinds i and j after collision

f_if_j = distribution functions of colliding particles of kind i and j before collision

g_{ij} = absolute value of the relative velocity of particles of kinds i and j before collision

b = distance of closest approach of particles of kinds i and j in absence of interparticle potential

The integral given by Eq. (2-19) represents the net change in particle density of particles of kind i due to collisions with particles of kind j which either deflect particles of kind i into or out of the control volume. The products of the distribution functions of particles of kinds i and j represent the most probable number of collisions occurring with the control volume in phase space, since both are evaluated at the same position $\mathbf{r}$. The specification of our problem is now complete. The difficult part is yet to come; it consists in solving Eq. (2-18) for the distribution functions f_i.

The Equations of Change. It is clear that, if Boltzmann's equation, Eq. (2-18), could be solved for the distribution functions f_i for all kinds of particles, the determination of the flow-field variables could be accomplished through Eqs. (2-10) and (2-14) to (2-17). The accomplishment of this objective, however, is no simple matter and, in fact, makes up a large portion of the subject matter of the books by Chapman and Cowling and Hirschfelder, Curtiss, and Bird cited earlier. We shall discuss that portion of the subject which is necessary to the understanding of the origin of the gas-dynamic equations and transport properties used in this book.

Let us first obtain the Maxwell-Enskog equation of change using Boltzmann's equation. Let us seek a differential equation which describes the variation of an average flow-field variable $\bar{\theta}_i$ associated with gas particles of kind i. We can develop such an equation by multiplying Eq. (2-18) by θ_i and integrating the resulting equation over all velocities $\mathbf{v}_i$; i.e.,

$$\int \theta_i \left(\frac{\partial f_i}{\partial t} + \mathbf{v}_i \cdot \frac{\partial f_i}{\partial \mathbf{r}} - \sum_j J_{ij} \right) d\mathbf{v}_i = 0 \qquad (2\text{-}20)$$

[1] Section 10-2 describes the conditions under which the gas-dynamic flow-field variables are adequately determined using the single-particle distribution function.

where
$$\frac{\partial}{\partial \mathbf{r}} = \frac{\partial}{\partial x}\,\mathbf{i} + \frac{\partial}{\partial y}\,\mathbf{j} + \frac{\partial}{\partial z}\,\mathbf{k} \qquad (2\text{-}21)$$

Now,
$$\int \theta_i \frac{\partial f_i}{\partial t}\,d\mathbf{v}_i = \frac{\partial}{\partial t}\left(\int \theta_i f_i\,d\mathbf{v}_i\right) - \int f_i \frac{\partial \theta_i}{\partial t}\,d\mathbf{v}_i$$

or in view of Eq. (2-7),

$$\int \theta_i \frac{\partial f_i}{\partial t}\,d\mathbf{v}_i = \frac{\partial}{\partial t}\,(n_i\bar{\theta}_i) - n_i \overline{\frac{\partial \theta_i}{\partial t}} \qquad (2\text{-}22)$$

also
$$\int \theta_i\left(\mathbf{v}_i\cdot\frac{\partial f_i}{\partial \mathbf{r}}\right) d\mathbf{v}_i = \frac{\partial}{\partial \mathbf{r}}\cdot\left(\int f_i\theta_i\mathbf{v}_i\,d\mathbf{v}_i\right) - \int f_i\mathbf{v}_i\cdot\frac{\partial\theta_i}{\partial\mathbf{r}}\,d\mathbf{v}_i$$

since $\mathbf{v}_i$ and $\mathbf{r}$ are independent variables in phase space. Or in view of Eq. (2-7),

$$\int \theta_i\left(\mathbf{v}_i\cdot\frac{\partial f_i}{\partial \mathbf{r}}\right) d\mathbf{v}_i = \frac{\partial}{\partial \mathbf{r}}\cdot(n_i\overline{\theta_i\mathbf{v}_i}) - n_i\left(\overline{\mathbf{v}_i\cdot\frac{\partial\theta_i}{\partial\mathbf{r}}}\right) \cdot \qquad (2\text{-}23)$$

and so Eq. (2-20) becomes Enskog's equation of change; viz.,

$$\frac{\partial n_i\bar{\theta}_i}{\partial t} + \frac{\partial}{\partial \mathbf{r}}\cdot(n_i\overline{\theta_i\mathbf{v}_i}) - n_i\left(\overline{\frac{\partial\theta_i}{\partial t}} + \overline{\mathbf{v}_i\cdot\frac{\partial\theta_i}{\partial\mathbf{r}}}\right) = \sum_j \int \theta_i J_{ij}\,d\mathbf{v}_i \qquad (2\text{-}24)$$

where J_{ij} is given by Eq. (2-19). Let us illustrate the use of the equation of change by letting $\theta_i = m_i$. Then, since m_i is constant,

$$\frac{\partial m_i}{\partial t} = 0 \qquad \text{and} \qquad \frac{\partial m_i}{\partial \mathbf{r}} = 0$$

and Eq. (2-24) becomes

$$\frac{\partial n_i m_i}{\partial t} + \frac{\partial}{\partial \mathbf{r}}\cdot(n_i m_i\bar{\mathbf{v}}_i) = m_i\sum_j \int J_{ij}\,d\mathbf{v}_i \qquad (2\text{-}25)$$

Now if we have a reacting gas mixture including chemical reactions involving particles of kind i, it is clear that the total number of particles of kind i must change upon collision of particles of kind i with other particles. That is,

$$\sum_j \int J_{ij}\,d\mathbf{v}_i = K_i \qquad (2\text{-}26)$$

where K_i is the total rate of change per unit volume of the number of particles of kind i due to chemical reaction upon collision with particles of kind j. Furthermore, since mass is conserved throughout such chemical reactions,

$$\sum_i K_i m_i = 0 \qquad (2\text{-}27)$$

Hence, making use of Eq. (2-26), Eq. (2-25) can be written

$$\frac{\partial n_i}{\partial t} + \frac{\partial}{\partial \mathbf{r}} \cdot (n_i \overline{\mathbf{v}}_i) = K_i \tag{2-28}$$

or in view of Eqs. (2-3) and (2-11), we have the alternate equations

$$\frac{\partial \rho_i}{\partial t} + \frac{\partial}{\partial \mathbf{r}} \cdot [\rho_i(\overline{\mathbf{V}}_i + \mathbf{v}_0)] = \dot{w}_i \tag{2-29}$$

and

$$\frac{\partial \rho C_i}{\partial t} + \frac{\partial}{\partial \mathbf{r}} \cdot [\rho C_i(\overline{\mathbf{V}}_i + \mathbf{v}_0)] = \dot{w}_i \tag{2-30}$$

and since $\Sigma K_i m_i = 0$, $\Sigma m_i n_i = \rho$, and $\Sigma m_i n_i \overline{\mathbf{V}}_i = 0$, we can obtain the equation for conservation of mass from Eq. (2-29) by summing over all species i.

$$\frac{\partial \rho}{\partial t} + \frac{\partial}{\partial \mathbf{r}} \cdot (\rho \mathbf{v}_0) = 0 \tag{2-31}$$

Furthermore, from Eqs. (2-30) and (2-31) we can obtain

$$\rho \frac{\partial C_i}{\partial t} + (\rho \mathbf{v}_0) \cdot \frac{\partial C_i}{\partial \mathbf{r}} = \dot{w}_i - \frac{\partial}{\partial \mathbf{r}} \cdot (\rho C_i \overline{\mathbf{V}}_i) \tag{2-32}$$

a form convenient to our later analyses. In the above $\dot{w}_i = m_i K_i =$ net mass rate of production of species i due to chemical reactions and C_i is the mass fraction of species i as defined before. Equation (2-32) is the species-conservation equation of gas dynamics which includes the diffusion velocity $\overline{\mathbf{V}}_i$ determined by the molecular distribution function as given by Eq. (2-12).

We can derive the equations for conservation of momentum and conservation of energy in a similar manner using the equation of change. Since these equations are not explicitly affected by the occurrence of chemical reactions, they will not be derived here.[1] The equation of change is useful in solving the Boltzmann equation as will be described next.

Solving Boltzmann's Equation. We outline here Enskog's method of solving Boltzmann's equation. We know that, when the gas mixture is in thermodynamic and chemical equilibrium, the distribution function is given by

$$f_i^{(0)} = n_i \left(\frac{m_i}{2\pi kT}\right)^{3/2} \exp\left(-\frac{m_i}{2kT}\overline{V}_i^2\right) \tag{2-33}$$

[1] They are derived in Hirschfelder, Curtiss, and Bird, *op. cit.*, pp. 459–464 and 496–498.

the classical Maxwell-Boltzmann distribution law. This can be derived using the methods of statistical mechanics and, in fact, is so derived as the combination of Eqs. (9-11) and (9-16). Let us suppose that our distribution function for our gas mixture not in equilibrium is given by the converging infinite series

$$f_i = f_i^{(0)} + \epsilon f_i^{(1)} + \epsilon^2 f_i^{(2)} + \cdots \tag{2-34}$$

and our Boltzmann equation is given by

$$\frac{\partial f_i}{\partial t} + \mathbf{v}_i \cdot \frac{\partial f_i}{\partial \mathbf{r}} = \frac{1}{\epsilon} \sum_j J_{ij} \tag{2-35}$$

Here $1/\epsilon$ is a measure of the frequency of collisions between particles of kind i and kind j. If $1/\epsilon$ is large, collisions are frequent, but according to Eq. (2-34), f_i will be close to $f_i^{(0)}$. In the limit as ϵ approaches zero, the gas mixture would then be in equilibrium. If we substitute Eq. (2-34) into Eq. (2-35), we obtain an infinite set of equations; viz.,

$$0 = \sum_j J_{ij}(f_i^{(0)}, f_j^{(0)}) \tag{2-36}$$

and

$$\frac{\partial f_i^{(0)}}{\partial t} + \mathbf{v}_i \cdot \frac{\partial f_i^{(0)}}{\partial \mathbf{r}} = \sum_j [J_{ij}(f_i^{(0)}, f_j^{(1)}) + J_{ij}(f_i^{(1)}, f_j^{(0)})] \tag{2-37}$$

etc. Direct substitution shows that Eq. (2-36) is satisfied by Eq. (2-33). Also, when Eq. (2-33) is substituted into the left-hand side of Eq. (2-37), we find that the right-hand side, which consists of integrals involving $f_i^{(1)}$ and $f_j^{(1)}$, is equated to an expression involving space and time derivatives of n_i, T, and $\mathbf{v}_0$. The time derivatives are eliminated using the equations of change for these dependent variables. It is found that the only expression for $f^{(1)}$ which when used with Eq. (2-34) satisfies Eqs. (2-10), (2-14), and (2-15) for $\mathbf{v}_0$, T, and ρ, respectively, and also satisfies integral equation (2-37) above is a linear combination of terms involving gradients of T, $\mathbf{v}_0$, p, and n_i. The coefficients of these terms are determined by a set of integral equations involving $\mathbf{V}_i$, n_i, and T. These equations to the first order in ϵ were solved by Chapman and Cowling[1] by expanding portions of the integrands of the integrals involved into converging infinite series of polynomials. Only a few terms were required to evaluate the integrals to reasonable accuracy. Explicit integrals for the transport coefficients D_{ij}, μ, and k result from this process, and the distribution functions other than $f_i^{(0)}$ are not solved for. The mathematical details are too cumbersome to be repeated here and

[1] Sidney Chapman and T. G. Cowling, "The Mathematical Theory of Nonuniform Gases," 2d ed., Cambridge University Press, New York, 1958.

can be found in the books by Chapman and Cowling and Hirsch-felder, Curtiss, and Bird referred to previously.

One result of these calculations is the equation for the diffusion velocity $\overline{\mathbf{V}}_i$, which is

$$\overline{\mathbf{V}}_i = \frac{n^2}{\rho n_i} \sum_j m_j D_{ij} \mathbf{d}_j - \frac{1}{n_i m_i} D_i^T \frac{\partial \log T}{\partial \mathbf{r}} \tag{2-38}$$

where, if $n = \sum_i n_i$,

$$\mathbf{d}_j = \frac{\partial}{\partial \mathbf{r}} \frac{n_j}{n} + \left(\frac{n_j}{n} - \frac{n_j m_j}{\rho} \right) \frac{\partial \log p}{\partial \mathbf{r}} \tag{2-39}$$

To a reasonable approximation for a binary mixture, the theory gives

$$D_{ij} = \frac{3}{16} \frac{[2\pi k(m_i + m_j)/m_i m_j]^{\frac{1}{2}}}{n\pi \sigma_{ij}^2 \Omega_{ij}^{(1,1)*}} \tag{2-40}$$

and, in Eq. (2-38), D_i^T is the coefficient of thermal diffusion. The collision integrals $\Omega_{ij}^{(l,s)*}$ and the collision diameter σ_{ij} will be discussed in Chap. 10. These collision integrals are found to be directly dependent upon the dynamics of particle collisions and hence the interparticle potential field. We note that Eq. (2-38) indicates that the diffusion velocity of any species i depends upon gradients of species, pressure, and temperature. In our applications of Eq. (2-38) we shall find the pressure and temperature gradient contributions to be of secondary importance.

Equations for thermal conductivity and viscosity in terms of the collision integrals $\Omega_{ij}^{(l,s)*}$ also result from the Chapman-Cowling theory. These equations are given and described in Chap. 10. It is most important to note at this point that the equations of gas dynamics are completely described by the molecular theory and what little empiricism that remains is involved in specifying inter-particle force fields. In principle even this element of empiricism can be removed by recourse to quantum theory, although the task is immense except for the simplest of particles.

The use of the molecular theory of gases as outlined here is the most rational approach to treating the flow of reacting gas mixtures. We make this statement for two reasons: (1) The method allows us properly to introduce terms into our flow equations which account for the effects of chemical reactions, and (2) the interparticle poten-tials, which are basically related to the forces between pairs of like and pairs of unlike particles in a gas mixture, are explicit in the equa-tions for the transport coefficients yielded by this theory. The species-conservation equation is the only gas-dynamics equation which explicitly contains a term due to chemical reactions in the form in which these equations are used in this book.

The Diffusion Velocity. The full equation for the diffusion velocity vector as given by the molecular theory of gases is given by Eqs. (2-38) and (2-39). It is possible to simplify these equations for our purposes by making several reasonable assumptions. First of all, most gas mixtures of interest to us will be essentially binary mixtures in the sense that all species present will fall into one of two classes: heavy particles and light particles. For example, dissociated air consists of the heavy particles O_2 and N_2 and the light particles O and N. If a graphite surface is reacting with an air mixture, the heavy particles of CO_2 and the light particles of O_2, N_2, and possibly CO will be present. If helium is used as a coolant gas in a transpiration cooling scheme, the heavy particles of O_2 and N_2 and the light particles of He will be present. One binary diffusion coefficient D_{12} will suffice to calculate the mass flux of O or N through O_2 and N_2, N_2 or O_2 through CO_2, or He through N_2 and O_2 to a reasonable approximation. For an effective two-species gas, then, Eq. (2-38) gives

$$\overline{\mathbf{V}}_1 = \frac{n^2 m_2}{\rho n_1} D_{12}\mathbf{d}_2 - \frac{1}{n_1 m_1} D_1^T \frac{\partial \log T}{\partial \mathbf{r}} \tag{2-41}$$

where $\mathbf{d}_2$ is given by Eq. (2-39) with $j = 2$, and a similar equation can be written for $\overline{\mathbf{V}}_2$. Now in a binary mixture

$$m_1 n_1 \overline{\mathbf{V}}_1 + m_2 n_2 \overline{\mathbf{V}}_2 = 0 \tag{2-42}$$

$$m_1 n_1 + m_2 n_2 = n\bar{m} \tag{2-43}$$

and
$$n_1 + n_2 = n \tag{2-44}$$

If Eq. (2-39) is used to write equations for $\mathbf{d}_1$ and $\mathbf{d}_2$ and use is made of Eqs. (2-43) and (2-44), it will be found that

$$\mathbf{d}_1 = -\mathbf{d}_2 \tag{2-45}$$

Furthermore, if Eq. (2-41) and the analogous equation for $\overline{\mathbf{V}}_2$ are used with Eqs. (2-42) and (2-45), it will be found that

$$D_{12} = D_{21} \tag{2-46}$$

and
$$D_1^T = -D_2^T \tag{2-47}$$

If Eq. (2-41) and the analogous equation for $\overline{\mathbf{V}}_2$ are used to obtain the difference $\overline{\mathbf{V}}_1 - \overline{\mathbf{V}}_2$ and Eqs. (2-45) through (2-47) are made use of, there results

$$\overline{\mathbf{V}}_1 - \overline{\mathbf{V}}_2 = -\frac{n^2}{n_1 n_2} D_{12}\left[\frac{\partial}{\partial \mathbf{r}} \frac{n_1}{n} + \left(\frac{n_1}{n} - \frac{n_1 m_1}{\rho}\right) \frac{\partial \log p}{\partial \mathbf{r}} + k_T \frac{\partial \log T}{\partial \mathbf{r}}\right] \tag{2-48}$$

where
$$k_T = \frac{D_1^T \rho}{D_{12} m_1 m_2 n^2} \qquad (2\text{-}49)$$

If, now, we combine Eq. (2-48) with Eq. (2-41) to obtain $\overline{\mathbf{V}}_1$, we obtain

$$\overline{\mathbf{V}}_1 = -\frac{n m_2}{n_1 \overline{m}} D_{12} \left[\frac{\partial}{\partial \mathbf{r}} \frac{n_1}{n} + \left(\frac{n_1}{n} - \frac{n_1 m_1}{\rho} \right) \frac{\partial \log p}{\partial \mathbf{r}} + k_T \frac{\partial \log T}{\partial \mathbf{r}} \right] \qquad (2\text{-}50)$$

It is convenient to introduce the mass fractions C_i where

$$C_i = \frac{n_i m_i}{n \overline{m}} \qquad (2\text{-}51)$$

and where
$$\overline{m} = \frac{\sum m_i n_i}{\sum n_i} = \left(\sum \frac{C_i}{m_i} \right)^{-1} \qquad (2\text{-}52)$$

thus
$$\frac{\partial}{\partial \mathbf{r}} \frac{n_1}{n} = \frac{\overline{m}^2}{m_1 m_2} \frac{\partial C_1}{\partial \mathbf{r}} \qquad (2\text{-}53)$$

Introducing Eqs. (2-51) to (2-53) into Eqs. (2-50) we obtain

$$\overline{\mathbf{V}}_1 = -D_{12} \left[\frac{\partial}{\partial \mathbf{r}} (\log C_1) + \frac{m_2 - m_1}{\overline{m}} C_2 \frac{\partial}{\partial \mathbf{r}} (\log p) + \frac{m_1 m_2}{\overline{m}^2 C_1} k_T \frac{\partial}{\partial \mathbf{r}} (\log T) \right]$$
$$(2\text{-}54)$$

Thus we see that gradients in concentration, pressure, and temperature all contribute to the mass diffusion flux.   In any given direction the relative ratios of the various terms contributing to a diffusion flux are in proportion as

$$\left| \frac{dC_1}{C_1} \right| : \left| \frac{dp}{p} \right| : \left| k_T \frac{dT}{T} \right|$$

where k_T is of the order of 10^{-1} for gas mixtures of interest to us. In many cases of interest to us it will be found that

$$\left| \frac{dC_1}{C_1} \right| \gg \left| \frac{dp}{p} \right| \qquad \left| k_T \frac{dT}{T} \right|$$

and to a reasonable approximation we can use

$$\overline{\mathbf{V}}_1 = -D_{12} \frac{\partial}{\partial \mathbf{r}} (\log C_1) \qquad \text{Fick's law} \qquad (2\text{-}55)$$

Thus, for example, Eq. (2-32) for the conservation of species becomes

$$\rho \frac{\partial C_i}{\partial t} + (\rho \mathbf{v}_0) \cdot \frac{\partial C_i}{\partial \mathbf{r}} = \dot{w}_i + \frac{\partial}{\partial \mathbf{r}} \cdot \left(\rho D_{12} \frac{\partial C_i}{\partial \mathbf{r}} \right) \qquad (2\text{-}56)$$

and this is the equation we shall use in that which follows. It should be apparent that the contribution to diffusion flux due to pressure gradients and the contribution to diffusion flux due to temperature gradients (Sorét effect) can be introduced using Eq. (2-54) if it is desirable to do so.

2-3. The Boundary-layer Equations. In this section we shall describe Prandtl's boundary-layer concept and illustrate how it can

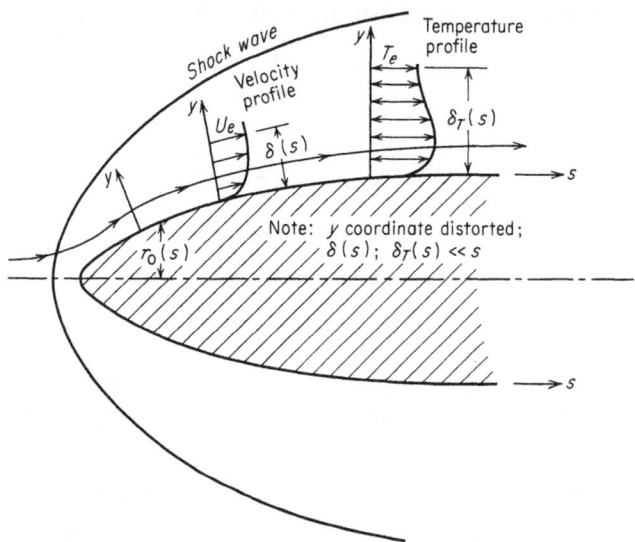

Fig. 2-1. Schematic drawing of boundary-layer flow.

be used to reduce the equations of gas dynamics to the boundary-layer equations. Following this, the boundary-layer equations as used throughout the remainder of this book will be presented.

Prandtl's Boundary-layer Concept. It is postulated that the effects of the transport phenomena in a dilute gas mixture are confined to a thin layer near the surface of the body over which the fluid flows. Figure 2-1 illustrates schematically the geometry of boundary-layer flow over a blunted object in flight at supersonic speeds. There will be, in general, at least three significant boundary-layer thicknesses. These are the velocity boundary-layer thickness δ (shown in Fig. 2-1), the temperature boundary-layer thickness δ_T (also shown in Fig. 2-1), and a species-concentration boundary-layer thickness δ_c. If the Prandtl number and Schmidt number of the gas mixture are both unity, these thicknesses are identical for the particular case of flow over a flat plate. If, as is true for most gas mixtures, the Prandtl number and Schmidt number are less than one,[1] then the temperature and concentration boundary-layer thicknesses are greater than the velocity boundary-layer thickness. These dimensionless transport parameters are defined as

$$P = \frac{C_{p_f}\mu}{k} \qquad \text{Prandtl number} \qquad (2\text{-}57)$$

and
$$S = \frac{\mu}{\rho D_{12}} \qquad \text{Schmidt number} \qquad (2\text{-}58)$$

and will appear repeatedly in the boundary-layer equations in that which follows. Their significance will be dealt with in Sec. 2-6. For most gas mixtures of interest they will be sufficiently close to unity that we can assume all boundary-layer thicknesses equal to the velocity boundary-layer thickness in the order-of-magnitude analysis to follow.

Prandtl's boundary-layer concept can be expressed by the inequality

$$\delta(s) \ll s$$

In fact, solutions to the boundary-layer equations will show that

$$\frac{\delta^2}{s} = O\left(\frac{1}{R_e}\right) \qquad R_e = \frac{\rho_e u_e s}{\mu_e} \qquad (2\text{-}59)$$

and we can use this order of magnitude for the boundary-layer thickness to reduce the equations of gas dynamics to the boundary-layer equations. Let us illustrate the procedure by reducing the s component of the Navier-Stokes equation (conservation of momentum) for a compressible fluid mixture to the s component of the boundary-layer equations where s and y are orthogonal coordinates as shown in Fig. 2-1. The s and y components of velocity are u and v, respectively, and we can write the s component of the Navier-Stokes equation as,[2] in the absence of external forces and for a steady flow,

$$\rho u \frac{\partial u}{\partial s} + \rho v \frac{\partial u}{\partial y} = -\frac{\partial p}{\partial s} + \frac{\partial}{\partial s}\left\{\mu\left[2\frac{\partial u}{\partial s} - \frac{2}{3}\left(\frac{\partial u}{\partial s} + \frac{\partial v}{\partial y}\right)\right]\right\} + \left[\frac{\partial}{\partial y}\mu\left(\frac{\partial u}{\partial y} + \frac{\partial v}{\partial s}\right)\right]$$

$$(2\text{-}60)$$

[1] Chapter 10 is devoted to transport properties of gas mixtures including those having Prandtl and Schmidt numbers other than unity.

[2] Here we assume that the bulk viscosity is equal to $-2/3$ as it is for a monatomic gas. For relaxing gas mixtures, this is not strictly correct, but the analysis used to obtain the boundary-layer momentum equation will result in Eq. (2-66) in either case. See E. V. Laitone, *J. Aeronaut. Sci.*, vol. 23, no. 9, pp. 846–854, September, 1956, for a discussion of the equations of motion for a compressible, viscous, relaxing gas.

and from Eqs. (2-21) and (2-31), the conservation of mass equation is, for a steady flow,

$$\frac{\partial \rho u}{\partial s} + \frac{\partial \rho v}{\partial y} = 0 \tag{2-61}$$

where in two dimensions

$$\mathbf{v}_0 = u\mathbf{i} + v\mathbf{j} \tag{2-62}$$

Now we seek to express these equations in dimensionless form such that all derivatives will be dimensionless and of the same order. This can be done by dividing all velocity components by a characteristic velocity, all coordinates by a characteristic length, and density and viscosity coefficients by a characteristic density and viscosity, respectively. If we denote the dimensionless quantities by an asterisk superscript (*), then we can state that, in the dimensionless coordinates, mass must still be conserved. That is,

$$\frac{\partial \rho^* u^*}{\partial s^*} + \frac{\partial \rho^* v^*}{\partial y^*} = 0 \tag{2-63}$$

and both terms of this equation must be of the same order lest the momentum flux in either direction be constant, a trivial solution. If we choose the characteristic lengths in the s and y directions as L and δ, then it is found that the proper dimensionless quantities are

$$u^* = \frac{u}{u_e} \qquad v^* = \frac{Lv}{\delta u_e} \tag{2-64a}$$

$$s^* = \frac{s}{L} \qquad y^* = \frac{y}{\delta} \tag{2-64b}$$

$$\rho^* = \frac{\rho}{\rho_e} \qquad \mu^* = \frac{\mu}{\mu_e} \qquad p^* = \frac{p}{\rho_e u_e^2} \tag{2-64c}$$

where the subscripts e denote local values at the edge of the boundary layer (see Fig. 2-1). The choice of normalizing parameters applied to v in Eq. (2-64a) can be inferred by using Eq. (2-63). When Eqs. (2-64a) through (2-64c) are used with Eq. (2-60), there results

$$\rho^* u^* \frac{\partial u^*}{\partial s^*} + \rho^* v^* \frac{\partial u^*}{\partial y^*} = -\frac{\partial p^*}{\partial s^*} + \frac{1}{R_e} \frac{\partial}{\partial s^*} \left\{ \mu^* \left[2\frac{\partial u^*}{\partial s^*} + \frac{2}{3}\left(\frac{\partial u^*}{\partial s^*} + \frac{\partial v^*}{\partial y^*} \right) \right] \right\}$$
$$+ \left(\frac{L}{\delta}\right)^2 \frac{1}{R_e} \frac{\partial}{\partial y^*} \left\{ \mu^* \left[\frac{\partial u^*}{\partial y^*} + \left(\frac{\delta}{L}\right)^2 \frac{\partial v^*}{\partial s^*} \right] \right\} \tag{2-65}$$

and if Prandtl's hypothesis applies, the terms crossed out are far smaller in magnitude than the remaining terms and can be neglected according to Eq. (2-59) if R_e is large. Note that all dimensionless derivatives are of the same order by virtue of the choice of normalizing characteristic lengths and velocities in Eqs. (2-64a) and (2-64b). Reverting to dimensional coordinates we have the s component of the boundary-layer momentum equation; viz.,

$$\rho u \frac{\partial u}{\partial s} + \rho v \frac{\partial u}{\partial y} = -\frac{\partial p}{\partial s} + \frac{\partial}{\partial y}\left(\mu \frac{\partial u}{\partial y}\right) \tag{2-66}$$

and the equation for conservation of mass is unchanged.

The equations for conservation of species and conservation of energy can be reduced in a similar manner to boundary-layer equations. The operation need not be repeated here.

The Boundary-layer Equations for a Reacting Gas. The boundary-layer equations resulting from the application to the gas-dynamic equations of the order-analysis technique described in the previous section will now be stated.

Some assumptions apply when these equations are used. These assumptions are:

1. All gas species considered behave as perfect-gas species.
2. The flow is in a steady state.
3. The effects of radiation from the gas to the surface and from the surface to the gas are neglected.
4. The flow is two-dimensional or axially symmetric.

The effects of radiation can often be taken into account in an additive fashion when they become significant, and they will not be further considered in this book.

For a perfect gas the equation of state is

$$p_i = \rho_i R_i T \tag{2-67}$$

where, from Dalton's law for partial pressures,

$$\sum_i p_i = p = T \sum_i \rho_i R_i = \rho \bar{R} T \tag{2-68a}$$

and the gas constant for the mixture is

$$\bar{R} = \sum_i C_i R_i \qquad C_i = \frac{\rho_i}{\rho} \tag{2-68b}$$

The equations for conservation of momentum are
s component:

$$\rho u \frac{\partial u}{\partial s} + \rho v \frac{\partial u}{\partial y} = -\frac{\partial p}{\partial s} + \frac{\partial}{\partial y}\left(\mu \frac{\partial u}{\partial y}\right) \tag{2-66}$$

y component:

$$0 = -\frac{\partial p}{\partial y} \qquad \text{or} \qquad p = p_e \tag{2-69}$$

The equation for conservation of mass is

$$\frac{\partial \rho u r_0^k}{\partial s} + \frac{\partial \rho v r_0^k}{\partial y} = 0 \tag{2-70}$$

where $k = 0$ for two-dimensional flow and $k = 1$ for axisymmetric flow. $r_0(s)$ is the radius of the body in a meridian plane for an axisymmetric shape. The equation for conservation of species is given by Eq. (2-56) and is

$$\rho u \frac{\partial C_i}{\partial s} + \rho v \frac{\partial C_i}{\partial y} = \frac{\partial}{\partial y}\left(\rho D_{12} \frac{\partial C_i}{\partial y}\right) + \dot{w}_i \tag{2-56}$$

where we have neglected thermal and pressure diffusion as being negligible relative to mass diffusion. Thus

$$\bar{V}_{iy} = -\frac{D_{12}}{C_i}\frac{\partial C_i}{\partial y} \qquad \text{Fick's law} \tag{2-71}$$

The equation for conservation of energy is

$$\rho u \frac{\partial I}{\partial s} + \rho v \frac{\partial I}{\partial y} = \frac{\partial}{\partial y}\left[\frac{\mu}{P}\frac{\partial I}{\partial y} + \mu\left(1 - \frac{1}{P}\right)\frac{1}{2}\frac{\partial u^2}{\partial y}\right]$$
$$- \frac{\partial}{\partial y}\left[\left(\frac{1}{L} - 1\right)\rho D_{12} \sum_i h_i \frac{\partial C_i}{\partial y}\right] \tag{2-72}$$

and where the enthalpy for any species i is given by

$$h_i = \int_0^T C_{p_i}\, dT + h_i^0 \tag{2-73a}$$

(h_i^0 is the heat of formation of species i)

$$h = \sum_i C_i h_i \tag{2-73b}$$

and

$$I = h + \frac{u^2}{2} \tag{2-73c}$$

Furthermore, we assume that the gas mixture is made up largely of light particles and heavy particles, so that one binary diffusion coefficient D_{12} applies for diffusion fluxes of all species. Equations (2-67) through (2-73c) constitute the set of boundary-layer equations for a reacting gas mixture.

2-4. Reduction of the Boundary-layer Equations to Ordinary Differential Equations. The boundary-layer equations as stated in Sec. 2-3 are nonlinear partial differential equations which are difficult to solve except under special circumstances where enough terms can be dropped to reduce the equations to ordinary differential equations (couette flow and pipe flow, for example). There are, fortunately, other circumstances under which the equations can be reduced to ordinary differential equations. These circumstances occur when there exists a natural coordinate system $(\bar{s},\eta)$, related to the cartesian coordinate system (s,y) through appropriate transformations, in which the derivatives of the dependent variables become separable and ordinary differential equations result. Let us seek such a coordinate system using our boundary-layer equations for a reacting gas mixture. The technique used here has been presented by Li and Nagamatsu.[1]

Introduce Eq. (2-68a) into Eq. (2-70). There results, making use of Eq. (2-69),

$$\frac{\partial}{\partial s}\frac{u}{\bar{R}T} + \frac{\partial}{\partial y}\frac{v}{\bar{R}T} = -\frac{u}{\bar{R}T}\frac{1}{p_e r_0^k}\frac{d(p_e r_0^k)}{ds} \qquad (2\text{-}74)$$

In Eq. (2-74) introduce the stream function

$$\psi(s,y) = \int_0^y \frac{u}{\bar{R}T}\,dy + \psi(s,0) \qquad (2\text{-}75)$$

Then Eq. (2-74) becomes

$$\frac{\partial}{\partial y}\frac{\partial \psi}{\partial s} + \frac{\partial}{\partial y}\frac{v}{\bar{R}T} = -\frac{1}{p_e r_0^k}\frac{d(p_e r_0^k)}{ds}\frac{\partial \psi}{\partial y} \qquad (2\text{-}76)$$

since

$$\frac{\partial}{\partial s}\frac{\partial \psi}{\partial y} = \frac{\partial}{\partial y}\frac{\partial \psi}{\partial s}$$

Integrating Eq. (2-76) from $y = 0$ to $y = y$ we obtain

$$\frac{v}{\bar{R}T} = -\frac{\partial \psi}{\partial s} - \frac{\psi}{p_e r_0^k}\frac{d(p_e r_0^k)}{ds} \qquad (2\text{-}77)$$

Thus, Eq. (2-66) can be written with the help of Eqs. (2-69), (2-76), and (2-77) as

$$\frac{\partial \psi}{\partial y}\frac{\partial u}{\partial s} - \left[\frac{\partial \psi}{\partial s} + \frac{\psi}{p_e r_0^k}\frac{d(p_e r_0^k)}{ds}\right]\frac{\partial u}{\partial y} = -\frac{1}{p_e}\frac{dp_e}{ds} + \frac{1}{p_e}\frac{\partial}{\partial y}\left(\mu\frac{\partial u}{\partial y}\right) \qquad (2\text{-}78)$$

Now let us seek a similar solution to Eq. (2-78) such that

$$\psi(s,\eta) = N(s)f(\eta) \qquad (2\text{-}79a)$$

and

$$u(s,\eta) = u_e(s)f'(\eta) \qquad (2\text{-}79b)$$

[1] Ting Yi Li and Henry T. Nagamatsu, *J. Aeronaut. Sci.*, vol. 22, no. 9, pp. 607–616, September, 1955.

The prime on $f'(\eta)$ in Eq. (2-79b) indicates differentiation with respect to η. η is a similarity variable $\eta(s,y)$ as yet undefined. We define it using Eqs. (2-75), (2-79a), and (2-79b). That is, from Eqs. (2-75) and (2-79a),

$$\frac{\partial \psi}{\partial y} = \frac{u}{\bar{R}T} = \frac{\partial \psi}{\partial \eta}\frac{\partial \eta}{\partial y} = N(s)f'(\eta)\frac{\partial \eta}{\partial y} \tag{2-80a}$$

and, from Eq. (2-79b),

$$\frac{u}{\bar{R}T} = \frac{u_e(s)f'(\eta)}{\bar{R}T} \tag{2-80b}$$

Hence, equating Eq. (2-80a) with Eq. (2-80b) we obtain

$$N(s)\frac{\partial \eta}{\partial y} = \frac{u_e(s)}{\bar{R}T}$$

and so

$$\eta = \frac{u_e(s)}{N(s)}\int_0^y \frac{dy}{\bar{R}T} \tag{2-81a}$$

or with the use of Eqs. (2-68a), (2-68b), and (2-69),

$$\eta = \frac{\rho_e u_e(s)}{p_e N(s)}\int_0^y \frac{\rho}{\rho_e}\,dy \tag{2-81b}$$

It remains to determine $N(s)$. This can be done by expressing Eq. (2-78) in terms of the independent variables s and η rather than s and y by using Eqs. (2-79a), (2-79b), and (2-81a). That is, using these equations, we find that

$$\frac{\partial \psi}{\partial y} = \frac{\rho u_e}{p_e N(s)}\frac{\partial \psi}{\partial \eta} = \frac{\rho u_e}{p_e}f' \tag{2-82a}$$

$$\frac{\partial u}{\partial s} = f'\frac{du_e}{ds} + u_e f''\frac{\partial \eta}{\partial s} \tag{2-82b}$$

$$\frac{\partial \psi}{\partial s} = f\frac{dN(s)}{ds} + N(s)f'\frac{\partial \eta}{\partial s} \tag{2-82c}$$

and

$$\mu\frac{\partial u}{\partial y} = \frac{\rho\mu u_e^2}{p_e N(s)}f'' = \frac{\rho_e \mu_e u_e^2}{p_e N(s)}Cf'' \tag{2-82d}$$

where

$$C = \frac{\rho\mu}{\rho_e \mu_e} \tag{2-83}$$

Substituting Eqs. (2-82a) through (2-82d) into Eq. (2-78) results in

$$\left[\frac{\rho u_e(f')^2}{p_e} - \frac{\rho_e u_e}{p_e}\right]\frac{du_e}{ds} - \left[f\frac{dN(s)}{ds} + \frac{fN(s)}{p_e r_0^k}\frac{d(p_e r_0^k)}{ds}\right]\frac{\rho u_e^2}{p_e N(s)}f''$$

$$= \frac{1}{p_e}\frac{\rho u_e}{p_e N(s)}\frac{\rho_e \mu_e u_e^2}{p_e N(s)}(Cf'')' \tag{2-84}$$

where, in obtaining Eq. (2-84), we have used the asymptotic value of Eq. (2-66); viz., as $y \to \infty$,

$$\rho_e u_e \frac{du_e}{ds} = -\frac{dp_e}{ds} \tag{2-85}$$

Now

$$f \frac{dN(s)}{ds} + \frac{N(s)f}{p_e r_0^k} \frac{d(p_e r_0^k)}{ds} \equiv \frac{f}{p_e r_0^k} \frac{d(Np_e r_0^k)}{ds} \tag{2-86}$$

hence, using Eq. (2-86) in Eq. (2-84), we obtain

$$(Cf'')' \frac{\rho_e \mu_e r_0^k u_e}{p_e N[d(Np_e r_0^k)/ds]} + ff'' + \frac{1}{u_e} \frac{du_e}{ds} \left\{ \frac{(\rho_e/\rho) - (f')^2}{(1/p_e r_0^k N)[d(Np_e r_0^k)/ds]} \right\} = 0 \tag{2-87}$$

Let us choose $N(s)$ such that

$$\frac{\rho_e \mu_e r_0^k u_e}{p_e N[d(Np_e r_0^k)/ds]} = 1 \tag{2-88}$$

Then

$$2\rho_e \mu_e r_0^{2k} u_e = \frac{d(Np_e r_0^k)^2}{ds}$$

whence, since $r_0^k = 0$ at $s = 0$,

$$(Np_e r_0^k)^2 = 2 \int_0^s \rho_e \mu_e r_0^{2k} u_e \, ds$$

and so

$$N(s) = \frac{\left(2 \int_0^s \rho_e \mu_e r_0^{2k} u_e \, ds \right)^{1/2}}{p_e r_0^k} \tag{2-89}$$

Combining Eqs. (2-81b) and (2-89) results in our desired independent variable transformations.   These are

$$\eta = \frac{\rho_e u_e r_0^k}{(2\bar{s})^{1/2}} \int_0^y \frac{\rho}{\rho_e} \, dy \tag{2-90}$$

and

$$\bar{s} = \int_0^s \rho_e \mu_e r_0^{2k} u_e \, ds \tag{2-91}$$

and hence, from Eqs. (2-89) and (2-91),

$$N = \frac{(2\bar{s})^{1/2}}{p_e r_0^k} \tag{2-92}$$

Thus, if

$$\psi = N(s)f(\eta) = \frac{(2\bar{s})^{1/2}}{p_e r_0^k} f(\eta) \tag{2-93}$$

$$u = u_e(s)f'(\eta) \tag{2-79b}$$

$$N = \frac{(2\bar{s})^{1/2}}{p_e r_0^k} \tag{2-92}$$

$$\eta = \frac{\rho_e u_e r_0^k}{(2\bar{s})^{1/2}} \int_0^y \frac{\rho}{\rho_e} \, dy \tag{2-90}$$

and
$$\bar{s} = \int_0^s \rho_e \mu_e r_0^{2k} u_e \, ds \qquad (2\text{-}91)$$

then Eq. (2-87) becomes

$$(Cf'')' + ff'' + \frac{2\bar{s}}{u_e} \frac{du_e}{d\bar{s}} \left[\frac{\rho_e}{\rho} - (f')^2 \right] = 0 \qquad (2\text{-}94)$$

We note here that transformation equations (2-90) and (2-91) result as a natural consequence of seeking a similarity solution to Eqs. (2-66) and (2-70). Transformation equations (2-90) and (2-91) contain several previously introduced transformations as special cases. For example,

1. When u_e, ρ_e, and μ_e are all constant and $k = 0$, then

$$\bar{s} = \rho_e \mu_e u_e s$$

and
$$\eta = \frac{y}{(2\mu_e s/\rho_e u_e)^{1/2}}$$

and these are the Blasius[1] transformation equations.

2. When u_e, ρ_e, and μ_e are all constant and $k = 1$, then

$$\bar{s} = \rho_e u_e \mu_e \int_0^s r_0^2 \, ds$$

$$\eta = \frac{y}{\left(2 \dfrac{\mu_e}{\rho_e u_e r_0^2} \displaystyle\int_0^s r_0^2 \, ds \right)^{1/2}}$$

and these are the Mangler[2] transformation equations.

3. When u_e, ρ_e, and μ_e vary with s and $k = 0$, then

$$\bar{s} = \int_0^s \rho_e u_e \mu_e \, ds$$

and
$$\eta = \frac{\rho_e u_e}{(2\bar{s})^{1/2}} \int_0^y \frac{\rho}{\rho_e} \, dy$$

and these are the Illingworth-Levy[3] transformation equations.

[1] H. Blasius, Z. Math. Physik, vol. 56, p. 1, 1908.
[2] See, for example, L. Howarth, "Modern Developments in Fluid Dynamics," pp. 382–386, Oxford University Press, New York, 1953.
[3] C. R. Illingworth, Proc. Roy. Soc. London, vol. 199, 1949; Solomon Levy, J. Aeronaut. Sci., vol. 21, no. 7, pp. 459–474, 1949.

4. For complete generality

$$\bar{s} = \int_0^s \rho_e u_e \mu_e r_0^{2k}\, ds \tag{2-91}$$

and

$$\eta = \frac{\rho_e u_e r_0^k}{(2\bar{s})^{1/2}} \int_0^y \frac{\rho}{\rho_e}\, dy \tag{2-90}$$

and these are the Lees-Dorodnitsyn[1] transformations.

Now that we have developed the proper transformation equations to reduce our partial differential equations to ordinary differential equations, let us apply transformation equations (2-79b) and (2-90) to (2-93) to the remaining boundary-layer equations (2-56) and (2-72). After some algebraic manipulation of the type already illustrated in deriving Eq. (2-94), there result the useful operators

$$\rho u \frac{\partial}{\partial s} + \rho v \frac{\partial}{\partial y} = \rho u_e^2 \rho_e \mu_e r_0^{2k}\left(f' \frac{\partial}{\partial \bar{s}} - f' \frac{\partial \eta}{\partial \bar{s}} \frac{\partial}{\partial \eta} - \frac{f}{2\bar{s}} \frac{\partial}{\partial \eta} \right) \tag{2-95}$$

and for any quantity $Q(s,y)$,

$$\frac{\partial}{\partial y}\left(Q \frac{\partial}{\partial y} \right) = \frac{\rho u_e^2 r_0^{2k}}{2\bar{s}} \frac{\partial}{\partial \eta}\left(\rho Q \frac{\partial}{\partial \eta} \right) \tag{2-96}$$

Applying operators (2-95) and (2-96) to Eqs. (2-56) and (2-72), where Eqs. (2-93) and (2-79b) are also used, there result

$$f' \frac{\partial C_i}{\partial \bar{s}} - f' \frac{\partial \eta}{\partial \bar{s}} \frac{\partial C_i}{\partial \eta} - \frac{f}{2\bar{s}} \frac{\partial C_i}{\partial \eta} = \frac{1}{2\bar{s}} \frac{\partial}{\partial \eta}\left(\frac{C}{S} \frac{\partial C_i}{\partial \eta} \right) + \frac{\dot{w}_i}{\rho_e \rho u_e^2 r_0^{2k} \mu_e} \tag{2-97}$$

and $$f' \frac{\partial I}{\partial \bar{s}} - f' \frac{\partial \eta}{\partial \bar{s}} \frac{\partial I}{\partial \eta} - \frac{f}{2\bar{s}} \frac{\partial I}{\partial \eta} = \frac{1}{2\bar{s}} \frac{\partial}{\partial \eta}\left[\frac{C}{P} \frac{\partial I}{\partial \eta} + C\left(1 - \frac{1}{P} \right)\frac{1}{2} \frac{\partial u^2}{\partial \eta} \right]$$

$$- \frac{1}{2\bar{s}} \frac{\partial}{\partial \eta}\left[\frac{C}{S}\left(\frac{1}{L} - 1 \right) \sum_i h_i \frac{\partial C_i}{\partial \eta} \right] \tag{2-98}$$

Now, let us assume that $C_i(s,\eta)$ and $I(s,\eta)$ have similar solutions also. That is, let

$$C_i = (C_i)_e z_i(\eta) \tag{2-99}$$

and

$$I = I_e g(\eta) \tag{2-100}$$

Then Eqs. (2-97) and (2-98) become

$$\left(\frac{C}{S} z_i' \right)' + f z_i' = \frac{2\bar{s} f' z_i}{(C_i)_e} \frac{d(C_i)_e}{d\bar{s}} - \frac{2\bar{s}\dot{w}_i}{\rho \rho_e u_e^2 \mu_e r_0^{2k}(C_i)_e} \tag{2-101}$$

---

[1] Lester Lees, *Jet Propulsion*, vol. 26, no. 4, 1956.

and the energy equation becomes

$$\left(\frac{C}{P}g'\right)' + fg' = \frac{2\bar{s}f'g}{I_e}\frac{dI_e}{d\bar{s}} + \left[\frac{C}{S}\left(\frac{1}{L} - 1\right)\sum_i \frac{h_i(C_i)_e}{I_e}z_i'\right]'$$

$$+ \frac{u_e^2}{I_e}\left[\left(\frac{1}{P} - 1\right)Cf'f''\right]' \quad (2\text{-}102)$$

and, for comparison, the momentum equation was shown to be

$$(Cf'')' + ff'' = \frac{2\bar{s}}{u_e}\frac{du_e}{d\bar{s}}\left[(f')^2 - \frac{\rho_e}{\rho}\right] \quad (2\text{-}94)$$

Boundary conditions for these equations are

At $y = 0, \eta = 0$ As $y \to \infty, \eta \to \infty$

$$z_i(0) = (z_i)_w = \frac{(c_i)_w}{(c_i)_e} \qquad z_i \to 1$$

$$f'(0) = 0 \qquad f(0) = f_w \qquad f' \to 1 \qquad f \to 0$$

$$g'(0) = g'_w \qquad\qquad g' \to 0$$

$$g(0) = g_w \qquad\qquad g \to 1$$

The transformed species-conservation, energy, and momentum equations are given by Eqs. (2-101), (2-102), and (2-94), respectively. Providing

$$\frac{\bar{s}}{(C_i)_e}\frac{d(C_i)_e}{d\bar{s}} = \text{const} \qquad \text{and} \qquad \frac{\bar{s}\dot{w}_i}{\rho_e u_e \mu_e r_0(C_i)_e} = \text{const}$$

$$\frac{\bar{s}}{I_e}\frac{dI_e}{d\bar{s}} = \text{const} \qquad \frac{u_e^2}{I_e} = \text{const} \qquad \frac{(h_i)_e(C_i)_e}{I_e} = \text{const}$$

and

$$\frac{2\bar{s}}{u_e}\frac{du_e}{d\bar{s}} = \text{const}$$

and both $h_i/(h_i)_e$ and ρ/ρ_e are functions of η only, we see that these equations are coupled ordinary differential equations with independent variable η. For a large number of useful cases these restrictions are satisfied, as we shall see in succeeding developments.

2-5. Some Particular Integrals of the Boundary-layer Equations. There are a number of important particular integrals which can be arrived at using Eqs. (2-101), (2-102), and (2-94) with their boundary conditions for the cases of $(z_i)_w$ and g_w constant. They are:

1. If $du_e/d\bar{s} = 0$, $dI_e/d\bar{s} = 0$, and $L = P = 1$, then it can be shown by direct substitution that

$$g = af' + b \qquad \text{Crocco's integral}[1]$$

or using the boundary conditions for g and f,

$$g = g_w + f'(1 - g_w) \tag{2-103}$$

2. If $S = P$, $d(C_i)_e/d\bar{s} = 0$, $dI_e/d\bar{s} = 0$, and $\dot{w}_i = 0$ (frozen flow), and if either $P = 1$ or $u_e = 0$, then it can be shown that

$$z_i = a_1 g + b_1 \qquad \text{Probstein's integral}[2]$$

or using the boundary conditions for g and z_i,

$$z_i = (z_i)_w + \frac{1 - (z_i)_w}{1 - g_w}(g - g_w) \tag{2-104}$$

3. If $S = 1$, $\dot{w}_i = 0$ (frozen flow), and $du_e/d\bar{s} = d(C_i)_e/d\bar{s} = 0$, then it can be shown that

$$z_i = a_2 f' + b_2 \qquad \text{Probstein's second integral}[2]$$

or using the boundary conditions for z_i and f',

$$z_i = (z_i)_w + [1 - (z_i)_w]f' \tag{2-105}$$

4. If $S = P = 1$, $\dot{w}_i = 0$ (frozen flow), and

$$\frac{dI_e}{d\bar{s}} = \frac{du_e}{d\bar{s}} = \frac{d(C_i)_e}{d\bar{s}} = 0$$

then Eqs. (2-104) and (2-105) apply and

$$g = a_3 f' + b_3$$

and, applying the boundary conditions for g and f',

$$g = g_w + f'(1 - g_w) \tag{2-106}$$

which is a more restricted statement of Crocco's integral.

5. If $\bar{s} = 0$, $u_e = 0$ (stagnation point), $L = 1$, $\dot{w}_i = 0$ (frozen flow), then

$$z_i = a_4 g + b_4 \qquad \text{Scala's integral}[3]$$

[1] L. Crocco, Sullo strato limite laminare nei gas lungo una lamina plana, *Rend. mat. Univ. Roma,* vol. 2, p. 138, 1941.

[2] Wallace D. Hayes and Ronald F. Probstein, "Hypersonic Flow Theory," p. 294, Academic Press, Inc., New York, 1959.

[3] Sinclaire M. Scala and Guido L. Vidale, *Intern. J. Heat and Mass Transfer,* vol. 1, no. 1, pp. 4–22, 1960.

and, applying the boundary conditions for z_i and g,

$$z_i = (z_i)_w + \frac{1 - (z_i)_w}{1 - g_w} (g - g_w) \tag{2-107}$$

which is the same as Probstein's second integral but shown by Scala to be applicable with the restrictions $dI_e/d\bar{s} = 0$ and $d(C_i)_e/d\bar{s} = 0$ removed.

6. If $d(C_k)_e/d\bar{s} = 0$, $dI_e/d\bar{s} = 0$, $L = 1$, and either $u_e = 0$ or $P = 1$, and if we combine

$$\bar{C}_i = \sum_k r_{i,k} C_k \tag{2-108}$$

with

$$\rho u \frac{\partial C_k}{\partial s} + \rho v \frac{\partial C_k}{\partial y} = \frac{\partial}{\partial y} \left(\rho D_{12} \frac{\partial C_k}{\partial y} \right) + \rho \frac{\partial C_k}{\partial t} \tag{2-109}$$

where $r_{i,k}$ = fraction of mass of species k which is contributed by element i

$\bar{C}_i$ = mass fraction of element i

then because elemental mass is conserved according to Eq. (2-27),

$$\frac{\partial \bar{C}_i}{\partial t} = \sum_k r_{i,k} \frac{\partial C_k}{\partial t} \equiv 0 \tag{2-110}$$

there results from Eqs. (2-108), (2-109), and (2-110)

$$\rho u \frac{\partial \bar{C}_i}{\partial s} + \rho v \frac{\partial \bar{C}_i}{\partial y} = \frac{\partial}{\partial y} \left(\rho D_{12} \frac{\partial \bar{C}_i}{\partial y} \right) \tag{2-111}$$

which, when transformed using Eqs. (2-95) and (2-96), becomes

$$\left(\frac{C}{S} \bar{z}_i' \right)' + f \bar{z}_i' = 0 \tag{2-112}$$

if $d(\bar{C}_i)_e/d\bar{s} = 0$ and if

$$\bar{C}_i = (\bar{C}_i)_e \bar{z}_i(\eta) \tag{2-113}$$

Thus, if $dI_e/d\bar{s} = 0$ in Eq. (2-102), it can be shown that, when

$$L = P = 1$$

$$\bar{z}_i = a_5 g + b_5 \qquad \text{Lees' integral}[1]$$

and applying the boundary conditions that $\bar{z}_i = (z_i)_w$ when $g = g_w$ and $\bar{z}_i = 1$ when $g = 1$, we obtain

$$\bar{z}_i = (\bar{z}_i)_w + \frac{1 - (\bar{z}_i)_w}{1 - g_w} (g - g_w) \tag{2-114}$$

7. If $d(C_i)_e/d\bar{s} = 0$, $dI_e/d\bar{s} = 0$, $L = P = 1$, and $u_e = 0$, then both Eqs. (2-103) and (2-114) apply and

$$\bar{z}_i = a_6 f' + b_6$$

[1] L. Lees, Convective Heat Transfer with Mass Addition and Chemical Reactions, in "Combustion and Propulsion Third AGARD Colloquium," pp. 451–498, Pergamon Press, New York, 1958.

or, applying boundary conditions for $\bar{z}_i$ and f',

$$\bar{z}_i = (\bar{z}_i)_w + [1 - (\bar{z}_i)_w]f' \tag{2-115}$$

Equations (2-103), (2-105), and (2-106) represent convenient relations among f', g, and z_i for frozen flow, and Eqs. (2-103), (2-114), and (2-110) represent convenient relations between f', g, and $\bar{z}_i$ for chemically reacting gas flows. We shall have occasion to use some of these valuable relations later. It should be apparent that Eqs. (2-114) and (2-115) are more general than Eqs. (2-105) and (2-106), since they contain Eqs. (2-105) and (2-106) as special cases when $\dot{w}_i = 0$.

All these useful integrals are tabulated in Table 2-1 along with the

TABLE 2-1. PARTICULAR INTEGRALS OF THE LAMINAR-BOUNDARY-LAYER EQUATIONS FOR MIXTURES OF REACTING AND NONREACTING SPECIES

Case	$\dfrac{du_e}{d\bar{s}}$	$\dfrac{dI_e}{d\bar{s}}$	$\dfrac{d(C_i)_e}{d\bar{s}}$	$\bar{s}$	u_e	$\dot{w}_i$	L	P	S	Integral†
1	0	0	‡	...	...	...	1	1	1	$g = af' + b$
2	...	0	0	...	...	0	1	S	P	$z_i = a_1g + b_1$
				0	0	...				
3	0	...	0	...	...	0	...	...	1	$z_i = a_2f' + b_2$
4	0	0	0	...	...	0	1	1	1	$g = a_3f' + b_3$
5	...	...	...	0	0	0	1	S	P	$z_i = a_4g + b_4$
6	...	0	0	...	0	...	1	S	P	$\bar{z}_i = a_5g + b_5$
					...	...	1	1	1	
7	...	0	0	...	0	...	1	1	1	$\bar{z}_i = a_6f' + b_6$

† All a_i and b_i are constants.
‡ Leaders mean that the variable or parameter can take any value.

conditions under which they are applicable. It should be understood that $(z_i)_w$ and/or g_w must be constant in order that these integrals be valid.

2-6. The Dimensionless Transport Parameters. Three dimensionless transport parameters appear in the boundary-layer equations as given in Sec. 2-4. They are

$$P = \frac{C_{p_f}\mu}{k} \qquad \text{Prandtl number} \tag{2-116}$$

$$S = \frac{\mu}{\rho D_{12}} \qquad \text{Schmidt number} \tag{2-117}$$

$$L = \frac{\rho D_{12}C_{p_f}}{k} = \frac{P}{S} \qquad \text{Lewis number} \tag{2-118}$$

where $\qquad C_{p_f} = \sum_i C_i C_{p_i} \qquad$ frozen specific heat

We shall explain the significance of these parameters in this section.

First of all, consider the terms on the right-hand side of the energy equation (2-72). We have

$$\frac{\partial}{\partial y}\left[\frac{\mu}{P}\frac{\partial I}{\partial y} + \mu\left(1 - \frac{1}{P}\right)\frac{1}{2}\frac{\partial u^2}{\partial y} - \left(\frac{1}{L} - 1\right)\rho D_{12}\sum_i h_i \frac{\partial C_i}{\partial y}\right] \quad (2\text{-}119)$$

Let us simplify this term by making use of Eqs. (2-73a) through (2-73c). We have

$$\frac{\partial I}{\partial y} = \frac{\partial}{\partial y}\left[\sum_i C_i\left(\int_0^T C_{p_i}\,dT + h_i^0\right) + \frac{u^2}{2}\right]$$

or, in view of Eqs. (2-73a) through (2-73c),

$$\frac{\partial I}{\partial y} = C_{p_f}\frac{\partial T}{\partial y} + \sum_i h_i \frac{\partial C_i}{\partial y} + \frac{1}{2}\frac{\partial u^2}{\partial y} \quad (2\text{-}120)$$

Substituting Eq. (2-120) into the relation (2-119) we obtain

$$\frac{\partial}{\partial y}\left(k\frac{\partial T}{\partial y} + \mu u \frac{\partial u}{\partial y} + \rho D_{12}\sum_i h_i \frac{\partial C_i}{\partial y}\right) \quad (2\text{-}121)$$

where the definitions for P, S, and L given by Eqs. (2-116) through (2-118) were used. The first term in relation (2-121) is recognized as the heat-conduction term, the second term as the shear work term, and the third term the transport of energy by diffusion. If we introduce the dimensionless variables and parameters below

$$u^* = \frac{u}{u_e} \qquad \mu^* = \frac{\mu}{\mu_e} \qquad \rho^* = \frac{\rho}{\rho_e} \qquad C_{p_e}T^* = \frac{C_{p_e}T}{h_e}$$

$$k^* = \frac{k}{k_e} \qquad y^* = \frac{y}{\delta_T} \qquad C_i^* = \frac{C_i}{(C_i)_e} \qquad h_i^* = \frac{(C_i)_e h_i}{h_e} \qquad (2\text{-}122)$$

into relation (2-121), we can show that

$$\frac{k(\partial T/\partial y)}{\mu u(\partial u/\partial y)} = \frac{\text{conduction}}{\text{shear work}} \propto \frac{1}{PE} \quad (2\text{-}123)$$

$$\frac{\rho D_{12}\sum_i h_i(\partial C_i/\partial y)}{\mu u(\partial u/\partial y)} = \frac{\text{diffusion}}{\text{shear work}} \propto \frac{1}{SE} \quad (2\text{-}124)$$

and $\quad \dfrac{\rho D_{12}\sum_i h_i(\partial C_i/\partial y)}{k(\partial T/\partial y)} = \dfrac{\text{diffusion}}{\text{conduction}} \propto \dfrac{P}{S} = L \quad (2\text{-}125)$

where $E = u_e^2/h_e$ is called Eckert's number.

We conclude from the above that, when E is large, shear work predominates over diffusion and conduction in determining the temperature profiles in the boundary layer. Furthermore, for a constant given Eckert number, if P is large, shear work predominates over conduction. If S is large, shear work predominates over diffusion. The converse conclusions apply if P and S are small. (They are both usually less than one for most gas mixtures.) If P is large and S is small, diffusion predominates over conduction in determining the temperature distribution, and the converse is true if S is large and P is small. If P and S are of the same magnitude and small, then both diffusion and conduction predominate over shear work providing E is small. For most practical cases of interest, P, S, and L are all close to unity and all terms must be considered for moderate values of Eckert's number. If Eckert's number is large, then the effects of diffusion and conduction are secondary in determining the temperature distribution in the boundary layer. We can show that

$$E = \frac{u_e^2}{h_e} \simeq \frac{\gamma_e \bar{R} u_e^2}{C_{p_e} \gamma_e \bar{R} T_e} = (\gamma_e - 1) M_e^2$$

since

$$a_e^2 = \gamma_e \bar{R} T_e$$

$$C_{p_e} - \bar{R} = C_{v_e}$$

and

$$\gamma_e = \frac{C_{p_e}}{C_{v_e}}$$

and so for $M_e = O(10)$ or greater, viscous dissipation will predominate in determining the temperature profiles through the boundary layer if P and S are of the order of one.

We can also interpret the transport parameters P and S in another way. Consider the energy equation (2-72). We have

$$\rho u \frac{\partial I}{\partial s} + \cdots = \frac{\partial}{\partial y} \frac{\mu}{P} \frac{\partial I}{\partial y} + \cdots$$

where all terms must be of the same order of magnitude if they are to be retained in the differential equation. Make the terms dimensionless by using the definitions of dimensionless parameters and variables given in Eqs. (2-122) with the additional definition that $I_e I^* = I$. We obtain

$$\rho^* u^* \frac{\partial I^*}{\partial s^*} + \cdots = \frac{1}{PR_e} \left(\frac{L}{\delta_T}\right)^2 \frac{\partial}{\partial y^*} \left(\mu^* \frac{\partial I^*}{\partial y^*}\right) + \cdots \qquad (2\text{-}126)$$

Thus, if both terms are to be the same order of magnitude,

$$\left(\frac{\delta_T}{L}\right)^2 \propto \frac{1}{PR_e} \qquad (2\text{-}127)$$

but in Sec. 2-3 we stated that

$$\left(\frac{\delta}{L}\right)^2 \propto \frac{1}{R_e} \tag{2-59}$$

Hence
$$\frac{\delta_T}{\delta} \propto \frac{1}{(P)^{1/2}} \tag{2-128}$$

Thus our previously stated conclusion that, if P is less than one, the thermal boundary layer is thicker than the velocity boundary layer is directly confirmed by relation (2-128).

Consider now the diffusion equation (2-56). We have

$$\rho u \frac{\partial C_i}{\partial s} + \cdots = \frac{\partial}{\partial y}\left(\rho D_{12} \frac{\partial C_i}{\partial y}\right) + \cdots$$

Make these terms dimensionless as before where we use the diffusion boundary-layer thickness δ_c instead of δ_T in transformation equations (2-122). We obtain

$$\rho^* u^* \frac{\partial C_i^*}{\partial s^*} + \cdots = \frac{1}{SR_e}\left(\frac{L}{\delta_c}\right)^2 \frac{\partial}{\partial y^*}\left(\rho^* \frac{\partial C_i^*}{\partial y^*}\right) + \cdots \tag{2-129}$$

Thus, if both terms are to be of the same order,

$$\left(\frac{\delta_c}{L}\right)^2 \propto \frac{1}{SR_e} \tag{2-130}$$

and making use of relation (2-59), we find that

$$\frac{\delta_c}{\delta} \propto \frac{1}{(S)^{1/2}} \tag{2-131}$$

and again our statement about the relative thicknesses of the diffusion and velocity boundary layers is verified.

2-7. Conclusions. The objective of this chapter was to introduce the boundary-layer equations for a reacting gas mixture which will be used as a point of departure for the work to be presented in succeeding chapters. The origins of the gas-dynamics equations, from which the boundary-layer equations are descended, were examined in order to trace the terms contributed by chemical reactions and in order to establish the basis for the transport property equations used in this book. The transport parameters L, S, and P were also examined in order to establish their significance in reacting-boundary-layer theory. We are now in a position to proceed to obtain solutions to the boundary-layer equations for various boundary conditions of practical significance.

3

Surface-material–Boundary-layer Interactions

3-1. Introduction. In Chaps. 4 through 8 we shall be concerned
with the details of the boundary-layer flow of reacting gas mixtures.
Before we plunge into the detailed treatment of various reacting-
boundary-layer problems it seems advisable to examine some of the
more complex problems with a somewhat simplified approach which,
while retaining essential features, isolates and illustrates some of
the effects of the gas-dynamic, thermodynamic, and chemical phenomena involved. Mass-transfer effects, melting surface layers,
ablation, and the response of solid material to high rates of boundary-layer heat transfer will be treated with this point of view in mind
in this chapter.

The present chapter will be primarily concerned with the interaction between the reacting, viscous gas layer and the surface over
which it flows. After some useful thermochemistry concepts have
been established, a simplified analysis of the heat transfer from a
reacting gas boundary layer to a cooler undersurface will be presented where the various terms contributing to heat transfer are
isolated. An analysis of the melting surface layer will be presented
next, followed by a simplified but definitive description of mass-transfer effects upon heat transfer from a viscous boundary layer. Heat
conduction in the solid state is then described. All the above concepts are illustrated with numerical examples.

The author believes that separating and independently examining
these different phenomena will sharpen the reader's appreciation of
their relative importance when dealing with problems in which
several of the separate effects discussed in this chapter occur concurrently. For example, it is entirely possible that practical problems arise in which both exothermic (combustion) and endothermic
(dissociation) chemical reactions occur in the gas layer, heterogeneous

reactions occur at the interface between the gas-layer species and the undersurface material over which the gas flows (burning at the interface), the undersurface is melting and vaporizing at the interface with consequent mass-transfer effects, and heat is being conducted to the solid-state interior. In fact, it is just this problem which an engineer must face when designing an ablating heat shield to protect an object which penetrates an atmosphere at hypervelocity. It is our purpose in this chapter to disassemble the parts of such a problem so that each contributing effect can be examined separately in order to assess and understand its relative importance in the overall problem. Of course, some sacrifice in exactness is involved in such a disassembly, since, in general, nonlinear interactions occur.

3-2. Some Thermochemistry. The reader who has come this far has no difficulty in concluding that, in addition to conventional gas-dynamic concepts, several fundamental concepts from physical chemistry and thermochemistry are likely to be required in dealing with problems involving reacting gas mixtures. The purpose of this section is to develop and present such concepts as are useful in this chapter. Chapter 9 presents a more thorough treatment of the thermochemistry used in analyses with which this book is concerned.

Common to most problems of interest to us here are simultaneous, competing chemical reactions involving gas-phase reactions (homogeneous reactions) and reactions between gas-layer species and liquid- or solid-surface species (heterogeneous reactions). Assume that no nuclear reactions occur; the mass of any element entering into these chemical reactions must be conserved during the reactions. That is, it is axiomatic and the *element* mass fractions remain fixed and unchanged during the chemical reactions. If we designate the element mass fractions by $\bar{C}_i$, where

$$\bar{C}_i = \frac{\bar{\rho}_i}{\rho} = \text{element mass fraction}$$

$$C_k = \frac{\rho_k}{\rho} = \text{species mass fraction}$$

$$\rho_k = \text{density of any species } k$$

$$\rho = \text{density of total gas mixture}$$

and $r_{i,k} = \text{fraction of mass of species } k \text{ contributed by element } i$

then the relation between element and species mass fractions is given by

$$\bar{C}_i = \sum_k r_{i,k} C_k \tag{3-1}$$

and, according to the above discussion, element mass is conserved during all chemical reactions; hence

$$d\bar{C}_i = 0 = \sum_k r_{i,k}\,dC_k \tag{3-2}$$

Equation (3-2) represents a useful relation among differential changes of species, mass fractions as subsequent developments will show.

Consider the case of a hot gas flowing over a surface the material of which (in either the liquid or solid state) is vaporizing to a significant extent. The surface material, assumed here to be an element with chemical symbol E for simplicity's sake, enters the hot gas boundary layer and is assumed to react with gas species present in the boundary layer. Assume that the hot gas layer is a mixture of nitrogen and oxygen molecules and atoms present in the proportions close to those of air at the same temperature and pressure. Several homogeneous chemical reactions can be going on simultaneously in the gas layer including the following:

$$\text{E} + \text{O} \rightleftharpoons \text{EO} \qquad \text{E} + 2\text{O} \rightleftharpoons \text{EO}_2 \qquad \text{E} + \text{N} \rightleftharpoons \text{EN}$$

$$2\text{E} + \text{O}_2 \rightleftharpoons 2\text{EO} \qquad \text{E} + \text{O}_2 \rightleftharpoons \text{EO}_2 \qquad 2\text{E} + \text{N}_2 \rightleftharpoons 2\text{EN}$$

$$\text{O} + \text{O} \rightleftharpoons \text{O}_2 \qquad\qquad\qquad\qquad\quad \text{N} + \text{N} \rightleftharpoons \text{N}_2$$

Then, according to Eq. (3-1), we can write for the elements E, N, and O

$$\bar{C}_{\text{E}} = C_{\text{E}} + r_{\text{E,EO}}C_{\text{EO}} + r_{\text{E,EO}_2}C_{\text{EO}_2} + r_{\text{E,EN}}C_{\text{EN}} \tag{3-3}$$

$$\bar{C}_{\text{N}} = C_{\text{N}} + C_{\text{N}_2} + r_{\text{N,EN}}C_{\text{EN}} \tag{3-4}$$

$$\bar{C}_{\text{O}} = C_{\text{O}} + C_{\text{O}_2} + r_{\text{O,EO}}C_{\text{EO}} + r_{\text{O,EO}_2}C_{\text{EO}_2} \tag{3-5}$$

and, according to Eq. (3-2), $d\bar{C}_{\text{E}} = 0$; hence solving Eq. (3-3) for dC_{E}, we obtain

$$dC_{\text{E}} = -r_{\text{E,EO}}\,dC_{\text{EO}} - r_{\text{E,EO}_2}\,dC_{\text{EO}_2} - r_{\text{E,EN}}\,dC_{\text{EN}} \tag{3-6}$$

Similar equations can be obtained for dC_{N} and dC_{O}. Equation (3-6) and similar equations for dC_{N} and dC_{O} represent useful relations among differential changes of species mass fractions which occur during the chemical reactions. We shall find occasions to refer to Eqs. (3-1) through (3-6) in that which follows.

None of the chemical reactions which we shall be concerned with will occur without a conversion of internal energy into heat (exothermic reaction) or the conversion of external translational energy into internal energy (endothermic reaction) concurrently with the formation of new species. The aspect of physical chemistry which deals with the heat changes accompanying chemical reactions is thermochemistry, and it is from thermochemistry that we borrow several established concepts for use in the work which follows.

The change in enthalpy which occurs accompanying a chemical reaction proceeding at constant pressure is characterized by ΔH, the heat of reaction. For example, for the reaction

$$C(s) + O_2(g) \rightarrow CO_2(g) \qquad \Delta H = -94.03 \text{ kcal}$$

That is, when 12 g of solid carbon (1 g mole C) and 32 g of gaseous oxygen (1 g mole O_2) react to form 44 g of gaseous carbon dioxide (1 g mole CO_2) at constant pressure, there is a decrease in enthalpy of 94,030 cal. This decrease in enthalpy shows up as 94,030 cal of heat evolved per 44 g CO_2 produced. This is so because energy must be conserved during this process.

Now it is not always useful to us in that which follows to express heat release in terms of calories per gram molecular weight of product of the reaction. Sometimes we desire to express our heat-release terms in terms of the specific heat release with units of calories per gram of reactant or calories per gram of product if desired (or Btu per pound, if English units are preferred). This can be taken care of as follows: Let

$-\Delta H = -$heat of reaction $=$ heat released during the reaction

Then for any chemical reaction described by the equation

$$\sum_i^m a_i A_i \rightarrow \sum_i^n b_i B_i$$

we can write

$$-\Delta H = \sum_i^m a_i H_{A_i} - \sum_i^n b_i H_{B_i} \qquad (3\text{-}7)$$

where A_i, B_i = chemical symbols of reactants and products, respectively

a_i, b_i = number of moles of reactants and products, respectively, participating in chemical reaction

H_{A_i}, H_{B_i} = molar enthalpy of reactants and products, respectively

Now define the specific enthalpies h_{A_i} and h_{B_i}, where

$$M_{A_i} h_{A_i} = H_{A_i}$$
$$M_{B_i} h_{B_i} = H_{B_i}$$

and M_{A_i} and M_{B_i} are the molecular weights of species A_i and B_i. Then Eq. (3-7) can be manipulated to obtain an equation for a specific-heat-release term. This is

$$\Delta Q = \frac{-\Delta H}{M_{A_1}} = a_1 h_{A_1} + \frac{1}{M_{A_1}} \left(\sum_{i=2}^m M_{A_i} a_i h_{A_i} - \sum_{i=1}^n M_{B_i} b_i h_{B_i} \right) \qquad (3\text{-}8a)$$

where the units of ΔQ are calories per gram of reactant A_1 or Btu per pound of reactant A_1 depending upon the units being used.[1] Apply Eq. (3-8a) to the chemical reaction

$$O_2 \rightarrow O + O$$

for example. Since the energy of dissociation must be supplied to a molecule of O_2 (presumably by a violent collision with another particle) before it will dissociate, it is plain that the enthalpy per unit mass of O at any temperature is greater than the enthalpy per unit mass of O_2 at the same temperature by an amount equal to the heat of dissociation per unit mass at that temperature. Symbolically, by Eq. (3-8a),

$$\Delta Q_O = -(h_O - h_{O_2}) \qquad |\text{heat/unit mass } O_2|$$

but

$$h_O = \int_0^T C_{p_O} \, dT + h_O^\circ$$

where h_O° is the specific heat of formation of O and

$$h_{O_2} = \int_0^T C_{p_{O_2}} \, dT$$

Thus $\quad \Delta Q_O = -h_O^\circ + \int_0^T (C_{p_{O_2}} - C_{p_O}) \, dT \simeq -h_O^\circ \qquad |\text{heat/unit mass } O_2|$

since $C_{p_{O_2}} \simeq C_{p_O}$ to all practical purposes. Thus the heat release for the reaction $O_2 \rightarrow O + O$ is *minus* and is equal to the heat of dissociation of O. The reaction is endothermic.

Applying Eq. (3-8a) further, we find that for $O + O \rightarrow O_2$

$$\Delta Q_{O_2} = 2(h_O - h_{O_2}) \simeq 2h_O^\circ \qquad |\text{heat/unit mass } O| \qquad (3\text{-}8b)$$

for $E + O \rightarrow EO$

$$\Delta Q_{EO} = h_O + \frac{r_{E,EO}}{r_{O,EO}} h_E - \frac{1}{r_{O,EO}} h_{EO} \qquad |\text{heat/unit mass } O| \qquad (3\text{-}8c)$$

since

$$r_{E,EO} = \frac{M_E}{M_{EO}}$$

$$r_{O,EO} = \frac{M_O}{M_{EO}}$$

[1] In Sec. 5-7 we derive an analogous equation for specific-heat release per unit mass of product. See Eq. (5-70).

In a similar way relations for specific-heat release can be derived for all the reactions mentioned earlier. They are

$$\Delta Q_{N_2} = 2(h_N - h_{N_2}) \simeq 2h_N^\circ \qquad (3\text{-}8d)$$

$$\Delta Q_{EN} = h_N + \frac{r_{E,EN}}{r_{N,EN}} h_E - \frac{1}{r_{N,EN}} h_{EN} \qquad (3\text{-}8e)$$

and

$$\Delta Q_{EO_2} = h_O + \frac{r_{E,EO_2}}{r_{O,EO_2}} h_E - \frac{1}{r_{O,EO_2}} h_{EO_2} \qquad (3\text{-}8f)$$

where here in all cases the specific-heat releases ΔQ_{J_2}, ΔQ_{EJ}, and ΔQ_{EJ_2} are expressed in terms of heat released per unit mass of reactant J.[1]

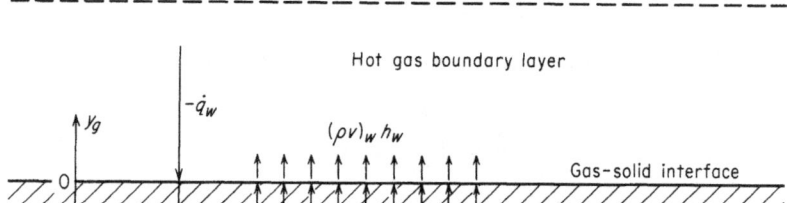

FIG. 3-1. Heat transfer from a reacting gas layer to a vaporizing surface.

We shall make use of Eqs. (3-8) in the next section, where we shall separate the various contributions to the heat transfer from a hot reacting gas boundary layer to a cool surface over which the gas layer flows.

3-3. Various Contributions to Reacting-boundary-layer Heat Transfer. Let us make use of the concepts described in Sec. 3-2 to examine the heat transfer from a reacting gas boundary layer to a vaporizing surface. Consider the situation depicted in Fig. 3-1. Neglecting radiation from the hot surface to the gas and from the hot gas to the surface (which can be accommodated when desired), we can write a heat-balance equation applicable at the interface between the gas layer and the solid. This is, if h is the enthalpy of the gas mixture,

$$\dot{q}_s = -\dot{q}_w - (\rho v)_w h_w + (\rho v)_w [h_E(s)]_w \qquad (3\text{-}9)$$

[1] See Eq. (5-70) for an equation for specific-heat release per unit mass of product.

but by definition, for $P = 1$, where P is the Prandtl number,

$$-\dot{q}_w = \rho_e u_e C_H \left(h_e - h_w + \frac{u_e^2}{2} \right) \tag{3-10}$$

and for the reaction $E(s) \rightleftharpoons E(g)$,

$$h_E(s) + L_v = h_E(g) \tag{3-11}$$

where L_v is the heat of vaporization of the surface material E and Eq. (3-10) constitutes a definition of the heat-transfer coefficient C_H. Now the enthalpy of a gas mixture is

$$h = \sum_i C_i h_i \tag{3-12}$$

and if we define a mass-transfer parameter B_3 as

$$B_3 = \frac{(\rho v)_w}{\rho_e u_e C_H} \tag{3-13}$$

then Eqs. (3-9) through (3-13) can be combined to obtain an equation for the heat transfer to the hot surface. Combine Eqs. (3-9) and (3-10) to get

$$\dot{q}_s = \rho_e u_e C_H \left(h_e - h_w + \frac{u_e^2}{2} \right) - (\rho v)_w h_w + (\rho v)_w [h_E(s)]_w$$

or, in view of Eqs. (3-11) and (3-13),

$$\dot{q}_s = \rho_e u_e C_H \left\{ h_e - h_w + \frac{u_e^2}{2} - B_3 h_w + B_3 [(h_E)_w(g) - L_v] \right\} \tag{3-14}$$

Equation (3-12) can now be used to write

$$h_e - h_w = \sum_i [(C_i)_e(h_i)_e - (C_i)_w(h_i)_w]$$

or $\quad h_e - h_w = \sum_i (C_i)_e[(h_i)_e - (h_i)_w] + \sum_i (h_i)_w[(C_i)_e - (C_i)_w] \tag{3-15}$

Thus, if $(C_E)_e = 0$, that is, the gas species E is confined to the boundary layer, then Eqs. (3-14) and (3-15) can be combined to obtain

$$\dot{q}_s = \rho_e u_e C_H \left\{ \sum_i (C_i)_e[(h_i)_e - (h_i)_w] + \frac{u_e^2}{2} \right.$$

$$+ \sum_{i \neq E} (h_i)_w[(C_i)_e - (1 + B_3)(C_i)_w]$$

$$\left. + (h_E)_w[B_3 - (1 + B_3)(C_E)_w] - B_3 L_v \right\} \tag{3-16}$$

Equation (3-16) can be further simplified by examining the boundary conditions. For example, at the surface, for all species except

E, the diffusion current for the element mass fractions must exactly equal the convection current created by the vaporization of E at the surface. That is, at the surface

$$\rho D_{12} \left(\frac{\partial \bar{C}_i}{\partial y} \right)_w = (\rho v)_w (\bar{C}_i)_w \qquad i \neq E \qquad (3\text{-}17)$$

or, assuming that $S = \mu / \rho D_{12} = 1$ for illustrative purposes, it can be shown that $\bar{C}_i = \bar{C}_i(u)$ and

$$\frac{\partial \bar{C}_i}{\partial y} = \frac{\partial \bar{C}_i}{\partial u} \frac{\partial u}{\partial y}$$

Then
$$\left(\frac{\partial \bar{C}_i}{\partial u} \right)_w = B_3 \frac{(\bar{C}_i)_w}{u_e} \qquad i \neq E \qquad (3\text{-}18a)$$

where, for a Prandtl number equal to one only,

$$\tau_w = \left(\mu \frac{\partial u}{\partial y} \right) = \rho_e u_e^2 C_{II} \qquad (3\text{-}18b)$$

has been used in deriving (3-18a) from (3-17). Similarly for the surface material E we write the mass flux balance at the interface as

$$(\rho v)_w + \rho D_{12} \left(\frac{\partial \bar{C}_E}{\partial y} \right)_w = (\rho v)_w (\bar{C}_E)_w \qquad (3\text{-}19)$$

which, combined with Eq. (3-18b), gives

$$\left(\frac{\partial \bar{C}_E}{\partial u} \right)_w = B_3 \frac{(\bar{C}_E)_w}{u_e} - \frac{B_3}{u_e} \qquad (3\text{-}20)$$

Now the species-conservation equation for the element mass fractions was derived in Sec. 2-5 and is

$$\rho u \frac{\partial \bar{C}_i}{\partial s} + \rho v \frac{\partial \bar{C}_i}{\partial y} = \frac{\partial}{\partial y} \left(\rho D_{12} \frac{\partial \bar{C}_i}{\partial y} \right) \qquad (2\text{-}111)$$

Furthermore, the s conservation-of-momentum equation was given in Sec. 2-3 as, if $\partial p / \partial s = 0$,

$$\rho u \frac{\partial u}{\partial s} + \rho v \frac{\partial u}{\partial y} = \frac{\partial}{\partial y} \left(\mu \frac{\partial u}{\partial y} \right) \qquad (2\text{-}66)$$

hence, if $S = \mu / \rho D_{12} = 1$, it can be verified by direct substitution in Eq. (2-111) that

$$\bar{C}_i = (\bar{C}_i)_w + [(\bar{C}_i)_e - (\bar{C}_i)_w] \frac{u}{u_e}$$

or
$$\left(\frac{\partial \bar{C}_i}{\partial u} \right)_w = \frac{(\bar{C}_i)_e - (\bar{C}_i)_w}{u_e} \qquad (3\text{-}21)$$

Hence, combining Eqs. (3-18a) and (3-21), we obtain

$$(\bar{C}_i)_e = (1 + B_3)(\bar{C}_i)_w \qquad i \neq E \tag{3-22}$$

and combining Eq. (3-20) with (3-21), we obtain

$$(\bar{C}_E)_e = (1 + B_3)(\bar{C}_E)_w - B_3 \tag{3-23}$$

Furthermore, when the element mass fractions in Eqs. (3-22) and (3-23) are expressed in terms of the species mass fractions by using Eqs. (3-3) through (3-5) developed previously, the resulting equations can be used to express the summation terms in Eq. (3-16) in terms of the heat-release terms of Eqs. (3-8). That is, Eqs. (3-22), (3-23), and (3-3) to (3-5) are used to obtain expressions for terms like

$$(h_i)_w[(C_i)_e - (1 + B_3)(C_i)_w]$$

Then it will be found that the summations in Eq. (3-16) become summations involving the ΔQ's of Eq. (3-8a) and the species mass fractions. The algebra involved is identical with that involved in deriving Eq. (5-86) and need not be presented here. The result is that Eq. (3-16) becomes

$$\dot{q}_s = \rho_e u_e C_H \left\{ \overbrace{\sum_i (C_i)_e[(h_i)_e - (h_i)_w]}^{①} + \overbrace{\frac{u_e^2}{2}}^{②} - \overbrace{B_3 L_v}^{③} \right.$$

$$\overbrace{+ \sum_{J=O,N} \Delta Q_{EJ}[(\bar{C}_J)_e - (1 + B_3)(\bar{C}_J - r_{J,EJ}C_{EJ})_w]}^{④}$$

$$\overbrace{+ \sum_{J=O} \Delta Q_{EJ_2}[(\bar{C}_J)_e - (1 + B_3)(\bar{C}_J - r_{J,EJ_2}C_{EJ_2})_w]}^{⑤}$$

$$\left. \overbrace{- \sum_{J=O,N} \Delta Q_{J_2}[(C_{J_2})_e - (1 + B_3)(C_{J_2})_w]}^{⑥} \right\} \tag{3-24}$$

where all specific heat-release terms are defined as heat per unit mass of *reactant* as given by equations (3-8d) to (3-8f) and where $(C_{EJ})_e = (C_{EJ_2})_e = 0$ for $J = O$, N; that is, all products of chemical reactions remain within the boundary layer. Equation (3-24) is a generalization of an equation first presented by Cohen et al.[1] and Bromberg and Lipkis[2] for a particular surface element E

[1] C. B. Cohen, R. Bromberg, and R. P. Lipkis, *Jet Propulsion*, vol. 28, pp. 659–668, 1958.

[2] R. Bromberg and R. P. Lipkis, *Jet Propulsion*, vol. 28, pp. 668–675, 1958.

(carbon) and extended by Lees.[1]   It is a most useful and remarkable equation which can be used to isolate and identify different contributions to the heat transfer from a reacting boundary layer to a cool surface.   While the equation was derived under the simplifying assumption that Schmidt number and Prandtl number (and hence Lewis number) are unity, its utility in pointing out the relative importance of various sources of energy transported to the surface is unimpaired.   For example,

The first term is the usual heat-convection term.

The second term is the term arising due to viscous dissipation within the boundary layer.

The third term is the heat dissipated in vaporizing the surface material.

The fourth and fifth terms are heat-release terms corresponding to the formation of compounds EJ and EJ_2, where $J = O$ or N in the former and $J = O$ in the latter.

The sixth term is the contribution due to dissociation of J_2, where $J = O$ or N.   Since ΔQ_{J_2} is the heat of dissociation, this represents a heat-sink term.

It can be seen from Eq. (3-24) that suppressing the formation of compounds EJ and EJ_2 (assuming that they are formed by exothermic reactions) will reduce the heat transfer to the solid undersurface as will increasing the heat of vaporization L_v or the amount of dissociation (an endothermic reaction).   It should be stated here that the effect of the reactions upon C_H through their effect on transport coefficients and velocity, composition, and density profiles through the boundary layer must be considered before all-inclusive conclusions can be drawn, although these effects will be shown to be secondary in Chap. 4.

For a typical example, assuming that the surface material is carbon and the only reaction is $C + O \rightarrow CO$, Lees[2] shows that the heat absorbed because of the vaporization of carbon exceeds the heat released because of the subsequent burning of C and formation of CO and that this fact, combined with the effect of reducing C_H resulting from mass transfer at the surface, can result in a reduction of overall heat transfer to the solid carbon surface.   It is apparent that Eq. (3-24) presents a useful means of calculating the relative importance of various reactions in the heat transfer from a reacting gas mixture to a cooler surface.

[1] Lester Lees, "Combustion and Propulsion, Third AGARD Colloquium," pp. 451–498, Pergamon Press, New York, 1959.
[2] *Ibid.*, p. 463.

3-4. The Melting Surface Layer. It has developed that a dependable lightweight heat-protection system for disposable hypersonic objects is made up of a material which melts and vaporizes.[1] The effective heat capacity of such a material includes the heat capacity of the material in the solid and liquid state plus the heats of fusion and/or vaporization of the material. An indirect benefit of a vaporizing liquid would be the effect of the vaporized material

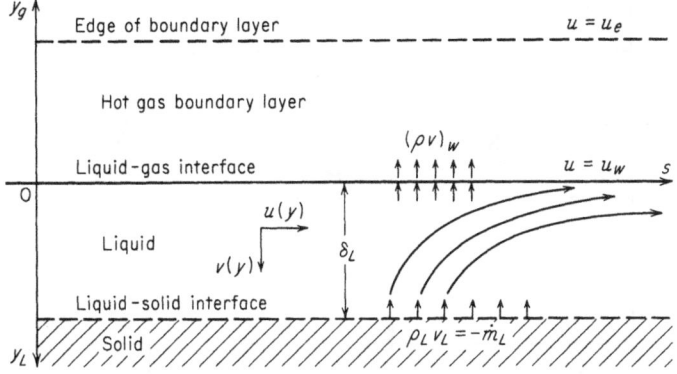

FIG. 3-2. Schematic of the interaction of a liquid layer with a reacting gas layer.

passing laterally into the boundary layer and thereby directly (favorably) influencing the velocity, diffusion, and temperature gradients at the surface. This latter aspect will be dealt with in detail in Sec. 3-5.

In order to treat the complete problem of the interaction of melting surface material with the reacting boundary layer, a description of the behavior of a melting layer is sought.

Figures 3-2 and 3-3 were prepared in anticipation of the solution sought in order to visualize the interaction between the liquid and gas layer and to illustrate the coordinate systems used.

We shall treat the case of a melting layer on a flat plate or circular cylinder exposed to a turbulent gas layer. The case of the three-dimensional stagnation point for a laminar boundary layer has been treated by Sutton,[2] Bethe and Adams,[3] and Lees.[4]

[1] L. Steg, *Am. Rocket Soc. Reprint* 836-59, 1959; Mac C. Adams, *ARS J.*, vol. 29, pp. 625–632, 1959; Lester Lees, in a paper presented to the Seventh Anglo-American Conference jointly sponsored by Royal Aeronautical Society and the Institute of the Aeronautical Sciences, October, 1959.

[2] G. Sutton, *J. Aerospace Sci.*, vol. 25, pp. 29–32, 1958.

[3] Hans A. Bethe and Mac C. Adams, *J. Aerospace Sci.*, vol. 26, pp. 321–328, 1959.

[4] Lester Lees, *ARS J.*, vol. 29, pp. 345–354, 1959.

For our purposes we shall present an approximate solution of the problem, which, however, retains the essential features of the exact solution. Our purpose here is to estimate the approximate magnitude of the heat absorbed by a melting layer and to determine the approximate thickness and flow velocity of the layer.

We shall assume that the physical properties of the liquid layer are constant. That is, the density ρ_L, the thermal conductivity k_L, the

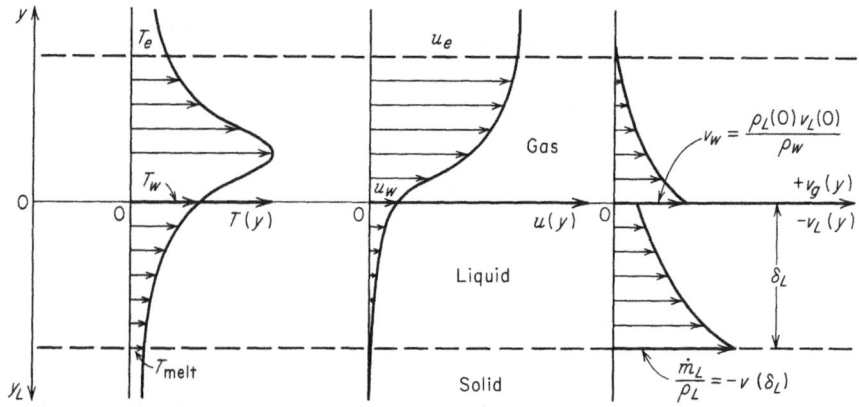

FIG. 3-3. Representative distributions of temperature, velocity, and transverse velocity for the situation depicted in Fig. 3-2.

heat capacity C_{P_L}, and the viscosity μ_L will all be assumed constant throughout the liquid layer to the order of approximation of this analysis. The equations we shall use in an analysis are the equation for conservation of mass

$$\frac{\partial u}{\partial s} + \frac{\partial v}{\partial y} = 0 \tag{3-25}$$

and, in lieu of the momentum equation, the approximate expression for the shear stress at the surface of the viscous liquid layer, which will be treated as a couette-flow-like layer,

$$\tau(0) = -\mu_L \frac{u_w}{\delta_L} \tag{3-26}$$

where u_w and δ_L are the interface velocity and the thickness, respectively, of the liquid layer. Figure 3-3 shows the coordinate system and symbols used. If we assume that the liquid layer is similar to couette flow, then the velocity distribution through the layer is

$$\frac{u}{u_w} = 1 - \frac{y}{\delta_L} \tag{3-27}$$

Furthermore, to the order of the approximation of this analysis, the heat-transfer rate to the surface of the liquid layer can be written

$$\dot{q}(0) = k_L \frac{T_w - T_m}{\delta_L} \tag{3-28}$$

Now the heat transferred to the solid undermaterial will be less than that transferred from the gas to the liquid by an amount equal to that heat absorbed by the melted layer. If the melted layer did not flow back along the surface, for our assumed constant physical properties the heat absorbed would be absorbed at the rate

$$\frac{1}{2} C_{p_L} \dot{m}_L (T_w - T_m)$$

where T_w is the melted surface temperature and T_m is the melting temperature at the melted-material–solid-material interface (see Fig. 3-3). However, the melted material does flow downstream and so absorbs more heat than it would if it remained in place. Rather than complicate the analysis at this point by precisely calculating the rate of absorption by the melted layer, we shall assume it to be

$$0.65 C_{p_L} \dot{m}_L (T_w - T_m)$$

and hence the rate of heat transfer to the solid material is, approximately,

$$\dot{q}_s = \dot{q}(0) - 0.65 C_{p_L} \dot{m}_L (T_w - T_m) \tag{3-29}$$

Now we need to relate $\tau(0)$ and $\dot{q}(0)$ to the gas-layer (boundary-layer) shear-stress and heat-transfer rates. Taking note of our coordinate system used in Fig. 3-3, we can write that

$$\tau_w + \tau(0) = 0$$

or

$$\tau(0) = -\tau_w \tag{3-30}$$

where τ_w is the shear stress exerted by the gas layer on the liquid-layer surface. Furthermore,

$$\dot{q}(0) = -\dot{q}_w \tag{3-31}$$

where $\dot{q}_w$ is the gas-layer heat-transfer rate. Additional relations found to be useful are the Reynolds analogy relation

$$C_H = \frac{C_f}{2} P^{-2/3} \tag{3-32}$$

where P is the gas-layer Prandtl number and C_H is defined by the equation

$$-\dot{q}_w = C_H \rho_e u_e (I_e - h_w) \tag{3-33}$$

and where ρ_e, u_e, and I_e, are the density, velocity, and total enthalpy of the gas layer at its outer edge and h_w is the gas-layer enthalpy evaluated at the melted surface temperature. C_f is defined by the equation

$$\tau_w = \frac{C_f}{2}\rho_e u_e^2 \tag{3-34}$$

Desired from the analysis are the temperature difference across the liquid layer $T_w - T_m$, the liquid-layer thickness δ_L, the liquid-layer surface velocity u_w, and the melting rate $\dot{m}_L$.

First, let us find the liquid-layer surface velocity. From Eq. (3-25) we can write

$$v(y) - v(0) = -\int_0^y \frac{\partial u}{\partial s}\, dy = -\frac{d}{ds}\left(\delta_L u_w \int_0^{y/\delta_L} \frac{u}{u_w}\frac{dy}{\delta_L}\right)$$

or, making use of Eq. (3-27),

$$v(y) - v(0) = -\frac{d}{ds}\left\{\delta_L u_w\left[\frac{y}{\delta_L} - \frac{1}{2}\left(\frac{y}{\delta_L}\right)^2\right]\right\}$$

Let $y = \delta_L$ to determine the melting velocity, and let $v(0) = 0$, since we assume no vaporization or gasification of the liquid layer. We obtain

$$v(\delta_L) = -\frac{\dot{m}_L}{\rho_L} = -\frac{d}{ds}\left(\frac{1}{2}\delta_L u_w\right) \tag{3-35}$$

Now, from Eqs. (3-26) and (3-30) we obtain

$$\delta_L = -\mu_L\frac{u_w}{\tau(0)} = \mu_L\frac{u_w}{\tau_w} \tag{3-36}$$

and from Eqs. (3-32) to (3-34) we obtain

$$\frac{1}{\tau_w} = -\frac{I_e - h_w}{u_e P^{2/3}\dot{q}_w} \tag{3-37}$$

Thus, combining Eqs. (3-36) and (3-37), we find that

$$\delta_L = -\mu_L\frac{u_w}{u_e}\frac{I_e - h_w}{P^{2/3}\dot{q}_w} \tag{3-38}$$

Equation (3-38) can now be inserted into Eq. (3-35) to obtain

$$\frac{d}{ds}\left[\left(\frac{u_w}{u_e}\right)^2\frac{1}{\dot{q}_w}\right] = \frac{-2\dot{m}_L P^{2/3}}{\rho_L \mu_L u_e(I_e - h_w)} \tag{3-39}$$

Now, let us define an effective heat capacity H of our melting material such that

$$H = \frac{\dot{q}(0)}{\dot{m}_L} = \text{const} \tag{3-40}$$

H, as defined by Eq. (3-40), is not a property of the melting material and is dependent upon the gas-layer characteristics. Since we are concerned in this example with the case of a turbulent gas boundary layer flowing over a melting flat plate or circular cylinder, we write

$$- \dot{q}_w = \dot{q}(0) = N s^{-1/5} \tag{3-41}$$

where N is a function of gas-layer properties which we assume are independent of s for our flat-plate solution here. Furthermore, from Eqs. (3-40) and (3-41)

$$\dot{m}_L = \frac{N s^{-1/5}}{H} \tag{3-42}$$

Inserting Eq. (3-42) into Eq. (3-39) we obtain the differential equation

$$d\left[\left(\frac{u_w}{u_e}\right)^2 \frac{1}{\dot{q}_w}\right] = - \frac{2N P^{2/3} s^{-1/5} \, ds}{\rho_L H \mu_L u_e (I_e - h_w)}$$

which can be integrated to obtain, where $\dot{q}_w$ at $s = 0$ is infinite,

$$\left(\frac{u_w}{u_e}\right)^2 = \frac{5}{2} \frac{P^{2/3}[\dot{q}(0)]^2 s}{\rho_L \mu_L (I_e - h_w) u_e H} \tag{3-43}$$

where Eqs. (3-41) and (3-42) have been used to eliminate N and $-\dot{q}_w$. Equation (3-43) gives the liquid-layer surface velocity u_w as a function of heat-transfer rate to the liquid layer $\dot{q}(0)$. We shall use this equation to obtain our other desired quantities.

We can now derive an equation for the effective heat capacity H of the melting material. By a heat balance at the liquid-layer solid interface $y = \delta_L$, we obtain

$$\dot{q}_s = \dot{q}(\delta_L) = \dot{m}_L [L_m + C_{p_s}(T_m - T_0)] \tag{3-44}$$

where L_m = heat of fusion
C_{p_s} = specific heat of solid
T_0 = initial temperature of solid

Equating Eq. (3-44) to Eq. (3-29), solving for $\dot{q}(0)$, substituting for $\dot{q}(0)$ into Eq. (3-40), and solving for H, we obtain the desired equation for the effective heat capacity of the melting material.

$$H = L_m + C_{p_s}(T_m - T_0) + 0.65 C_{p_L}(T_w - T_m) \tag{3-45}$$

H can be determined once the temperature rise across the melt is obtained. We shall develop an equation for the temperature rise next. From Eq. (3-28)

$$\frac{T_w - T_m}{T_m} = \frac{\dot{q}(0) \, \delta_L}{k_L T_m}$$

or, in view of Eqs. (3-26), (3-30), and (3-31),

$$\frac{T_w - T_m}{T_m} = -\frac{\dot{q}_w \, \mu_L u_w}{\tau_w \, T_m k_L}$$

Making use of Eq. (3-37),

$$\frac{T_w - T_m}{T_m} = \frac{(I_e - h_w)\mu_L}{P^{2/3} T_m k_L} \frac{u_w}{u_e}$$

Equation (3-43) can be used to eliminate u_w/u_e to obtain

$$\frac{T_w - T_m}{T_m} = \left(\frac{5}{2}\right)^{1/2} \left\{ \frac{(I_e - h_w)\mu_L s [\dot{q}(0)]^2}{P^{2/3} \rho_L T_m^2 k_L^2 u_e H} \right\}^{1/2}$$

or, introducing the relation between $\dot{q}(0)$ and C_f obtained from Eqs. (3-31) to (3-33), i.e.,

$$\frac{C_f}{2} \rho_e u_e^2 = \frac{\dot{q}(0) u_e P^{2/3}}{I_e - h_w}$$

we can obtain the desired equation for the temperature rise across the liquid layer.

$$\frac{T_w - T_m}{T_m} = \left(\frac{5}{2}\right)^{1/2} \frac{C_f}{2} (R_e)^{1/2} \left[\frac{C_{p_L}}{k_L} \left(\frac{\rho_e}{\rho_L}\right)^{1/2} \frac{(\mu_L \mu_e)^{1/2}}{P} \left(\frac{I_e - h_w}{H}\right)^{1/2} \frac{I_e - h_w}{C_{p_L} T_m} \right] \tag{3-46}$$

where ρ_e and μ_e are the density and viscosity, respectively, of the gas layer evaluated at its outer edge and where

$$R_e = \frac{\rho_e u_e s}{\mu_e}$$

is the gas-layer Reynolds number.

Equation (3-46) directly relates temperature rise to liquid-layer and gas-layer physical properties and gas-layer heat-transfer rate. It is apparent that temperature rise is proportional to liquid-layer viscosity raised to the half power, for example.

Simple expressions for R_L, the Reynolds number of the liquid layer, and δ_L can be derived using several of the above equations in a straightforward manner. These expressions are

$$R_L = \frac{\rho_L u_w \delta_L}{(\mu_L)_w} = \frac{5}{2} \frac{\dot{q}(0)s}{(\mu_L)_w H} \tag{3-47}$$

and

$$\delta_L = \frac{5}{2} \frac{\dot{q}(0)s}{H \rho_L u_w} \tag{3-48}$$

Equations similar to Eqs. (3-29), (3-43), and (3-46) have been derived by Lees[1] for the laminar boundary layer over axisymmetric bodies. The present example, while analogous to Lees' treatment, is appropriate to turbulent gas layers flowing over a melting flat plate or circular cylinder.

In order to calculate $\dot{q}(\delta_L)$, the rate of heat transfer into the solid,

Table 3-1. Properties of Corning No. 7740 Glass

$T_m = 3000°\text{R}$	$\rho_L = 131 \text{ lb/ft}^3$
$C_{p_L} = 1/3 \text{ Btu/lb} - °\text{F}$	$k_L = 1.71 \times 10^{-3} \text{ Btu/ft} - °\text{R-sec}$
$L_m + C_{p_s}(T_m - T_0) \simeq 750 \text{ Btu/lb}$	$u_L = 0.0672 \exp\left(\dfrac{8720}{T_w}\right)^{1.612}, T_w \text{ in } °\text{R}$

Table 3-2. Typical Turbulent Gas-layer Conditions during Ablation
$(u_e = 20{,}000 \text{ ft/sec, alt} = 60{,}000 \text{ ft, flat plate})$

$R_e = \frac{2}{5} \times 10^7$	$\mu_e = 9.52 \times 10^{-6} \text{ lb/ft-sec}$	$P = 0.72$
$\dfrac{C_f}{2} = 2.4 \times 10^{-4}$	$I_e - h_w = 6500 \text{ Btu/lb}$	$\rho_e = 0.00722 \text{ lb/ft}^3$
$-\dot{q}_w = \dot{q}(0) = 280 \text{ Btu/ft}^2\text{-sec}$	$u_e = 20{,}000 \text{ ft/sec}$	$T_e = 392°\text{R}$

Table 3-3. Results Obtained Using Data Given in Tables 3-1 and 3-2 and Eqs. (3-40), (3-43), and (3-45) to (3-48)

$H = 876 \text{ Btu/lb}$	$R_L = 0.159$
$T_w = 3550°\text{R}$	$\dfrac{u_w}{u_e} = 4.61 \times 10^{-5}$
$\dot{m}_L = 0.319 \text{ lb/ft}^2\text{-sec}$	$u_w = 0.922 \text{ ft/sec}$
$\delta_L = 6.61 \times 10^{-3} \text{ ft}$	

one uses Eqs. (3-45) and (3-46) to determine H and the temperature rise $T_w - T_m$ across the layer. Once H is thus determined, the melting rate is found using Eq. (3-40). The heat-transfer rate into the solid material follows from Eq. (3-29).

It is appropriate to illustrate the use of these equations through an example. Let us use the properties of Corning No. 7740 glass,[2] and let us also choose representative gas-layer conditions (determined by flight velocity, altitude, and the geometry of the body) shown in Table 3-2.

If the values tabulated in Tables 3-1 and 3-2 are used with Eqs. (3-40), (3-43), and (3-45) to (3-48), the results given in Table 3-3 are obtained.

[1] Lees, *op. cit.*

[2] Sutton, *op. cit.*

Several conclusions can be drawn when the results listed in Table 3-3 are examined.

1. The liquid-layer Reynolds number is very small and assures that this layer will be laminar.

2. The velocity of the liquid at the interface is quite low and justifies the assumption that equations for solid-surface gas-layer heat transfer can be used.

3. The liquid layer is thin.

4. The heat capacity of the liquid-solid material is increased by the amount of heat absorbed by the liquid layer. This amount is quite modest and represents about 13.7 per cent of the total heat absorbed for the example shown.

5. For the input conditions shown in Table 3-2, the heat-transfer rate is relatively low and leads to a modest temperature rise of 550°R across the liquid layer. Higher heat-transfer rates would result in larger temperature rises across the liquid layer and larger liquid-layer heat capacities without invalidating conclusions 1, 2, and 3 above.

6. The gain in effective heat capacity contributed by the melted material (about 126 Btu/lb in the present case) is relatively modest, and the benefits from considering such materials are limited chiefly to their capacity to absorb high heat-transfer rates. They do not effectively reduce the weight of a thermal protection system, since their effective heat capacity is low compared with materials which melt *and* vaporize, as Secs. 3-5 and 3-6 will show.

According to Eq. (3-45), the effective heat capacity of the melted layer depends upon the temperature rise $T_w - T_m$ across the layer. Since a high effective heat capacity H is desirable and attainable through high values of gas-liquid interface temperature, Eq. (3-46) tells us that effective heat capacity will be high if the liquid viscosity is high and/or the liquid conductivity and density are low, all other factors being constant. Furthermore, liquid-layer temperature rise and, hence, effective heat capacity H depend directly upon gas-layer heat-transfer rate through the driving enthalpy potential $(I_e - h_w)$ of the gas layer.

It should be pointed out that an additional contribution to blocking the flow of heat from the hot gas layer to the solid material is realized if the liquid layer vaporizes at the gas-liquid interface or otherwise gasifies. This blocking effect will be discussed in the section following.

3-5. Mass-transfer Effects. In Sec. 3-3 we saw how endothermic and exothermic chemical reactions affect the gas-layer heat-transfer potential. Section 3-4 discussed the interaction of the hot gas layer with a melting undersurface.   It was suggested that, if the melt vaporizes, an additional heat-flow blocking effect might result. The purpose of this section is to examine, to a first approximation, this possible heat-flow blockage effect commonly referred to as the "mass-transfer effect."

There are, of course, many ways to create mass transfer at the gas-layer–liquid-layer or solid-wall interface.   Early work in this field dealt with the possibilities of introducing a gas or liquid into the hot gas layer through a porous wall.   Later approaches to the problem of a thermal protection system involved vaporization or gasification of a liquid layer or the use of sublimators which vaporize in the absence of liquefaction (such as graphite).   As far as the gas layer is concerned, these three approaches can be treated within the framework of a single mathematical description of the problem.   The phase changes involved in the vaporization of the liquid or solid undersurface both result in a mass transfer at the boundary of the gas layer.   Of course, some vaporized species (such as the noble gases) may not react with the gas layer, and others, such as carbon, will react.   The effect of these reactions has been discussed briefly in Sec. 3-3 and will be dealt with further in Chaps. 4 through 8.   We concern ourselves here with the effect of mass addition at the liquid or solid subsurface upon gas-layer skin friction and heat transfer in the absence of gas-phase reactions.

Intuitively, one can arrive at the results to be expected.   For example, introducing gas into the boundary layer will thicken the boundary layer.   This will decrease skin friction and heat transfer, other factors being constant.   Furthermore, even if one assumes that the boundary layer was not thickened, the transverse mass flux (blowing) must carry low-velocity boundary-layer gas into regions of higher velocity gas, much as does a turbulent fluctuation, and thereby reduce the velocity gradient near the surface and hence the shear stress upon the liquid or solid undersurface.   An analogous effect on heat transfer is easily envisioned.   These effects are indicated schematically in Fig. 3-4.

Theory and experiment support intuition.   As will be shown, quantitative expressions for the reduction in skin friction or heat transfer with blowing (or their increase with sucking) can be derived using boundary-layer theory.   Chapters 5 and 8 will present some of the theories in detail.   It is in keeping with the purpose of this chapter to treat the mass-transfer effects to a first approximation, leaving the detailed treatment to later chapters.

In order to develop a familiarity with the order of magnitude of the effect of mass transfer upon gas-layer-produced surface shear and heat transfer, consider the thin film approximation to the problem. Consider that we have a thin film which contains all the mass introduced into the gas layer. Assume that this thin layer has couette-flow properties, that is, $|\partial/\partial s| \ll |\partial/\partial y|$, and its thickness is unaffected by the mass introduced. Consider the transverse mass flux ρv to be

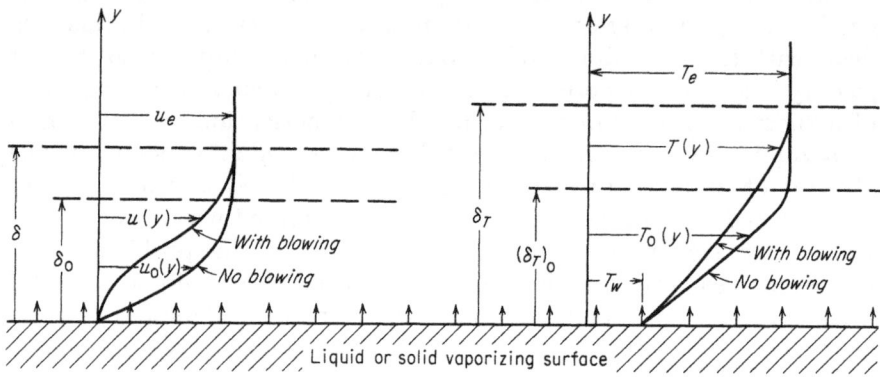

Note:
1. Boundary layer thickens with blowing: $\delta > \delta_0$; $\delta_T > (\delta_T)_0$
2. du/dy and dT/dy decrease with blowing.
3. Since $P < 1$, then $\delta_T > \delta$ as shown.

Fig. 3-4. The effects of mass transfer (blowing) upon boundary-layer characteristics.

constant through this layer and equal to its value at the surface. Then according to these assumptions,

$$\rho v = (\rho v)_w \tag{3-49}$$

and the s momentum equation becomes

$$(\rho v)_w \frac{du}{dy} = \frac{d}{dy}\left(\mu \frac{du}{dy}\right) \tag{3-50}$$

Introduce a coordinate transformation

$$Y = \int_0^y \frac{1}{\mu}\, dy \tag{3-51}$$

Then Eq. (3-50) becomes

$$(\rho v)_w \frac{du}{dY} = \frac{d^2u}{dY^2} \tag{3-52}$$

with boundary conditions $u = 0$ when $Y = 0$, $du/dY = \tau_w$ when $Y = 0$, and $u = u_e$ when $Y = Y_\delta$. The solution to Eq. (3-52) is

$$Y_\delta = \frac{1}{(\rho v)_w} \ln \frac{\tau_w + (\rho v)_w u_e}{\tau_w} \tag{3-53}$$

Now when $(\rho v)_w = 0$, the solution to Eq. (3-52) is

$$Y_{\delta_0} = \frac{u_e}{\tau_{w_0}} \qquad (3\text{-}54)$$

where the subscript 0 means the value at zero mass transfer. Consistent with our assumptions, if $Y_\delta = Y_{\delta_0}$, then equating Eqs. (3-53) and (3-54) produces

$$\psi_1 = \frac{C_f}{C_{f_0}} = \frac{B_1}{\exp B_1 - 1} \qquad (3\text{-}55)$$

where
$$B_1 = \frac{2(\rho v)_w}{\rho_e u_e C_{f_0}} \qquad (3\text{-}56)$$

In a completely analogous manner Lees[1] has shown that

$$\psi_2 = \frac{C_H}{C_{H_0}} = \frac{B_2}{\exp B_2 - 1} \qquad (3\text{-}57)$$

where
$$B_2 = \frac{(\rho v)_w}{\rho_e u_e C_{H_0}} \qquad (3\text{-}58)$$

In the special case where $C_f/2 = C_H$ ($P = 1$), then $B_1 = B_2$ and $\psi_1 = \psi_2 = \psi$. Both Eqs. (3-55) and (3-57) indicate that increasing blowing ($B_1 > 0$) will produce a marked decrease in C_H or C_f.

Some notion of the accuracy of Eq. (3-57) can be obtained by comparing it with experimental measurements and more exact theory. Figure 3-5 presents a comparison of Eq. (3-57) with measurements of heat transfer from a turbulent layer flowing over a flat plate, reported by Bartle and Leadon.[2] Also shown is the theoretical estimate by Low[3] for the laminar boundary layer over a flat plate. It seems reasonable to assume that the laminar boundary-layer theory with mass transfer present should closely approximate measurements, since this is true in the absence of mass transfer. It is clear that Eq. (3-57), which is independent of Mach number, overestimates the effect upon turbulent-boundary-layer heat transfer and underestimates the effect upon laminar-boundary-layer heat-transfer rates but gives the effect of mass transfer within an order of magnitude. Equation (3-57) also ignores the effects of gas-phase reactions involving vaporized material. The more exact solutions for mass-transfer effects upon heat transfer will be discussed in later chapters.

[1] Lester Lees, "Combustion and Propulsion, Third AGARD Colloquium," pp. 451–498, Pergamon Press, New York, 1959.

[2] E. Roy Bartle and Bernard M. Leadon, *J. Aerospace Sci.*, vol. 27, pp. 78–80, 1960.

[3] George M. Low, *NACA TN* 3404, March, 1955.

The results of this section and the preceding one can be used to derive an expression for the effective heat capacity of a melting and vaporizing surface material which includes the blocking effect due to mass transfer discussed in this section. From Eqs. (3-40) and (3-45)

$$\dot{q}(0) = \dot{m}_L[L_m + C_{p_s}(T_m - T_0) + 0.65\,C_{p_L}(T_w - T_m)] \qquad (3\text{-}59)$$

but if vaporization occurs so that vaporized material appears only at the gas-liquid or solid interface $[(C_E)_w = 1;\ (C_i)_w = 0,\ \text{for } i \neq E]$, then from Eqs. (3-10) and (3-16)

$$\dot{q}_s = \dot{q}(0) = -\dot{q}_w - L_v(\rho v)_w \qquad (3\text{-}60)$$

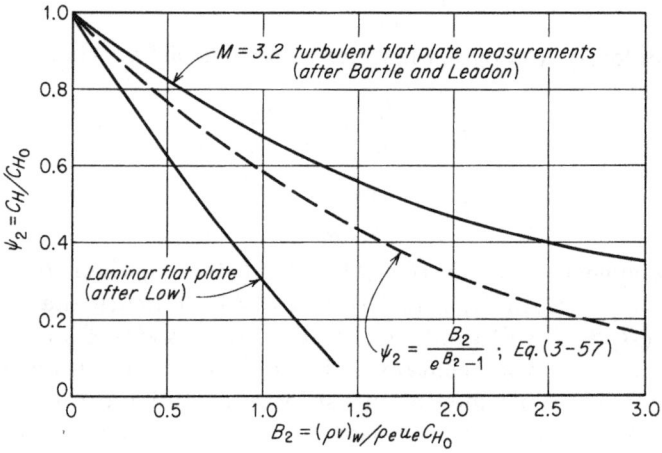

Fig. 3-5. Effect of mass transfer (blowing) upon heat-transfer coefficient.

where L_v is the heat of vaporization of the liquid surface material. Thus combining Eqs. (3-59) and (3-60)

$$-(\dot{q}_w)_0\psi_2 = \dot{m}_L[L_m + C_{p_s}(T_m - T_0) + 0.65(T_w - T_m) + L_v f] \quad (3\text{-}61)$$

where ψ_2 is defined by Eq. (3-57) and

$$f = \frac{(\rho v)_w}{\dot{m}_L}$$

Thus

$$H_{\text{eff}} = \frac{-(\dot{q}_w)_0}{\dot{m}_L} = \frac{H + fL_v}{\psi_2} \qquad (3\text{-}62)$$

where H_{eff} is the effective heat capacity of our melting *and* vaporizing material. Note that, if $f = 0$, $\psi_2 = 1$ and $H_{\text{eff}} = H$. For a vaporizing surface material $f > 0$ and $\psi_2 < 1$, and thus H_{eff} is increased significantly over H, both through the additional heat capacity of the heat of vaporization and through the mass-transfer effect which results in ψ_2 being less than one. The number 0.65 appearing in Eq. (3-61) depends upon the geometry of the melting layer. For example, for the melting layer at the stagnation point of an axisymmetric body, Lees[1] shows that 0.65 should be replaced by 0.60.

Equation (3-57) can be expanded to obtain

$$\psi_2 = 1 - \frac{B_2}{2} + \frac{1}{12} (B_2)^2 - \cdots \qquad B_2 < 2 \qquad (3\text{-}63)$$

where

$$B_2 = \frac{(\rho v)_w}{\rho_e u_e C_{H_0}} = \frac{(\rho v)_w}{(-\dot{q}_w)_0} (I_e - h_w)_0 \qquad (3\text{-}64)$$

since, by definition, if $P = 1$, $I_r = I_e$, and from Eq. (3-10)

$$C_{H_0} = \frac{-(\dot{q}_w)_0}{\rho_e u_e (I_e - h_w)_0} \qquad (3\text{-}10)$$

This suggests that, for small B_2, to the first order in B_2

$$\psi_2 \simeq 1 - \beta \frac{(\rho v)_w (I_e - h_w)_0}{(-\dot{q}_w)_0} \qquad (3\text{-}65)$$

where, according to experiment for turbulent boundary layers and exact calculations for laminar boundary layers, $\beta = \frac{1}{6}$ to $\frac{2}{3}$ and depends upon composition of the vaporized gas.[2] If Eq. (3-65) is used with Eqs. (3-61) and (3-62), there results

$$H_{\text{eff}} = H + f[L_v + \beta(I_e - h_w)_0] \qquad (3\text{-}66)$$

where

$$H = L_m + C_{p_s}(T_m - T_0) + 0.65(T_w - T_m) \qquad (3\text{-}45)$$

as before. Equation (3-66) reveals that H_{eff} depends directly upon enthalpy difference across the gas layer in the absence of ablation, the heat of vaporization, and the fraction of the melted material which vaporizes as well as upon the solid undersurface and melting layer heat capacities. The parameter β also influences the value of H_{eff} and is dependent upon composition of the vaporizing surface material and the state of the gas layer (laminar or turbulent). In general, for high flight velocities ($u_e \geq 15{,}000$ ft/sec) the second term in Eq. (3-66) is the dominant term, thus bearing out the supposition that vaporization enhances effective heat capacity significantly.

The fraction f of melted material which vaporizes depends upon liquid-layer–melted-layer interface temperature T_w and the vapor pressure of the vaporizing material. Its determination, using theory,

[1] Lester Lees, *ARS J.*, vol. 29, pp. 345–354, 1959.
[2] See, for example, Mac C. Adams, *ARS J.*, vol. 29, pp. 625–632, 1959.

could be a tedious process. Fortunately, in practice a composite material is used, such as a plastic suspended in a fiber glass matrix, which vaporizes in a definite proportion. That is, the plastic vaporizes completely and the glass melts but does not vaporize. The value of f is, then, simply the fraction by mass of the plastic in the original composite material. The interface temperature T_w can be calculated using the methods of Sec. 3-4 once physical properties of the plastic and glass are known. (Some suitable averaging process is obviously required for practicality.)

Adams[1] has presented calculations and measurements of H_{eff} for

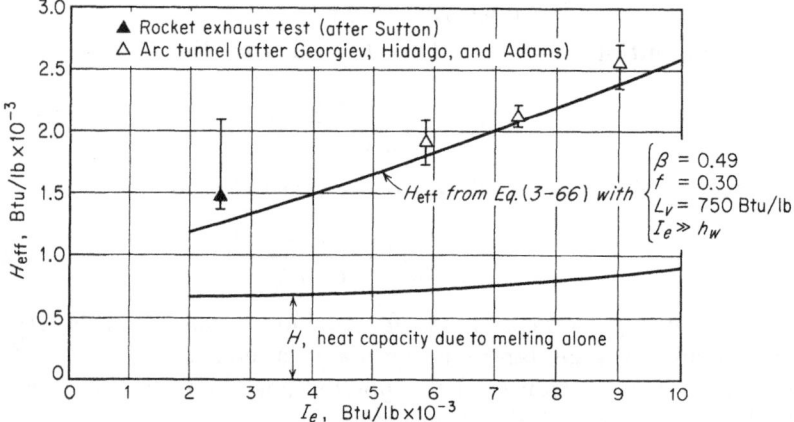

FIG. 3-6. Stagnation-point heat of ablation for a composite material of 30 per cent phenolic plastic and 70 per cent fiber glass. (*After Mac C. Adams, ARS J., vol. 29, pp. 625–632, 1959.*)

such a composite material for several values of $(I_e - h_w)_0$ appropriate to stagnation-point flow. A comparison with his measurements and the theory [Eq. (3-66) with $\beta = 0.49$, $f = 0.30$, $L_v = 750$ Btu/lb, and 0.65 replaced by 0.60] is shown in Fig. 3-6. It is apparent that effective heat capacities above those of the purely melting layer are realized when vaporization occurs.

3-6. Heat Conduction in the Solid State. There remains a reservoir for heat storage which has not yet been treated, i.e., the inherent capability of the unmelted material to absorb and store heat in the solid state. It is entirely possible that the heat-transfer rate to the solid material remains appreciable even with the intervening phase changes and their effects which occur between the solid interface and the source of heat generation, the hot gas boundary layer. In fact, Eq. (3-29) gives the rate of heat transfer to the solid material

[1] *Ibid.*

from the liquid layer. Of course, if the heat-transfer rates are comparatively low, no material phase changes occur, and the heat transfer to the solid is equal to the heat-transfer rate of the hot gas layer.

The theory for heat conduction within a solid material is expounded in several treatises on the subject. The reader is referred to such books as McAdams,[1] Jakob,[2] and Carslaw and Jaeger[3] among others for an exhaustive treatment of this subject. The purpose of this section is to illustrate in a relatively simple way how the unmelted or unvaporized material of a hypervelocity object responds to surface heat-transfer rates generated by the hot gas boundary layer.

The history of heat-transfer rate to which a hypervelocity object is exposed depends directly upon its trajectory within an atmosphere. Given equal initial and final velocities, to a first approximation, according to Eq. (1-10), objects of a similar size, shape, and weight will experience equivalent total amounts of heat transfer to their surfaces, differing, however, in their heating cycle. Objects plunging steeply into an atmosphere will experience a heat-transfer-rate pulse of short duration and high amplitude. Objects spending their kinetic energy in a relatively leisurely fashion by penetrating the atmosphere at a glancing angle will experience a low-amplitude long-duration heat pulse. While the amount of heat transferred to each of these objects may be of the same order of magnitude, the response of the skin material of the objects will differ greatly, since any material can transfer heat to its interior at a finite rate limited by its physical properties. If heat is transferred to the surface of the object at a rate exceeding the rate with which the material can transfer it further into the cooler portions of the skin, the surface material will begin to undergo phase changes such as we have discussed in Secs. 3-4 and 3-5. Modern technology has shown that, far from being catastrophic, these phase changes and their effects can work to the advantage of a designer of disposable hypervelocity objects by minimizing the weight required for thermal protection of the interior payload, as we have seen.

The differential equation governing the flow of heat into the interior of a solid is

$$\frac{\partial}{\partial x}\left(k\,\frac{\partial T}{\partial x}\right) + \frac{\partial}{\partial y}\left(k\,\frac{\partial T}{\partial y}\right) + \frac{\partial}{\partial z}\left(k\,\frac{\partial T}{\partial z}\right) = \rho\,\frac{\partial(C_p T)}{\partial t} \qquad (3\text{-}67)$$

[1] W. H. McAdams, "Heat Transmission," 3d ed., McGraw-Hill Book Company, Inc., New York, 1954.

[2] M. Jakob, "Heat Transfer," John Wiley & Sons, Inc., New York, 1949.

[3] H. S. Carslaw and J. C. Jaeger, "Conduction of Heat in Solids," 2d ed., Oxford University Press, New York, 1948.

in the absence of any internal heat sources or sinks. k, ρ, and C_p are the thermal conductivity, density, and specific heat, respectively, of the solid material. Boundary conditions appropriate to hypervelocity objects would be stated as

$$T(x,y,z,t) = T_0 \qquad\qquad t = 0$$

$$-k\left(\frac{\partial T}{\partial n}\right)_w = \dot{q}_w(x_w,y_w,z_w,t) \qquad t \geq 0$$

where n is in a direction normal to the surface and $T(x,y,z,t)$ is the temperature sought. The references cited previously in this section

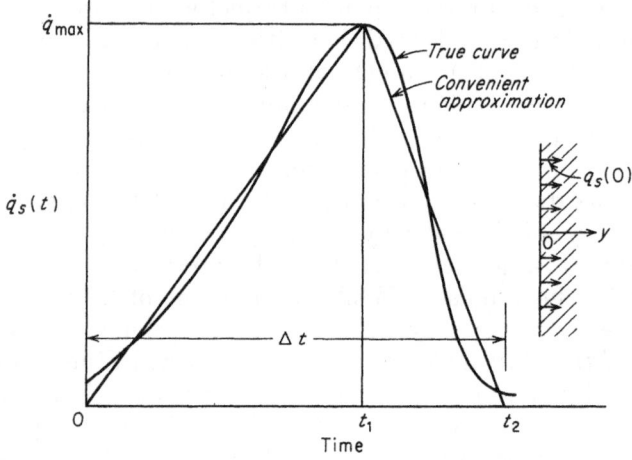

FIG. 3-7. A representative heat pulse for a hypervelocity atmospheric entry object.

present numerous example solutions for this boundary-value problem and techniques for obtaining solutions.

Let us consider the problem presented by an object penetrating an atmosphere at a sharp angle, such as might a long-range ballistic missile. It is known that the heat pulse takes the form shown in Fig. 3-7.[1] We shall find it convenient in that which follows to approximate the true curve with the triangle shown. Let us consider the response of a semi-infinite slab exposed to such a heat-transfer-rate cycle. Equation (3-67) simplifies in this case to

$$k\frac{\partial^2 T}{\partial y^2} = \rho C_p \frac{\partial T}{\partial t} \tag{3-68}$$

[1] See, for example, Leo Steg, *Am. Rocket Soc. Reprint* 836-59, 1959.

if k and C_p are assumed to be constant. Boundary conditions are

$$T(x) = T_0 \qquad \text{at } t = 0$$

$$-k\left(\frac{\partial T}{\partial y}\right)_{0,t} = \dot{q}_s(t) \qquad \text{at } t > 0 \tag{3-69}$$

and for our triangular heat pulse,

$$\dot{q}_s(t) = \dot{q}_{\max} \frac{t}{t_1} \qquad\qquad 0 \le t \le t_1 \tag{3-70a}$$

$$\dot{q}_s(t) = \dot{q}_{\max} \frac{t_2 - t}{t_2 - t_1} \qquad t_1 \le t \le t_2 \tag{3-70b}$$

Now the solution to Eq. (3-68) is known to be,[1] for boundary conditions given in Eq. (3-69),

$$T(y,t) = T_0 + \frac{1}{k}\left(\frac{\alpha}{\pi}\right)^{1/2} \int_0^t \theta^{-1/2} \dot{q}_s(t - \theta) \exp\left(\frac{-y^2}{4\alpha\theta}\right) d\theta \tag{3-71}$$

where $\alpha = k/C_p\rho$ = thermal diffusivity.

Furthermore, since it is the surface which melts first, let us examine the surface temperature $T(0,t)$ for various materials using Eqs. (3-71), (3-70a), and (3-70b); i.e.,

$$T(0,t) = T_0 + \frac{1}{k}\left(\frac{\alpha}{\pi}\right)^{1/2} \int_0^t \theta^{-1/2} \dot{q}_s(t - \theta) \, d\theta \tag{3-72}$$

where $\dot{q}_s(t - \theta)$ for our example is given by

$$\dot{q}_s(t - \theta) = \dot{q}_{\max} \frac{t - \theta}{t_1} \qquad\qquad 0 \le t \le t_1 \tag{3-73}$$

and $\quad \dot{q}_s(t - \theta) = \dot{q}_{\max} \frac{t_2 - t + \theta}{t_2 - t_1} \qquad t_1 \le t \le t_2 \tag{3-74}$

Since the maximum amount of heat has been transferred to the surface at $t = t_2$, let us evaluate the integral to obtain $T(0,t_2)$. We obtain

$$\frac{\pi^{1/2}[T(0,t_2) - T_0]k}{\dot{q}_{\max}(\alpha)^{1/2}} = \frac{4}{3}(t_1)^{1/2} + \frac{2}{3}\frac{t_2^{3/2} - t_1^{3/2}}{t_2 - t_1} \tag{3-57}$$

Now, for simplicity's sake and to illustrate our point, assume that

$$t_2 = \Delta t$$

$$t_1 = \frac{\Delta t}{2}$$

---

[1] Carslaw and Jaeger, *op. cit.*

Then Eq. (3-75) becomes

$$\dot{q}_{max} = 0.97 \left(\frac{C_p \rho k}{\Delta t}\right)^{1/2} (T_m - T_0) \tag{3-76}$$

where the value of $\dot{q}_{max}$ is that value of $\dot{q}_{max}$ which must not be exceeded for a heating cycle of the type shown on Fig. 3-7 if the material is to remain unmelted where the maximum surface temperature occurs at time t_2. Equation (3-76) tells us many facts about the desirable properties of nonmelting heat-sink materials. Assume that a high-performance nonmelting object requires a high allowable maximum heat-transfer rate. This quantity is maximized if

1. The product $(C_p \rho k)^{1/2} T_m$ is high.
2. Δt, the duration of the heat pulse, is low.

Table 3-4 gives some representative values of $(C_p \rho k)^{1/2}(T_m - T_0)$

Table 3-4. Representative Values of the Parameter $(C_p \rho k)^{1/2}(T_m - T_0)$ for Several Heat-sink Materials $(T_0 = 530°R)$

Material	Approximate value of $(C_p \rho k)^{1/2}(T - T_0)$, in Btu/ft^2 − (hr)$^{1/2}$	Approximate heat capacity in Btu/lb†
Copper	1.68×10^5	150–200
Steel................	2×10^4	90
Graphite	1.88×10^5	1600
Titanium carbide	1.86×10^5	1100
Magnesium oxide	2.47×10^4	1650
Silicon carbide	3.52×10^4	350
Beryllium oxide ..,....	4.16×10^4	2750

† Calculated using average values of the physical properties over the temperature range from T_0 to T_m.

for several materials. The higher the value of this parameter, the more desirable is this material for use as a nonmelting heat sink, other considerations such as ease of fabrication and ductility being equal.

It is apparent that copper compares well with refractory materials on the basis of the maximum heat-transfer rate it can withstand before melting, because of its excellent thermal conductivity. Graphite and titanium carbide compare well with copper on this basis but lack the workability which copper possesses. Of course, if melting and/or vaporization are presumed to be allowable, other considerations enter into the choice of the material to use in a heat-protection system, as Secs. 3-3 to 3-5 point out. For example, a desirable ablator should exhibit some properties just the opposite to those exhibited by a good solid heat sink.   That is, the thermal conductivity and vaporization temperature should be low for a good ablation material, other properties being equal.

The simple solution presented here is inadequate, of course, for detailed design purposes.   Considerations of geometry of the hypervelocity object, the variation of heat-transfer rate with geometry, and the variation of physical properties with temperature all indicate that Eq. (3-67) should be solved for the particular problem at hand. Our purpose here is to bring out in a straightforward manner the relationship between choice of heat-sink material and the heating cycle to which the surface material is exposed.

3-7. Conclusions.   We have described how the hot gas layer interacts with the solid or liquid sublayer and how the complicated problems of many simultaneously occurring phenomena can be treated.   Section 3-3 discussed the effects of chemical reactions in the gas layer upon heat transfer to the solid or liquid subsurface.   Section 3-4 treated the response of a melting subsurface to high heat-transfer rates of the gas layer.   Section 3-5 described how the vaporizing subsurface in turn affects the gas-layer heat transfer and surface shear.   Section 3-6 then indicated how the heat flow which penetrates an intermediate melting and vaporizing layer or which flows directly into a solid surface can be treated.   The elements are described in such a way that almost any eventuality can be handled by one or more of the techniques described, at least in an approximate way. We are now prepared to improve upon these approximations in succeeding chapters.

There are, of course, some valuable conclusions to be drawn at this point.   Among them are the following:

1. It is apparent from the results presented in Sec. 3-3 that exothermic reactions should be inhibited in the boundary layer if heat-transfer rates to the subsurface are to be minimized.   Thus, it is desirable to choose (if possible) noncombustible vaporizing ablators when ablation is relied upon for thermal protection.   Inorganic materials seem to be called for.

2. Section 3-4 made it clear that other desirable characteristics of melting and vaporizing ablating materials are high heat of vaporization, high viscosity (so as to encourage heat storage in the liquid state), and low conductivity in the liquid state (so as to encourage storage of heat in the melt, inhibit heat flow to the solid undersurface, and encourage vaporization).

3. The mass-transfer effect upon gas-layer heat transfer described in Sec. 3-5 is significant and lends support to the desirability of choosing a heat-protection system of the vaporizing ablation type.

4. It is apparent from considering Sec. 3-6 that nonmelting heat-sink thermal-protection systems cannot compete with ablating systems on a weight basis. The heat capacity of the best of the nonmelting heat-sink materials shown in Table 3-4, copper, is about 150 to 200 Btu/lb, whereas an ablating system possesses an effective heat capacity of greater than 1000 Btu/lb.

5. The simple, dimensionally correct, approximate expressions derived in Sec. 3-4 serve notice that many parameters must be matched in laboratory testing of melting heat-protection systems. Equation (3-46) serves to illustrate this.

In closing this chapter is is appropriate to remind the reader that several approximations were made in deriving the equations presented here in order to bring out what the author believes to be the controlling factors in the problem. Some of these approximations will be removed in succeeding chapters at the cost of clarity in concept. It is believed that the approaches given here will lend some insight into methods of treating similar problems in this complex science.

4

The Dissociated Laminar Boundary Layer

4-1. Introduction. Chapter 3 presented several simplified analyses most of which applied equally well to reacting laminar or turbulent boundary layers. In order to improve on the accuracy of such simplified analyses, we must include features of the boundary layer in our analyses which depend upon its physical state, that is, upon whether it is a laminar or a turbulent boundary layer. Chapters 4 and 5 will present some of the more accurate theories for the reacting laminar boundary layer. Chapters 7 and 8 will be concerned with the theory of the reacting turbulent boundary layer.

In the present chapter we are concerned with the dissociated laminar boundary layer, including the effects of body shape and pressure gradient. Beginning with a statement of the appropriate boundary-layer equations for the dissociated boundary layer, we shall proceed from some simplified solutions to the most precise calculations available. The effects of dissociation upon heat transfer from a dissociated boundary layer to a cooler body will be described and accounted for in these analyses.

As the boundary-layer theory for viscous hypersonic flow developed, it became apparent that two classes of boundary-layer flow over bodies could be distinguished: that of flow over blunt objects and that of flow over objects having a relatively "sharp" leading edge. The laminar-boundary-layer flow over sharp-leading-edge objects received early and extensive treatment because of the interesting interaction phenomena which occurred owing to a self-induced pressure gradient imposed upon the boundary layer because of the displacement of the external gas stream, particularly close to the leading edge or nose of the body. These effects will be treated in Chap. 6 of this book.

While the problem of leading-edge interaction remains an important one in hypersonic boundary-layer theory, particularly for objects constrained to fly at hypersonic Mach numbers and low Reynolds numbers, it became increasingly apparent that hypersonic flight in the region of high density or high Reynolds number precluded, for practical reasons, the use of sharp leading edges or noses. These practical reasons include the requirement of minimizing heat-transfer rates at the stagnation point and, furthermore, the necessity of somehow providing enough heat capacity at the stagnation point to accept the high heat-transfer rates present there. Because blunt noses entail highly heated gases which have passed through strong nose shock waves, it is this latter situation which is most influenced by the gas-dissociation effects treated in the present chapter.

4-2. The Boundary-layer Equations. The boundary-layer equations for a reacting gas mixture were derived in Chap. 2. They will be stated here as a point of departure for the work to be presented in

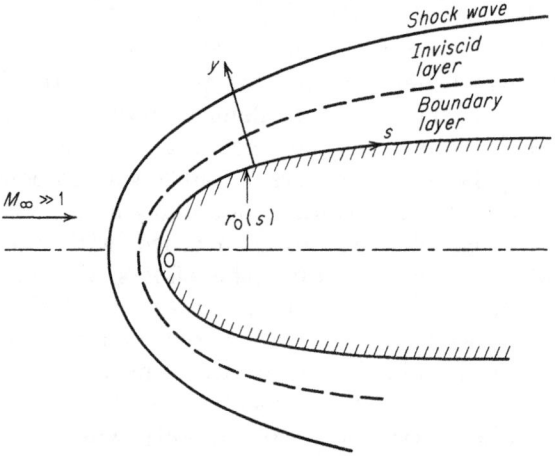

FIG. 4-1. Coordinate system used in analyzing the boundary-layer flows over objects in flight at hypersonic speeds.

this chapter. The coordinate system used is illustrated in Fig. 4-1.

From Eq. (2-67) the equation of state is

$$p_i = \rho_i \frac{k}{m_i} T \qquad (4\text{-}1a)$$

where k is Boltzmann's constant and m_i is the mass of one molecule of species i. Furthermore, from Eq. (2-68a),

$$p = \sum_i p_i = \rho \bar{R} T \qquad (4\text{-}1b)$$

where, from Eq. (2-68b),

$$\bar{R} = \sum_i C_i R_i \qquad C_i = \frac{\rho_i}{\rho} \qquad (4\text{-}1c)$$

The equation for conservation of species is given by Eq. (2-56) and is

$$\rho u \frac{\partial C_i}{\partial s} + \rho v \frac{\partial C_i}{\partial y} = \frac{\partial}{\partial y}\left(\rho D_{12} \frac{\partial C_i}{\partial y}\right) + \dot{w}_i \tag{4-2}$$

The equation of continuity is given by Eq. (2-70) and is

$$\frac{\partial \rho\, u r_0^k}{\partial s} + \frac{\partial \rho\, v r_0^k}{\partial y} = 0 \tag{4-3}$$

where $r_0(s)$ is the radius of a body of revolution in a meridian plane (see Fig. 4-1), $k = 0$ for a planar body, and $k = 1$ for a body of revolution.

The equations for conservation of momentum are given by Eqs. (2-66) and (2-69) and are

$$\rho u \frac{\partial u}{\partial s} + \rho v \frac{\partial u}{\partial y} = -\frac{\partial p}{\partial s} + \frac{\partial}{\partial y}\left(\mu \frac{\partial u}{\partial y}\right) \tag{4-4a}$$

and

$$0 = -\frac{\partial p}{\partial y} \qquad p = p_e \tag{4-4b}$$

The equation for conservation of energy is given by Eq. (2-72) and is

$$\rho u \frac{\partial I}{\partial s} + \rho v \frac{\partial I}{\partial y} = \frac{\partial}{\partial y}\left[\frac{\mu}{P} \frac{\partial I}{\partial y} + \mu\left(1 - \frac{1}{P}\right)\frac{1}{2}\frac{\partial u^2}{\partial y}\right]$$

$$- \frac{\partial}{\partial y}\left[\left(\frac{1}{L} - 1\right)\rho D_{12} \sum_i h_i \frac{\partial C_i}{\partial y}\right] \tag{4-5}$$

where

$$I = h + \frac{u^2}{2} \tag{4-6}$$

$$h = \sum_i C_i h_i \tag{4-7}$$

and

$$h_i = \int_0^T C_{p_i}\, dT + h_i^0 \tag{4-8}$$

An alternative form of the energy equation can be written. From Eq. (4-6)

$$\rho u \frac{\partial I}{\partial s} + \rho v \frac{\partial I}{\partial y} = \rho u \frac{\partial h}{\partial s} + \rho v \frac{\partial h}{\partial y} + u\left(\rho u \frac{\partial u}{\partial s} + \rho v \frac{\partial u}{\partial y}\right)$$

or, using Eq. (4-4a),

$$\rho u \frac{\partial I}{\partial s} + \rho v \frac{\partial I}{\partial y} = \rho u \frac{\partial h}{\partial s} + \rho v \frac{\partial h}{\partial y} + u \frac{\partial}{\partial y}\left(\mu \frac{\partial u}{\partial y}\right) - u \frac{\partial p}{\partial s} \tag{4-9}$$

however,
$$u \frac{\partial}{\partial y}\left(\mu \frac{\partial u}{\partial y}\right) = \frac{\partial}{\partial y}\left(\mu \frac{1}{2}\frac{\partial u^2}{\partial y}\right) - \mu\left(\frac{\partial u}{\partial y}\right)^2 \qquad (4\text{-}10)$$

Also, from Eq. (4-6),
$$\frac{\mu}{P}\frac{\partial I}{\partial y} = \frac{\mu}{P}\frac{\partial h}{\partial y} + \frac{\mu}{P}\left(\frac{1}{2}\frac{\partial u^2}{\partial y}\right) \qquad (4\text{-}11)$$

and from Eq. (4-7),
$$\frac{\mu}{P}\frac{\partial h}{\partial y} = \frac{\mu}{P}\left(\sum_i h_i \frac{\partial C_i}{\partial y}\right) + \frac{C_{p_f}\mu}{P}\frac{\partial T}{\partial y} \qquad (4\text{-}12)$$

where
$$C_{p_f} = \sum_i C_i \frac{dh_i}{dT} = \sum_i C_i C_{p_i} \qquad (4\text{-}13)$$

When Eq. (4-10) is substituted into Eq. (4-9), Eq. (4-12) is substituted into Eq. (4-11), and the resulting two equations substituted into Eq. (4-5), the alternative form of the energy equation results:

$$\rho u \frac{\partial h}{\partial s} + \rho v \frac{\partial h}{\partial y} = u \frac{\partial p}{\partial s} + \frac{\partial}{\partial y}\left(k\frac{\partial T}{\partial y} + \rho D_{12}\sum_i h_i \frac{\partial C_i}{\partial y}\right) + \mu\left(\frac{\partial u}{\partial y}\right)^2 \qquad (4\text{-}14)$$

where the definitions for Prandtl number P and Lewis number L given below are used.

$$P = \frac{C_{p_f}\mu}{k} \qquad (4\text{-}15)$$

and
$$L = \frac{\rho D_{12}C_{p_f}}{k} \qquad (4\text{-}16)$$

Boundary conditions for the above equations are

At $y = 0$	As $y \to \infty$
$\rho = \rho_w$	$\rho \to \rho_e$
$T = T_w$	$T \to T_e$
$h = h_w$	$h \to h_e$
$I = I_w$	$I \to I_e$
$u = 0$	$u \to u_e$
$v = 0$	$v \to 0$
$C_i = (C_i)_w$	$C_i \to (C_i)_e$

also $\mu = \mu(C_i,T)$, $k = k(C_i,T)$, $D_{12} = D_{12}(T)$, and $p = p_e(s)$.

In stating the above equations for our reacting gas mixture, it has been assumed that we are dealing with a steady flow system.

4-3. Boundary-layer Equations for an Ideal Dissociating Gas. It was Lighthill[1] who first introduced the concept of an ideal

[1] M. J. Lighthill, *J. Fluid Mech.*, vol. 2, part 1, pp. 1–32, 1957.

dissociating gas when dealing with dissociating gas mixtures.   Lighthill simplified the study of a dissociating gas mixture such as air by replacing it with the study of a simple ideal dissociating gas mixture the chemistry of which is represented by the reaction equation

$$A_2 \rightleftharpoons 2A \qquad (4\text{-}17)$$

That is, we assume only two species present in a dissociating gas mixture: molecules A_2 and atoms A.   The thermodynamic and chemical properties are then determined to be some suitable average value for the atoms and molecules of the real gas mixture.   The concept has proved very useful in the study of inviscid reacting flow systems and, as we shall demonstrate, will be useful in studying the dissociating boundary layer.

Accepting the approximation of the ideal dissociating gas, we can write

$$\frac{\rho_A}{\rho} = \alpha \qquad \text{mass fraction of atoms} \qquad (4\text{-}18)$$

and

$$\frac{\rho_M}{\rho} = 1 - \alpha \qquad \text{mass fraction of molecules} \qquad (4\text{-}19)$$

We shall now use Eqs. (4-18) and (4-19) in deriving the boundary-layer equations for an ideal dissociating gas.   The equation of state, Eq. (4-1a), becomes

$$p_A = \rho_A \frac{k}{m_A} T \qquad (4\text{-}20)$$

and

$$p_M = \rho_M \frac{k}{2m_A} T \qquad (4\text{-}21)$$

Thus, using Eqs. (4-20) and (4-21) in Eq. (4-1b), we obtain

$$p = \rho \bar{R} T \qquad (4\text{-}1b)$$

where, from Eq. (4-1c),

$$\bar{R} = (1 + \alpha) \frac{k}{2m_A} \qquad (4\text{-}22)$$

Furthermore, using Eq. (4-18) in Eq. (4-2), we obtain

$$\rho u \frac{\partial \alpha}{\partial s} + \rho v \frac{\partial \alpha}{\partial y} = \frac{\partial}{\partial y}\left(\rho D_{12} \frac{\partial \alpha}{\partial y} \right) + \dot{w}_A \qquad (4\text{-}23)$$

Equations (4-3), (4-4a), and (4-4b), the equations of continuity and conservation of momentum, respectively, remain unchanged.   The equations for conservation of energy, Eqs. (4-5) and (4-14), become, using Eqs. (4-18) and (4-19),

$$\rho u \frac{\partial I}{\partial s} + \rho v \frac{\partial I}{\partial y} = \frac{\partial}{\partial y}\left[\frac{\mu}{P}\frac{\partial I}{\partial y} + \mu\left(1 - \frac{1}{P}\right)\frac{1}{2}\frac{\partial u^2}{\partial y}\right]$$

$$- \frac{\partial}{\partial y}\left[\left(\frac{1}{L} - 1\right)\rho D_{12}(h_A - h_M)\frac{\partial \alpha}{\partial y}\right] \quad (4\text{-}24)$$

and

$$\rho u \frac{\partial h}{\partial s} + \rho v \frac{\partial h}{\partial y} = u \frac{\partial p}{\partial s} + \frac{\partial}{\partial y}\left[k\frac{\partial T}{\partial y} + \rho D_{12}(h_A - h_M)\frac{\partial \alpha}{\partial y}\right] + \mu\left(\frac{\partial u}{\partial y}\right)^2 \quad (4\text{-}25)$$

where

$$h = \alpha h_A + (1 - \alpha)h_M \quad (4\text{-}26)$$

$$h_A = \int_0^T C_{p_A}\,dT + h_A^0 \quad (4\text{-}27)$$

and

$$h_M = \int_0^T C_{p_M}\,dT \quad (4\text{-}28)$$

since $h_M^0 = 0$. $h_A^0 = D/2m_A$, where D is the dissociation energy per molecule for the ideal dissociating gas molecules.

Boundary conditions are

At $y = 0$	As $y \to \infty$
$\rho = \rho_w$	$\rho \to \rho_e$
$T = T_w$	$T \to T_e$
$h = h_w$	$h \to h_e$
$I = I_w$	$I \to I_e$
$u = 0$	$u \to u_e$
$v = 0$	$v \to 0$
$\alpha = \alpha_w$	$\alpha \to \alpha_e$

and $\mu = \mu(\alpha, T)$, $k = k(\alpha, T)$, $D_{12} = D_{12}(T)$, and $p = p_e(s)$ as before. These simplified equations are found to be sufficient in treating the dissociated boundary layer, as we shall demonstrate.

4-4. Effects of Dissociation: The Reaction Conductivity. Let us see if we can anticipate what the effects of boundary-layer gas dissociation will be upon heat transfer to the surface. In the absence of gas dissociation the energy equation (4-14) becomes, for a two-dimensional flat plate,

$$\rho u \frac{\partial h}{\partial s} + \rho v \frac{\partial h}{\partial y} = \frac{\partial}{\partial y}\left(k\frac{\partial T}{\partial y}\right) + \mu\left(\frac{\partial u}{\partial y}\right)^2 \quad (4\text{-}29)$$

The nonreacting gas, constant-property (ρ, μ, k, C_p all constant) solutions to Eq. (4-29) and the momentum equation (4-4a) give[1]

$$-\dot{q}_w = C_H \rho_e u_e (I_r - h_w) \qquad (4\text{-}30)$$

where

$$I_r = h_e + r\frac{u_e^2}{2}$$

$$r = (P)^{1/2}$$

and

$$C_H = \frac{C_f}{2}(P)^{-2/3} \qquad (4\text{-}31)$$

P is the Prandtl number for the nonreacting, constant-property gas mixture, and it is understood that for this case

$$-\dot{q} = k\frac{\partial T}{\partial y} \qquad (4\text{-}32)$$

When dissociation occurs and two species, atoms and molecules, are present, then the heat flux includes a term due to diffusion of the atoms and molecules through the boundary layer. For example, if the surface is cooler than the external stream, then atoms will diffuse toward the surface to be recombined near or at the surface and molecules will diffuse away from this region toward the external stream, where they will be dissociated. The heat-flux term then becomes for the ideal dissociated gas

$$-\dot{q} = k\frac{\partial T}{\partial y} + \rho D_{12}(h_A - h_M)\frac{\partial \alpha}{\partial y} \qquad (4\text{-}33)$$

and the energy equation is given by Eq. (4-25).

$$\rho u \frac{\partial h}{\partial s} + \rho v \frac{\partial h}{\partial y} = \frac{\partial}{\partial y}\left[k\frac{\partial T}{\partial y} + \rho D_{12}(h_A - h_M)\frac{\partial \alpha}{\partial y} \right] + \mu \left(\frac{\partial u}{\partial y}\right)^2 \qquad (4\text{-}34)$$

Now, from Eqs. (4-27) and (4-28),

$$h_A - h_M = h_A^0 + \int_0^T (C_{p_A} - C_{p_M})\, dT$$

but

$$h_A^0 \gg \int_0^T (C_{p_A} - C_{p_M})\, dT$$

hence our heat-flux term, Eq. (4-33), becomes

$$-\dot{q} = k\frac{\partial T}{\partial y} + \rho D_{12}h_A^0 \frac{\partial \alpha}{\partial y} \qquad (4\text{-}35)$$

[1] See, for example, Hermann Schlichting, "Boundary Layer Theory," 4th ed., Eq. (7.35a), p. 120, for $C_f/2$ and Eq. (14.68), p. 318, for $-\dot{q}_w$, McGraw-Hill Book Company, Inc., New York, 1960.

Now, if our ideal dissociating gas was everywhere in local chemical equilibrium, then

$$\alpha = \alpha(T,p)$$

Hence, since, for our flat-plate example, $p = \text{const} = p_e$,

$$\frac{\partial \alpha}{\partial y} = \frac{d\alpha}{dT}\frac{\partial T}{\partial y}$$

and we find that

$$-\dot{q} = (k + k_r)\frac{\partial T}{\partial y} \tag{4-36}$$

where

$$k_r = \rho D_{12} h_A^0 \frac{d\alpha}{dT} \tag{4-37}$$

is the "reaction conductivity" or thermal conductivity due to chemical reactions. The sum of k and k_r will be the apparent thermal conductivity of an equilibrium reacting mixture. Some notion of the relative magnitude of k_r can be gained by assuming, for example, that our ideal dissociating gas is oxygen. That is, for

$$O_2 \rightleftharpoons 2O$$

we have at equilibrium

$$\frac{4\alpha^2}{1 - \alpha^2} = \frac{K(T)}{p} \tag{4-38}$$

where[1]

$$K(T) \propto \exp\left(-\frac{D}{kT}\right) \tag{4-39}$$

where D is the dissociation energy of oxygen per molecule O_2. Equations (4-38) and (4-39) can be used with Eqs. (4-1b), (4-22), and (4-37) to obtain

$$k_r = \frac{pD_{12}}{2T}\left(\frac{D}{kT}\right)^2 \alpha(1 - \alpha) \tag{4-40}$$

It is apparent that k_r will be at its maximum value somewhere close to $\alpha = \frac{1}{2}$. The value of k_r has been calculated by Butler and Brokaw[2] and Hirschfelder[3] for O_2, and it is found that k_r can be of the order of ten times k at its maximum value.

When a reaction conductivity is defined in this manner, the explicit appearance of diffusion energy transport terms can be eliminated from the energy equation. However, in that which follows this section we shall not use the concept of reaction conductivity because its

[1] See, for example, Eq. (9-128).

[2] James N. Butler and Richard S. Brokaw, *J. Chem. Phys.*, vol. 26, no. 6, pp. 1636–1643, 1957.

[3] Joseph O. Hirschfelder, *J. Chem. Phys.*, vol. 26, no. 2, pp. 271–273, 1957.

definition requires that the flow be in local chemical equilibrium and because there is no intrinsic savings in labor inasmuch as k_r depends upon local concentrations and temperature. We can use the concept to anticipate the effects of dissociation, however.

Now, from Eq. (4-26), we find that

$$\frac{dh}{dT} = \frac{d\alpha}{dT} h_A^0 + C_{pf}$$

where for our ideal dissociating gas

$$C_{pf} = \sum_i C_i \frac{dh_i}{dT} = \alpha \frac{dh_A}{dT} + (1 - \alpha) \frac{dh_M}{dT}$$

Thus
$$\frac{d\alpha}{dT} = \frac{(dh/dT) - C_{pf}}{h_A^0} \tag{4-41}$$

Combining Eqs. (4-37) and (4-41) we obtain

$$k_r = \rho D_{12}(C_p - C_{pf}) \tag{4-42}$$

and so Eq. (4-36) becomes

$$-\dot{q} = k\left[1 + L\left(\frac{C_p}{C_{pf}} - 1\right)\right]\frac{\partial T}{\partial y}$$

where
$$L = \frac{\rho D_{12} C_{pf}}{k}$$

Define two convenient Prandtl numbers. The first depends upon both the flow field and the gas mixture and is

$$P_{eq} = \frac{C_p \mu}{k + k_r} \tag{4-43}$$

and, as previously defined, a Prandtl number dependent upon the gas mixture alone is

$$P = \frac{C_{pf}\mu}{k} \tag{4-44}$$

Note that $P_{eq} = P$ for a nonreacting gas mixture. From the above it follows that

$$\frac{P_{eq}}{P} = \frac{C_p}{C_{pf}} \frac{k}{k + k_r}\left(\frac{C_{pf}}{C_p} + L - L\frac{C_{pf}}{C_p}\right)^{-1} \tag{4-45}$$

since from Eq. (4-42)

$$\frac{k}{k + k_r} = \left(1 + L\frac{C_p}{C_{pf}}\right)^{-1}$$

Now by analogy with (4-30) our heat transfer from an equilibrium ideal dissociating gas boundary layer will be

$$-\dot{q}_w = \frac{C_f}{2}\,\rho_e u_e P_{eq}^{-2/3}(I_r - h_w)$$

or, making use of Eq. (4-45),

$$-\dot{q}_w = \frac{C_f}{2}\,\rho_e u_e P^{-2/3}(I_r - h_w)\left[\frac{(C_{p_f})_{av}}{(C_p)_{av}}\,(1 - L) + L\right]^{2/3} \tag{4-46}$$

now let

$$(C_p)_{av} \simeq \frac{I_r - h_w}{T_r - T_w}$$

and

$$(C_{p_f})_{av} \simeq \frac{I_r - h_w - (\alpha_e - \alpha_w)h_{\mathrm{A}}^0}{T_r - T_w}$$

Substituting the above equations for $(C_p)_{av}$ and $(C_{p_f})_{av}$ into Eq. (4-46) we obtain

$$-\dot{q}_w = \frac{C_f}{2}\,\rho_e u_e P^{-2/3}(I_r - h_w)\left[1 + (L - 1)\frac{(\alpha_e - \alpha_w)h_{\mathrm{A}}^0}{I_r - h_w}\right]^{2/3} \tag{4-47}$$

and thus the explicit effects of boundary-layer gas dissociation appear in the last factor of Eq. (4-47). Note that, if either $L = 1$ or $\alpha_e = \alpha_w$, there will be no explicit effect of dissociation upon heat transfer to the surface. If $L = 1$, then, according to Sec. 2-6, heat is diffused to the surface at a rate equal to the rate at which it is conducted and convected to the surface and it will make no difference which mode of heat transfer is used providing the surface is catalytic to recombination. If $\alpha_e = \alpha_w$, the boundary layer is essentially a uniform mixture of atoms and molecules and no diffusion currents will be present at the surface. The implicit effects of dissociation appear primarily in the enthalpy potential $I_r - h_w$, where, using Eqs. (4-6) and (4-26) to (4-28), we can write

$$I_r - h_w = (h_{\mathrm{A}} - h_{\mathrm{M}})_e\alpha_e + (h_{\mathrm{M}})_e + r\frac{u_e^2}{2} - (h_{\mathrm{A}} - h_{\mathrm{M}})_w\alpha_w + (h_{\mathrm{M}})_w$$

but we have already shown that

$$(h_{\mathrm{A}} - h_{\mathrm{M}})_e \simeq h_{\mathrm{A}}^0 \simeq (h_{\mathrm{A}} - h_{\mathrm{M}})_w$$

and therefore

$$I_r - h_w \simeq h_{\mathrm{A}}^0(\alpha_e - \alpha_w) + (h_{\mathrm{M}})_e - (h_{\mathrm{M}})_w + r\frac{u_e^2}{2} \tag{4-48}$$

and dissociation enters in through the first term of this equation. For boundary-layer conditions where $h_{\mathrm{A}}^0(\alpha_e - \alpha_w)$ is of the order of

50 per cent of $I_r - h_w$, this latter effect of dissociation is the principal effect, since L is usually close to one for gas mixtures like air. The recovery factor r is relatively insensitive to dissociation.

Another implicit effect of dissociation is the effect upon the transport properties of the dissociating gas mixture, which, however, is secondary to the effects described above as later developments will show.

An equation analogous to Eq. (4-47) applicable to stagnation-point heat transfer was first presented by Rosner,[1] who showed that the last term of Eq. (4-47) differed from the results of correlations of more exact calculations by less than 3 per cent for a wide range of variation of the factor

$$\frac{(\alpha_e - \alpha_w)h_A^0}{I_e - h_w}$$

for values of Lewis number equal to 1.4, a value appropriate to a dissociated air mixture.

4-5. The Frozen-boundary-layer Approximation. The previous section presented an analysis of the effect of dissociation upon the flat-plate boundary-layer heat transfer. The remainder of this chapter will be devoted to more precise treatments of the dissociating laminar boundary layer flowing over blunt axisymmetric bodies. The adaptation of such solutions to the flat-plate problem will be discussed in Sec. 5-11. We devote more attention to the blunt-body problem because this model is more appropriate to problems of hypersonic-body heat transfer.

When the gas dynamicist is faced with the problem of exactly determining the heat transfer from a dissociating gas boundary layer, he must solve the boundary-layer equations for the atom concentrations as well as for the variables of state and the flow velocity. For an ideal dissociating gas, six equations, Eqs. (4-1b), (4-3), (4-4a), (4-4b), (4-23), and (4-24) or (4-25), must be solved for six unknowns, u, v, ρ, T, p, and α. p is usually given as equal to $p_e(s)$ and is determined by the inviscid flow external to the boundary layer. As such, it represents a link between the external flow and the boundary-layer flow.

Let us now use the coordinate transformation developed in Sec. 2-4 to reduce our nonlinear partial differential equations to ordinary differential equations if possible. It was shown in Sec. 2-4 that, if the substitutions and transformations listed below are used with Eqs. (4-3), (4-4a), (4-4b), (4-23), and (4-24), differential equations will

[1] Daniel E. Rosner, *ARS J.*, vol. 30, no. 1, pp. 114–115, 1960.

result which will be ordinary differential equations under certain specified conditions. That is, if

$$\frac{u}{u_e(s)} = f'(\eta) = \frac{df}{d\eta}$$

$$\psi = \frac{2\bar{s}}{p_e r_0^k} f(\eta)$$

$$\eta = \frac{\rho_e u_e r_0^k}{(2\bar{s})^{1/2}} \int_0^y \frac{\rho}{\rho_e} \, dy \qquad (4\text{-}49)$$

and

$$\bar{s} = \int_0^s \rho_e \mu_e r_0^{2k} u_e \, ds \qquad (4\text{-}50)$$

then Eq. (4-3) is satisfied directly; Eq. (4-23), the conservation of species equation, becomes

$$\left(\frac{C}{S} z_A'\right)' + f z_A' = \frac{2\bar{s} f' z_A}{\alpha_e} \frac{d\alpha_e}{d\bar{s}} - \frac{2\bar{s}\dot{w}_A}{\rho \rho_e u_e^2 \mu_e r_0^{2k} \alpha_e} \qquad (4\text{-}51)$$

Eq. (4-4a), the momentum equation, becomes

$$(Cf'')' + ff'' = \frac{2\bar{s}}{u_e} \frac{du_e}{d\bar{s}} \left[(f')^2 - \frac{\rho_e}{\rho} \right] \qquad (4\text{-}52)$$

and the energy equation (4-24) becomes

$$\left(\frac{C}{P} g'\right)' + f g' = \frac{2\bar{s} f' g}{I_e} \frac{dI_e}{d\bar{s}} + \left[\frac{u_e^2}{I_e} \left(1 - \frac{1}{P}\right) Cf'f'' \right]'$$

$$+ \left[\frac{C}{S} \left(\frac{1}{L} - 1\right) \frac{(h_A - h_M)\alpha_e}{I_e} z_A' \right]' \qquad (4\text{-}53)$$

where

$$z_A(\eta) = \frac{\alpha}{\alpha_e} \qquad (4\text{-}54)$$

$$z_M(\eta) = \frac{1 - \alpha}{1 - \alpha_e} \qquad (4\text{-}55)$$

$$f'(\eta) = \frac{u}{u_e} \qquad (4\text{-}56)$$

$$g(\eta) = \frac{I}{I_e} \qquad (4\text{-}57)$$

and

$$C(\eta) = \frac{\rho\mu}{\rho_e\mu_e} \qquad (4\text{-}58)$$

Several terms appear in Eqs. (4-51) through (4-53) which could vitiate any advantage gained by our transformations because they could be dependent upon the coordinate $\bar{s}$. Fortunately, there are several solutions to Eqs. (4-51) through (4-53) which satisfy the requirements of independence of all terms from $\bar{s}$ and which can be used to illustrate the effects of dissociation upon laminar-boundary-layer skin friction and heat transfer.

The solution for heat transfer to a stagnation point is one such useful solution. In this case $d\alpha_e/d\bar{s} = 0$ in Eq. (4-51), $\dfrac{\bar{s}}{u_e}\dfrac{du_e}{d\bar{s}} = \text{const}$ in Eq. (4-52), and $I_e = \text{const}$ and $u_e = 0$ in Eq. (4-53). Providing the last term in Eq. (4-53) can be expressed as a function of η alone, the system of equations for flow at the stagnation point becomes a system of ordinary differential equations with η as the independent variable. The boundary conditions for this case are

$$
\begin{aligned}
z_{\mathrm{A}}(0) &= \frac{\alpha_w}{\alpha_e} & z_{\mathrm{A}}(\infty) &= 1 \\
f'(0) &= f(0) = 0 & f'(\infty) &= 1 \\
g'(0) &= g'_w & g'(\infty) &= 0 \\
g(0) &= g_w & g(\infty) &= 1
\end{aligned}
\tag{4-59}
$$

Solutions to Eqs. (4-51) through (4-53) are sought for various assumed implicit variations of C, P, and S with η and for a variety of values of reaction-rate constants which determine $\dot{w}_{\mathrm{A}}$.

It is instructive, first, to treat one of the simpler special cases for heat transfer to a stagnation point, namely, the case of frozen flow in the gas layer where $\dot{z}_{\mathrm{A}} = 0$ in Eq. (4-51). Consider Eq. (4-35)

$$
-\dot{q}_w = \left(k\frac{\partial T}{\partial y} + \rho D_{12} h_{\mathrm{A}}^0 \frac{\partial \alpha}{\partial y} \right)_w
\tag{4-35}
$$

where we introduce the reasonable simplification that $h_{\mathrm{A}} - h_{\mathrm{M}} \simeq h_{\mathrm{A}}^0$. Now Eq. (4-35) can be used to show, if we define a "partial enthalpy" I_f, where

$$
I_f = I - \sum_i C_i h_i^0
\tag{4-60}
$$

that is, for a binary mixture

$$
I_f = I - \alpha h_{\mathrm{A}}^0
\tag{4-61}
$$

then, in Eq. (4-35) the conduction term is, making use of Eq. (4-11),

$$
\left(k\frac{\partial T}{\partial y} \right)_w = \frac{1}{C_{p_f}}\frac{\partial I_f}{\partial y} = -(\dot{q}_w)_c
\tag{4-62}
$$

and the diffusion term in Eq. (4-35) is

$$\left(\rho D_{12} h_A^0 \frac{\partial \alpha}{\partial y}\right)_w = -(\dot{q}_w)_d \tag{4-63}$$

and the total heat transfer to the wall is $-\dot{q}_w = -(\dot{q}_w)_c - (\dot{q}_w)_d$. This breaking down of the heat-transfer rate into two terms, one due to conduction, the other due to diffusion, results in a very useful simplification in the case of a frozen binary-mixture boundary layer as will be demonstrated. Proceed as follows: Define

$$g_f = \frac{I_f}{(I_f)_e} \tag{4-64a}$$

and

$$z_A = \frac{\alpha}{\alpha_e} \tag{4-64b}$$

then, from Eqs. (4-57), (4-61), and (4-64a)

$$g = \frac{(I_f)_e}{I_e} g_f + \frac{\alpha_e h_A^0}{I_e} z_A \tag{4-65}$$

Substitute Eq. (4-65) into Eq. (4-53) and (4-64b) into Eqs. (4-51) and (4-52), and subtract the equation which results from the latter substitution from the equation which results from the former substitution. There result the following equations, remembering that $\dot{w}_A = 0$ for frozen flow:

$$\left(\frac{C}{S} z_A'\right)' + f z_A' = 0 \tag{4-66}$$

$$(Cf'')' + ff'' = 0 \tag{4-67}$$

$$\left(\frac{C}{P} g_f'\right)' + f g_f' = 0 \tag{4-68}$$

where Eq. (4-67) has implicit within it that the term in the momentum equation accounting for the effects of pressure gradient is negligible relative to the remaining terms. That is,

$$\frac{p_e}{\rho} - (f')^2 \simeq 0$$

a reasonable assumption through most of the boundary layer for a cool wall.[1] The boundary conditions on Eqs. (4-66) through (4-68) are

$$
\begin{array}{ll}
z(0) = z_w & z(\infty) = 1 \\
f'(0) = f(0) = 0 & f'(\infty) = 1 \\
g_f(0) = (g_f)_w & g_f(\infty) = 1
\end{array}
\tag{4-69}
$$

[1] Equation (4-67) is strictly correct only for flat-plate flow. We neglect the pressure-gradient term on the right-hand side of Eq. (4-52) as a matter of convenience only, since we are here concerned with an approximate solution for the stagnation point.

Furthermore, using Eqs. (4-35), (4-49), (4-50), (4-62), (4-63,) (4-64a), and (4-64b), it can be shown that

$$-\dot{q}_w = \frac{(I_f)_e}{P} \frac{\rho_w \mu_w}{(\rho_e \mu_e)^{1/2}} (2)^{k/2} \left[\left(\frac{du_e}{ds} \right)_0 \right]^{1/2} \left[g_f'(0) + L \frac{h_A^0 \alpha_e z_A'(0)}{(I_f)_e} \right] \quad (4\text{-}70)$$

where, near a stagnation point,

$$r_0 = s$$

$$u_e = s \left(\frac{du_e}{ds} \right)_0$$

and $k = 0$ for a two-dimensional body and 1 for a body of revolution.

Expressing the heat-transfer rate in the form of Nusselt's number, where

$$Nu = \frac{-\dot{q}_w s C_{p_w}}{k_w (I_e - h_w)} \quad (4\text{-}71)$$

and

$$R_w = \frac{\rho_w u_e s}{\mu_w}$$

then it follows that

$$\frac{Nu}{(R_w)^{1/2}} = \frac{2^{k/2}}{I_e - h_w} [(I_f)_e g_f'(0) + L h_A^0 \alpha_e z_A'(0)] \quad (4\text{-}72)$$

where Eqs. (4-66) through (4-68) must be solved simultaneously in conjunction with boundary conditions (4-69) to obtain the values of $g_f'(0)$ and $z_A'(0)$. Solutions to Eqs. (4-66) through (4-68) subject to boundary conditions (4-69) exist in the literature for special cases. Li and Nagamatsu[1] provide solutions to Eq. (4-67) coupled with Eq. (4-68) or (4-66) for $C = 1$ and P or $S = 1$; Levy[2] provides solutions for $C = 1$ and P or $S = 1.0$ or 0.7; Cohen and Reshotko[3] provide solutions for $C = 1$ and P or $S = 1.0$. Admitting that these solutions are special cases (P and S will vary weakly through the boundary layer, and C will vary more strongly between limits of 1.0 and about 0.20), nevertheless these solutions are possible cases and can be used here to help us anticipate the appearance and form of any exact solution obtained using the full equations with no approximations. Their results indicate that

$$z_A'(0) \simeq 0.47 S^{1/3} [1 - z_A(0)] \quad (4\text{-}73)$$

and

$$g_f'(0) \simeq 0.47 P^{1/3} [1 - g_f(0)] \quad (4\text{-}74)$$

[1] Ting Y. Li and Henry T. Nagamatsu, *J. Aeronaut. Sci.*, vol. 22, pp. 607–616, 1955.

[2] Solomon Levy, *J. Aeronaut. Sci.*, vol. 21, pp. 459–474, 1954.

[3] Clarence B. Cohen and Eli Reshotko, *NACA Rept.* 1293, 1956.

within ± 5 per cent accuracy. That $z'(0)$ and $g'(0)$ take this form can be anticipated from the solutions to Eqs. (4-66), (4-67), and (4-68), which are by quadrature

$$z'_A(0) = \left(\frac{S}{C}\right)_w \frac{1 - z_A(0)}{G(\infty;S)}$$

$$f''(0) = \left(\frac{1}{C}\right)_w \frac{1}{G(\infty;1)}$$

and

$$g'_f(0) = \left(\frac{P}{C}\right)_w \frac{1 - g_f(0)}{G(\infty;P)}$$

where, if we assume that $C = \text{const} = C_w$,

$$G(\infty,Z) = \frac{Z}{C} \int_0^\infty \exp\left[-\int_0^{\eta'} \frac{Z}{C} f \, d\eta''\right] d\eta' \qquad (4\text{-}75)$$

where $Z = S$, 1, or P. Furthermore, from the boundary conditions for the momentum equation, $f(0) = f'(0) = 0$. Thus, by Taylor's series expansion

$$f(\eta) = f''(0) \frac{\eta^2}{2} + \cdots$$

where, as can be shown by solving for $f''(0)$,

$$f''(0) = \frac{0.47}{C^{1/2}}$$

whence Eq. (4-75) becomes

$$G(\infty;Z) \simeq \frac{Z}{C} \int_0^\infty \exp\left[-\frac{Z}{C^{3/2}}(0.0783)\eta^3\right] d\eta \qquad (4\text{-}76)$$

Now call

$$\xi = 0.0783\eta^3 \frac{Z}{C^{3/2}}$$

hence

$$d\eta = \frac{0.785 C^{1/2} \xi}{Z^{1/3} \xi^{2/3}} \qquad (4\text{-}77)$$

and so, using (4-77) in Eq. (4-76), we obtain

$$G(\infty;Z) \simeq 0.785 \frac{Z^{2/3}}{C^{1/2}} \int_0^\infty \exp(-\xi) \frac{d\alpha}{\xi^{2/3}}$$

or

$$G(\infty;Z) \simeq 0.785 \frac{Z^{2/3}}{C^{1/2}} \Gamma\left(\frac{1}{3}\right) = 0.785 \frac{Z^{2/3}}{C^{1/2}} 3\Gamma\left(\frac{4}{3}\right)$$

and evaluating the gamma function we obtain, to a first approximation,

$$G(\infty;Z) \simeq 2.11 \frac{Z^{2/3}}{C^{1/2}} \tag{4-78}$$

Thus, for example,

$$f''(0) \simeq \left(\frac{1}{C}\right)_w \frac{C_w^{1/2}}{2.11} = \frac{0.47}{C^{1/2}}$$

$$z'_A(0) \simeq \frac{0.47 S^{1/2}}{C^{1/2}} [1 - z_A(0)]$$

and

$$g'_f(0) \simeq \frac{0.47 P^{1/3}}{C^{1/2}} [1 - g_f(0)]$$

and the appearance of Eqs. (4-73) and (4-74) is explained, since $C \simeq 1$ for many cases. The solution of Eqs. (4-66) and (4-68) is coupled to the solution of Eq. (4-67), since the solution of Eq. (4-67), $f(\eta)$, appears explicitly in $G(\infty;Z)$. A series of iterations is involved in obtaining the various solutions desired.

Substituting Eqs. (4-73) and (4-74) into Eqs. (4-70) and (4-72) along with making use of Eq. (4-65) results in

$$-\dot{q}_w = 0.66 P^{-2/3} (\rho_e \mu_e)^{1/2} \left[\left(\frac{du_e}{ds}\right)_0 \right]^{1/2} (I_e - h_w) \left[1 + (L^{2/3} - 1) \frac{h_c}{I_e - h_w} \right] \tag{4-79}$$

and

$$\frac{Nu}{(R_w)^{1/2}} = 0.66 P^{1/3} \left[1 + (L^{2/3} - 1) \frac{h_c}{I_e - h_w} \right] \tag{4-80}$$

where

$$k = 1 \qquad \text{axisymmetric flow}$$

$$C = 1 = \frac{\rho_e \mu_e}{\rho_w \mu_w}$$

and

$$h_c = h_A^0 (\alpha_e - \alpha_w)$$

While Eqs. (4-79) and (4-80) as derived are strictly applicable to frozen-boundary-layer flow with $C = 1$, nevertheless, and as events will bear out, they anticipate the solution for the general case to a remarkable degree of accuracy, as was first pointed out by Lees[1] and as will be shown later. Note the similarity of Eq. (4-79) to the approximate Eq. (4-47) derived in Sec. 4-4.

4-6. Stagnation-point Solutions. Now that we have examined and gained some insight into the nature of the results to be expected for the heat transfer to the stagnation point from a dissociating binary mixture from the solution for the special case of frozen flow with constant P, S, and C, let us turn to the exact solution for this

---

[1] Lester Lees, *Jet Propulsion*, vol. 26, no. 4, pp. 259–269, 1956.

case for complete generality in the variations of P, S, and C with η and arbitrary reaction rates. Equations (4-51) through (4-53) apply with boundary conditions (4-59). Because we are dealing with stagnation-point flow, $\bar{s} = 0$ and near the stagnation point

$$r_0 = s$$

$$u_e = s \left(\frac{du_e}{ds} \right)_0$$

thus

$$\frac{2\bar{s}}{u_e} \frac{du_e}{d\bar{s}} = \frac{1}{k+1} = 2^{-k} \qquad k = 0 \text{ or } 1$$

Also, the last term in Eq. (4-51) becomes, in view of the above,

$$\frac{2\bar{s}\dot{w}_A}{\rho \rho_e u_e^2 \mu_e r_0^{2k} \alpha_e} = \frac{\dot{z}_A}{(k+1)(du_e/ds)_0}$$

Fay and Riddell,[1] using plausible assumptions regarding the kinetics of the reaction

$$A_2 + \begin{Bmatrix} A \\ A_2 \end{Bmatrix} \underset{k_R}{\overset{k_D}{\rightleftharpoons}} A + A + \begin{Bmatrix} A \\ A_2 \end{Bmatrix}$$

show that

$$\frac{\dot{z}_A}{(k+1)(du_e/ds)_0} = C_1 \theta^{-3.5} \frac{z_A^2 - (z_A^*)^2}{1 + \alpha_e z_A} \alpha_e$$

where

$$\theta = \theta(\eta) = \frac{T}{T_e} (\eta)$$

$$\alpha^* = \text{equilibrium value of } \alpha = \alpha^*(\eta)$$

and

$$C_1 = K_1 p_0^2 T_0^{-3.5} \left[\left(\frac{du_e}{ds} \right)_0 \right]^{-1} \qquad (4\text{-}81)$$

and p_0 and T_0 are the stagnation pressures and temperatures of the external stream. K_1 is directly related to k_R.[2]

The parameter C_1 can be thought of as the ratio of a characteristic flow time to a characteristic time of reaction (dissociation). If C_1 is large, the flow is near equilibrium. If C_1 is small, diffusion predominantly determines the atom concentration profiles in the boundary layer. Our system of equations now becomes

$$\left(\frac{C}{S} z_A' \right)' + f z_A' + C_1 \theta^{-3.5} \alpha_e \frac{z_A^2 - (z_A^*)^2}{1 + \alpha} = 0 \qquad (4\text{-}82)$$

$$(Cf'')' + ff'' + \frac{1}{k+1} \left[\frac{\rho_e}{\rho} - (f')^2 \right] = 0 \qquad (4\text{-}83)$$

$$\left(\frac{C}{P} g' \right)' + fg' + \frac{1}{I_e} \left[\frac{C}{P} (L-1)(h_A - h_M)\alpha_e z_A' \right]' = 0 \qquad (4\text{-}84)$$

[1] J. A. Fay and F. R. Riddell, *J. Aeronaut. Sci.*, vol. 25, no. 2, pp. 73–85, 1958.
[2] The above equations are derived in detail in Sec. 5-9.

where an auxiliary equation relating T, g, f', and z_A is required. Equations (4-6), (4-18), (4-19), (4-26) through (4-27), and (4-45) reveal that, for a binary mixture,

$$T(\eta) = \frac{g(\eta)I_e - \alpha_e z_A(\eta)h_A^0 - \frac{1}{2}[f'(\eta)]^2}{\alpha_e z_A(\eta)C_{p_A} + [1 - \alpha_e z_A(\eta)]C_{p_M}} \qquad (4\text{-}85)$$

is the required relation between T and g, f', and z. C_{p_A} and C_{p_M} are the specific heats of the atoms and molecules, respectively. Equation (4-85) can be used with Eqs. (4-82), (4-83), and (4-84) to obtain an explicit differential equation for $T(\eta)$.[1] Boundary conditions are

$$z(0) = z_w \qquad z(\infty) = 1$$
$$f(0) = f'(0) = 0 \qquad f'(\infty) = 1$$
$$g(0) = g_w \qquad g(\infty) = 1 \qquad g'(\infty) = 0$$

Fay and Riddell solved these equations for values of C_1 ranging from 0 to ∞ for constant values of $P = 0.71$ and $L = 1.0$, 1.4, and 2.0 ($P = SL$). Scala and Baulknight[2] solved these equations for $C_1 = 0$ and ∞ (frozen flow and equilibrium flow) and with S, P, and L taking values which varied through the boundary layer according to the local thermodynamic state variables and species concentrations. Scala and Baulknight also included terms in Eqs. (4-82) and (4-84) which account for thermal diffusion, an effect which is of second order through most, but not all, of the boundary layer. Both sets of authors assume that $C = C(\eta)$ through the boundary layer and determine the density ratio according to the equation

$$p_w = \rho_w(1 + \alpha_w)\frac{k}{2m_A}T_w = p = \rho(1 + \alpha)\frac{k}{2m_A}T$$

whence
$$\frac{\rho}{\rho_w} = \frac{(1 + \alpha_w)T_w}{(1 + \alpha)T} = \frac{[1 + \alpha_e(z_A)_w]T_w}{(1 + \alpha_e z_A)T} \qquad (4\text{-}86)$$

The viscosity laws used differed in the two cases. Fay and Riddell used Sutherland's formula; viz.,

$$\frac{\mu}{\mu_w} = \frac{T_w + 216}{T + 216}\left(\frac{T}{T_w}\right)^{1.5} \qquad (4\text{-}87)$$

---

[1] Fay and Riddell base the transformation involving the independent variable $\bar{s}$, Eq. (4-50), upon $\rho_w\mu_w$ rather than $\rho_e\mu_e$ as in the present case and also use $C = \rho\mu/\rho_w\mu_w$ instead of $C = \rho\mu/\rho_e\mu_e$. This results in the identical system of equations to be solved and identical boundary conditions.

[2] Sinclaire M. Scala and Charles W. Baulknight, *ARS J.*, vol. 30, no. 4, pp. 329–336, 1960. Scala and Baulknight also use $\rho_w\mu_w$ instead of $\rho_e\mu_e$ in transformation equation (4-50) as well as $C = \rho\mu/\rho_w\mu_w$ instead of $C = \rho\mu/\rho_e\mu_e$ as in the present case. The equations to be solved and their boundary conditions are identical with the present system.

Scala and Baulknight[1] used calculations of viscosity for equilibrium binary mixtures with various degrees of dissociation and various temperatures. Their results, then, depend upon pressure and hence altitude for any given flow velocity.

The results of numerical integrations of Eqs. (4-82) through (4-84) can be summarized by Fay's and Riddell's correlation formulas for the cases of a catalytic wall, that is, a surface catalytic to atom recombination there. These correlation formulas are as follows:

1. For a frozen boundary layer ($C_1 \equiv 0$), where $h_w \ll I_e$,

$$\frac{(Nu)_f}{(R_w)^{1/2}} = 0.763P^{0.4}\left[1 + (L^{0.63} - 1)\frac{h_c}{I_e}\right] \tag{4-88}$$

or

$$(-\dot{q}_w)_f = 0.763P^{-0.6}(\rho_e\mu_e)^{1/2}\left[\left(\frac{du_e}{ds}\right)_0\right]^{1/2}\left(\frac{\rho_w\mu_w}{\rho_e\mu_e}\right)^{0.1}$$
$$\times (I_e - h_w)\left[1 + (L^{0.63} - 1)\frac{h_c}{I_e}\right] \tag{4-89}$$

The anticipated resemblance between Eqs. (4-88) and (4-89) and the earlier Eqs. (4-80) and (4-79) is borne out. Equation (4-88) reduces to the result of Sibulkin[2] when $L = 1$ and $\rho\mu = \rho_w\mu_w = \rho_e\mu_e$.

2. For an equilibrium boundary layer ($C_1 \to \infty$), where $h_w \ll I_e$,

$$\frac{(Nu)_{eq}}{(R_w)^{1/2}} = 0.763P^{0.4}\left(\frac{\rho_e\mu_e}{\rho_w\mu_w}\right)^{0.4}\left\{1 + [L^{0.52} - 1]\frac{h_c}{I_e}\right\} \tag{4-90}$$

and

$$(-\dot{q}_w)_{eq} = 0.763P^{-0.6}(\rho_e\mu_e)^{1/2}\left[\left(\frac{du_e}{ds}\right)_0\right]^{1/2}\left(\frac{\rho_w\mu_w}{\rho_e\mu_e}\right)^{0.1}$$
$$\times (I_e - h_w)\left[1 + (L^{0.52} - 1)\frac{h_c}{I_e}\right] \tag{4-91}$$

The variation of $Nu/(R_w)^{1/2}$ with C_1 is shown in Fig. 4-2 for a typical set of hypersonic flight conditions. Two conclusions can be drawn from Fig. 4-2.

1. There is no significant variation of heat-transfer rate with reaction-rate parameter C_1 if the wall is catalytic to recombination.

2. If the wall is noncatalytic, that is, inhibits recombination at the wall, as some glassy substances are apt to do, then significant reduction in heat-transfer rate can be achieved providing the gas-layer reaction rate is low enough. We shall return to this point in Sec. 4-7.

It is readily apparent from inspection of Eqs. (4-79), (4-89), and

[1] Scala and Baulknight, *ARS J.* vol. 29, no. 1, pp. 39–45, 1959.
[2] M. Sibulkin, *J. Aeronaut. Sci.*, vol. 19, no. 8, pp. 570–571, 1952.

(4-91) that $-\dot{q}_w$ is proportional to $(\mu_e)^n$ where $n = 0.5$ or 0.4. Since it is at the high temperatures at which the viscosity is less accurately determined, the uncertainty in μ_e is probably the main uncertainty in calculating stagnation-point laminar-boundary-layer heat using the equations developed herein. Some notion as to the uncertainty in knowledge of μ_e for high values of T_e can be obtained by examining

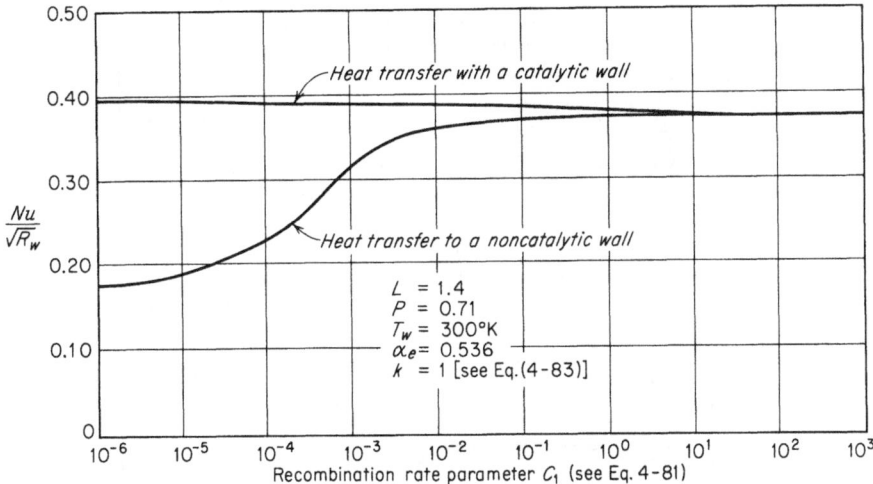

FIG. 4-2. Variation of stagnation-point heat-transfer parameter with a dissociation reaction-rate parameter. (*After J. A. Fay and F. R. Riddell, J. Aeronaut. Sci., vol. 25, no. 2, pp. 73–85, 1958.*)

Fig. 4-3, where several different calculations of high-temperature viscosity variations are presented. The calculation methods are explained in Chap. 10.

Curve 1 is the variation with temperature for pure O_2 calculated, assuming that the Lennard-Jones interaction potential applies throughout the temperature range of interest. Curve 2 is a calculation of viscosity for a binary mixture in equilibrium according to methods outlined by Hansen[1] for a pressure $p_e = 0.034$ atm. Curve 3 is calculated using Sutherland's formula, Eq. (4-85). Curve 4 results from calculations by Scala and Baulknight of the viscosity of pure O, assuming a rigid-sphere interaction potential throughout the temperature range of interest. Curve 5 results from calculations by Scala and Baulknight of the viscosity of an equilibrium mixture of O, N, O_2, and N_2 according to the methods described in Chap. 10 for a pressure $p_e = 0.034$ atm. It is the difference between curves

[1] C. F. Hansen, *NACA TN* 4150, March, 1958.

2, 3, and 5 at high temperatures which serves to indicate the uncertainty in knowledge of μ_e. An uncertainty in heat-transfer rates of the order of 10 to 20 per cent results because of the uncertainty in μ_e at the higher temperatures.

Scala and Baulknight compared the results of their calculations with results obtained using Eqs. (4-79) and (4-91) and with measurements by Vitale, Kaegi, Diaconis, and Warren.[1] The measure-

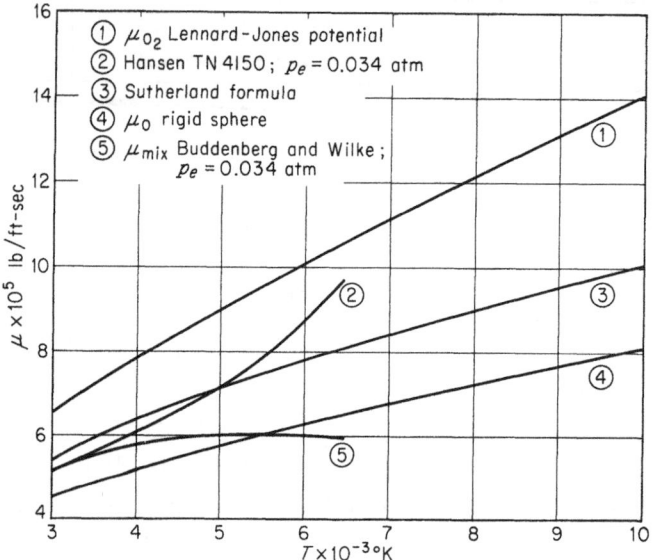

Fig. 4-3. Comparison of various calculations of viscosity at high temperature. (*After Sinclaire M. Scala and Charles W. Baulknight, ARS J., vol. 30, no. 4, pp. 329–336, 1960.*)

ments were made in a hypersonic shock tunnel over ranges of flow Mach number from 7 to 12 and stagnation temperatures T_e from 1500 to 5300°K. The test conditions represented flight at an altitude between 200,000 and 240,000 ft at velocities between 10,000 and about 20,000 ft/sec. Under these conditions, the effects of dissociation of the air in the region of the stagnation point should begin to make their appearance. Figure 4-4 shows the results of the comparison. Based on these test results it can be concluded for these test conditions that:

1. As might be expected from the approximate nature of the theory, Eq. (4-79) is least accurate.

[1] A. E. Vitale, E. M. Kaegi, N. S. Diaconis, and W. R. Warren, *Proc. Heat Transfer and Fluid Mech. Inst.*, pp. 204–275, Stanford University Press, Stanford, Calif., 1958.

2. The experimental scatter is such as to make the direct determination of viscosity by such experiments impractical, at least until techniques improve to reduce the experimental scatter.

Shock-tube measurements of stagnation-point heat transfer reported by Rose and Stark[1] substantiate these conclusions.

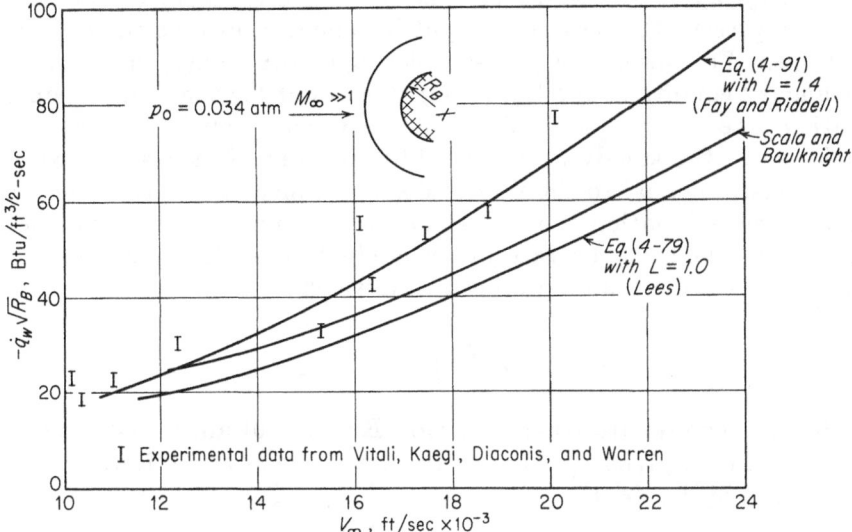

FIG. 4-4. A comparison among various theoretical calculations of stagnation-point heat-transfer rate and experimental measurements. Calculations assume that the surface is catalytic to recombination.

4-7. The Effects of Surface Recombination. Figure 4-2 aptly illustrates the advantages to be gained by inhibiting or preventing atom recombination at the cool surface. A reduction in heat-transfer rate to the stagnation point of the order of 50 per cent might thereby be realized at high flight velocities ($u_\infty > 16,000$ ft/sec). How reasonable is it to expect to achieve a noncatalytic surface to atomic oxygen or nitrogen recombination? In essence, the calculations discussed to this point have shown us that the reaction rates in the gas layer have very little effect upon heat transfer to the surface *providing the recombination rates are high at the surface* (see Fig. 4-2). If the recombination rates are "slow" both in the gas layer and at the surface, then a diffusion-inhibiting blanket ·of unrecombined atoms can pile up near the surface and thereby reduce the heat transfer to the

[1] P. H. Rose and W. I. Stark, *J. Aerospace Sci.*, vol. 25, no. 2, pp. 86–97, February, 1958.

wall by diffusion, which could result in a reduction of heat transfer to the surface.

Since the possible beneficial effects of a noncatalytic wall are limited to the cases where relatively slow reaction rates prevail in the gas layer, it seems appropriate to assess the effects of a noncatalytic wall for the most extreme (and most simple) case, the frozen boundary layer. Certainly calculations made for a frozen boundary layer represent the best that might be achieved using a noncatalytic surface. For this case Eqs. (4-66) through (4-68) apply, with modified boundary conditions for Eq. (4-66) to account for the finite recombination rate at the wall. This boundary condition is arrived at as follows: In the steady state, the diffusion flux of atoms toward the wall must be equal to the rate of disappearance of atoms at the wall due to recombination there. At low wall temperatures, it has been demonstrated that the rate of recombination at the wall follows a simple first-order reaction process and we can write

$$\rho_w D_{12}\left(\frac{\partial \alpha}{\partial y}\right)_w = (k_R \rho \alpha)_w \qquad (4\text{-}92)$$

Using our coordinate transformation Eqs. (4-49) and (4-50) and the definition for z given by Eq. (4-64b), we arrive at the following transformation of Eq. (4-92):

$$z'_A(0) = \left[\frac{\rho_e \mu_e}{2(du_e/ds)_0}\right]^{1/2} \frac{(k_R)_w}{\rho_w D_{12}} z_A(0) \qquad (4\text{-}93)$$

but, from Eq. (4-73),

$$z'_A(0) = 0.47 S^{1/3}[1 - z_A(0)] \qquad (4\text{-}73)$$

Therefore, combining Eqs. (4-93) and (4-73) we arrive at our new boundary condition for Eq. (4-66):

$$z_A(0) = \frac{1}{1 + \left[\dfrac{\rho_e \mu_e}{2(du_e/ds)_0}\right]^{1/2} \dfrac{(k_R)_w}{0.47 S^{1/3} \rho_w D_{12}}} \qquad (4\text{-}94)$$

Clearly when $(k_R)_w = 0$, then $\alpha_w = \alpha_e$ or

$$z_A(0) = 1 \qquad \text{when } (k_R)_w = 0$$

and when k_R is infinite then for a cool wall, $\alpha_w = 0$ or

$$z_A(0) = 0 \qquad \text{when } (k_R)_w = \infty$$

and it was the latter case which we dealt with previously in deriving Eqs. (4-79) and (4-80). Also using Eq. (4-94) with Eq. (4-73) we arrive at

$$z'(0) = 0.47 S^{1/3} \Phi \qquad (4\text{-}95)$$

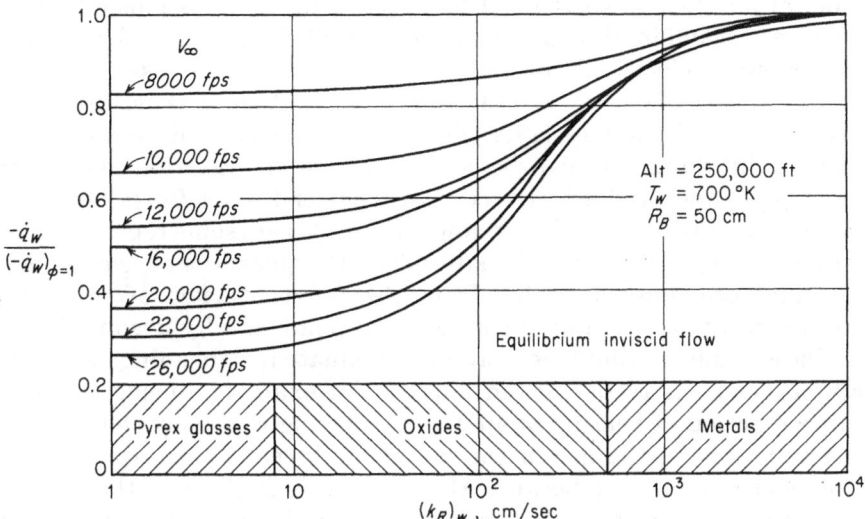

FIG. 4-5. Effect of surface-recombination-rate coefficient upon stagnation-point heat transfer from a frozen dissociated boundary layer. (*After R. Goulard, Jet Propulsion, vol. 28, no. 11, pp. 737–745, 1958.*)

where

$$\Phi = \left\{ 1 + \frac{0.47S^{-2/3}[2(du_e/ds)_0\rho_e\mu_e]^{1/2}}{\rho_w(k_R)_w} \right\}^{-1} \tag{4-96}$$

and $\rho_w\mu_w = \rho_e\mu_e$ as was assumed before. Using Eq. (4-95) with Eq. (4-72) we arrive at

$$\frac{Nu}{(R_w)^{1/2}} = 0.66P^{1/3}\left[1 + (L^{2/3}\Phi - 1)\frac{h_c}{I_e - h_w} \right] \tag{4-97}$$

or

$$-\dot{q}_w = 0.66P^{-2/3}(\rho_e\mu_e)^{1/2}\left[\left(\frac{du_e}{ds}\right)_0 \right]^{1/2}(I_e - h_w)\left[1 + (L^{2/3}\Phi - 1)\frac{h_c}{I_e - h_w} \right] \tag{4-98}$$

Goulard[1] derived Eqs. (4-97) and (4-98) and has computed variations of heat-transfer rate with free-stream velocity u_∞ for different values of $\Phi(k_R)$. It was shown that, as $(k_R)_w$ decreases, the heat transfer to the stagnation point also decreases as was anticipated. Because of the structure of Eq. (4-96), it is seen that reducing density (increasing altitude) or decreasing nose radius [increasing $(du_e/ds)_0$] will have a similar effect upon heat transfer from a boundary layer.

Figure 4-5 presents some results of using Eq. (4-98) calculations.

---

[1] R. Goulard, *Jet Propulsion*, vol. 28, no. 11, pp. 737–745, 1958.

The ratio of $(-\dot{q}_w)$ with $\Phi \leq 1$ to $(-\dot{q}_w)$ with $\Phi = 1$ is plotted versus surface recombination reaction-rate coefficient $(k_R)_w$ for several hypersonic velocities. In these calculations, it is assumed that $L = 1.4$, $S = 0.485$, $P = 0.71$, and $C = 1$. Also, it is assumed that the gas at the boundary-layer edge is in thermodynamic and dissociation equilibrium. As can be seen, reducing the reaction-rate coefficient $(k_R)_w$ does, indeed, reduce heat transfer from a *frozen* boundary layer at hypersonic speeds. Shown in Fig. 4-5 are some representative values of $(k_R)_w$ for several materials. It appears that glasses and ceramics are most desirable from the point of view of inhibiting surface recombination at the stagnation point of a hypersonic body.[1]

These results should be carefully evaluated, depending upon the applications the reader has in mind. For example, during the most intense period of heating of an object entering relatively sharply into the atmosphere, the gas in the boundary layer is most apt to be in chemical equilibrium because the density is high, and the present results would have little bearing upon the problem of heat-protection-system design. For an object entering the atmosphere at a slight angle or gliding at great altitude at hypersonic speeds, the present results have some significance.

We have now completed the analysis of the effects of dissociation on the heat transfer from a laminar boundary layer to the stagnation point of a blunt body. In the next section we shall treat the calculation of laminar-boundary-layer heat transfer in regions downstream from the stagnation point.

4-8. Effect of Pressure Gradient and Curvature. Sections 4-6 and 4-7 dealt with the solution to the laminar-boundary-layer equations for a dissociating binary mixture near the stagnation point. The present section deals with the extension of such solutions to regions away from the stagnation point on blunt bodies. Such solutions are required to treat the heat transfer to objects in flight at hypersonic speeds in regions away from the stagnation point when the boundary layer (hopefully) remains in the laminar-flow regime. As in the preceding text, we shall lead up to the more exact treatment of this problem by proceeding from approximate solutions to more exact solutions.

Let us begin our analyses by seeking the solution for the laminar-boundary-layer heat transfer to a body of revolution under the special circumstances where $L = 1$, the hot gas layer is in thermodynamic

---

[1] P. M. Chung has presented an analysis of the effects of air injection at the stagnation point upon surface recombination and heat transfer to a blunt body with finite surface catalytic activity in *NASA TN* D-27, 1959. As might be anticipated, air injection is effective in reducing heat transfer.

equilibrium, and the surface temperature is much less than the external stream temperature. For most hot gas mixtures of interest to the hypersonic gas dynamicist, $L \simeq 1.0$, and the approximation involved in assuming that $L = 1$ is a reasonable one. For $L = 1$ and $I_e = \text{const}$, Eqs. (4-52) and (4-53) become

$$(Cf'')' + ff'' + \frac{2\bar{s}}{u_e}\frac{du_e}{d\bar{s}}\left[\frac{\rho_e}{\rho} - (f')^2\right] = 0 \qquad (4\text{-}99)$$

and

$$\left(\frac{C}{P}g'\right)' + fg' + \frac{u_e^2}{2I_e}\left[2C\left(1 - \frac{1}{P}\right)f'f''\right]' = 0 \qquad (4\text{-}100)$$

with boundary conditions

$$\begin{array}{ll}
f(0) = f'(0) = 0 & f'(\infty) = 1 \\
g(0) = g_w(\bar{s}) \quad \text{or} \quad g'(0) = 0 & g(\infty) = 1
\end{array} \qquad (4\text{-}101)$$

where the conditions that Eqs. (4-99) and (4-100) be ordinary differential equations with independent variable η are

$$C = C(\eta) \text{ or const}$$

$$\frac{2\bar{s}}{u_e}\frac{du_e}{d\bar{s}}\left[\frac{\rho_e}{\rho} - (f')^2\right] = \text{a function of } \eta \text{ or const}$$

$$P = 1 \text{ or } \frac{u_e^2}{2I_e} = \text{const and } P = P(\eta)$$

$$g_w = \text{const}$$

If the above conditions are not satisfied, the usefulness of Eqs. (4-99) and (4-100) is impaired, since they then become nonlinear equations. In general, these conditions are unlikely to be satisfied away from the stagnation point. However, let us see if we can obtain a useful approximate solution for this case. Examine, first, the solutions to Eqs. (4-99) and (4-100) for the case where they become ordinary differential equations. Consider, for example, the special use where $P = 1$ and $C = 1$. Then, since $\rho_e/\rho = T/T_e$, Crocco's integral, Eq. (2-106), gives

$$\frac{\rho_e}{\rho} = g + \frac{\gamma - 1}{2}M_e^2[g - (f')^2] \qquad (4\text{-}102)$$

and Eqs. (4-99) and (4-100) become

$$f''' + ff'' + \beta[g - (f')^2] = 0 \qquad (4\text{-}103)$$

and

$$g'' + fg' = 0 \qquad (4\text{-}104)$$

where Eq. (4-104) is an ordinary differential equation if

$$\beta = \frac{2\bar{s}}{M_e} \frac{dM_e}{d\bar{s}} = \text{const} \tag{4-105}$$

or
$$M_e = \text{const } (\bar{s})^{\beta/2} \tag{4-106}$$

Equations (4-103) and (4-104) have been solved subject to boundary conditions (4-101) by Cohen and Reshotko,[1] Levy,[2] and Li and Nagamatsu[3] among others. Their results show that, providing $g(0)$ is small, the solutions for $g'(0)$ and $f''(0)$ are insensitive to variations in the pressure-gradient parameter β and, to a first approximation, might be taken to be constant. The implications of this finding in the present case are as follows: In the hypersonic case $g(0) \approx T_w/T_e$ is less than one for blunt bodies near the stagnation point. Also, $\rho\mu = \rho_e\mu_e$ or $C = 1$ over a large portion of the boundary layer. Furthermore, $P \simeq 1$ for dissociated air at the temperatures of interest to us here. It seems reasonable that solutions to Eqs. (4-103) and (4-104) obtained for $\beta = \text{const}$ might be applied to a case where β varies slowly along the body, and if so, we would have a method for calculating heat transfer in regions away from the stagnation point of a body. It has been shown that solutions to Eq. (4-103) with $\beta = \beta_{\text{local}} = \beta(s)$ do, indeed, approximate *measured* heat-transfer rates,[4] and thus the concept of "local similarity" is experimentally demonstrated to be acceptable. Let us develop a method of calculating heat transfer using this concept.

For equilibrium flow with $L = 1$, making use of Eq. (4-56) and our coordinate transformation equations (4-49) and (4-50), we have

$$\tau_w = \mu_w \frac{\partial u}{\partial y} = \frac{\rho_w \mu_w u_e^2 r_0^{2k}}{(2\bar{s})^{1/2}} f''(0) \tag{4-107}$$

and
$$-\dot{q}_w = \left(\frac{k}{C_p}\right)_w \left(\frac{\partial I}{\partial y}\right)_w = \frac{k_w}{C_{p_w}} \frac{\rho_w u_e}{(2\bar{s})^{1/2}} I_e g'(0) \tag{4-108}$$

For the initial approximation we shall use the values of $f''(0)$ and $g'(0)$ obtained for $\beta = 0$. It is shown in Sec. 4-5 that, if $\beta = 0$,

$$f''(0) \simeq 0.47 \tag{4-109}$$

and
$$g'(0) \simeq 0.47 P^{1/3}[1 - g(0)] \tag{4-110}$$

[1] Clarence B. Cohen and Eli Reshotko, *NACA Rept.* 1293, 1956.

[2] Solomon Levy, *J. Aeronaut. Sci.*, vol. 21, pp. 459–474, 1954.

[3] Ting Yi Li and Henry T. Nagamatsu, *J. Aeronaut. Sci.*, vol. 22, pp. 607–616, 1955.

[4] See, for example, Howard A. Stine and Kent Wanlass, *NACA TN* 3344, 1954.

Now define

$$\omega_e = \frac{\mu_e}{R_e T_e} \tag{4-111}$$

and

$$F(s) = \frac{\dfrac{p}{p_e} \dfrac{\omega_e}{(\omega_e)_0} \dfrac{r_0^k}{(2)^{1/2}} \dfrac{u_e}{u_\infty}}{\left[\displaystyle\int_0^s \frac{p}{p_0} \frac{u_e}{u_\infty} \frac{\omega_e}{(\omega_e)_0} r_0^{2k}\, ds\right]^{1/2}} \tag{4-112}$$

where p_0 is the stagnation pressure at the stagnation point, u_∞ is the flight velocity of our blunt body, and the subscript 0 denotes values at the stagnation point. Then it can be shown, substituting (4-110) and (4-112) into (4-108), that

$$-\dot{q}_w(s) = 0.47 P^{-2/3}(I_e - h_w)[(\rho_e\mu_e)_0 u_\infty]^{1/2} F(s) \tag{4-113}$$

Furthermore, at the stagnation point,

$$F(0) = 2^{k/2}\left[\frac{1}{u_\infty}\left(\frac{du_e}{ds}\right)_0\right]^{1/2}$$

and $\omega_e = (\omega_e)_0$; hence

$$\frac{-\dot{q}_w(s)}{-\dot{q}_w(0)} = \frac{F(s)}{2^{k/2}[(1/u_\infty)(du_e/ds)_0]^{1/2}} \tag{4-114}$$

where $F(s)$ is given by Eq. (4-112). Equation (4-114) was first derived by Lees.[1] Implicit within its derivation are the assumptions of $\beta = 0$, $g(0) = g_w \ll 1$ (a cool wall), $\rho_e\mu_e = \rho_w\mu_w$, $L = 1$, and $P \simeq 1$, and the dissociating binary mixture is in chemical equilibrium at every point in the boundary layer. The utility of this equation will be demonstrated later when it is compared with more exact solutions and with experiment.

Kemp, Rose, and Detra,[2] in an attempt to improve upon Eq. (4-113), used the local similarity concept and took into account the variation of β with $\bar{s}$ and the variation of C with η and assumed that $P = 0.71$ through the boundary layer. In essence they solved Eqs. (4-99) and (4-100) subject to boundary conditions (4-101) assuming local values of u_e and $\dfrac{2\bar{s}}{u_e}\dfrac{du_e}{d\bar{s}}$, which vary "slowly" with $\bar{s}$. They further stated that Eq. (4-89) or (4-91) rather than Eq. (4-79) should be used at the stagnation point, since the former equations result

---

[1] Lester Lees, *Jet Propulsion*, vol. 26, no. 4, pp. 259–269, 1956.

[2] Nelson H. Kemp, Peter H. Rose, and Ralph W. Detra, *J. Aerospace Sci.*, vol. 26, no. 7, pp. 421–430, July, 1959.

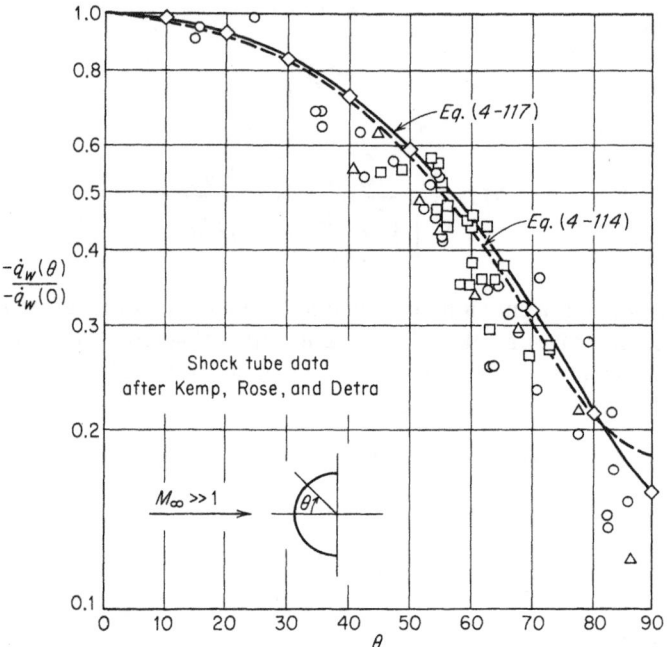

FIG. 4-6. Heat-transfer distribution on a hemisphere.

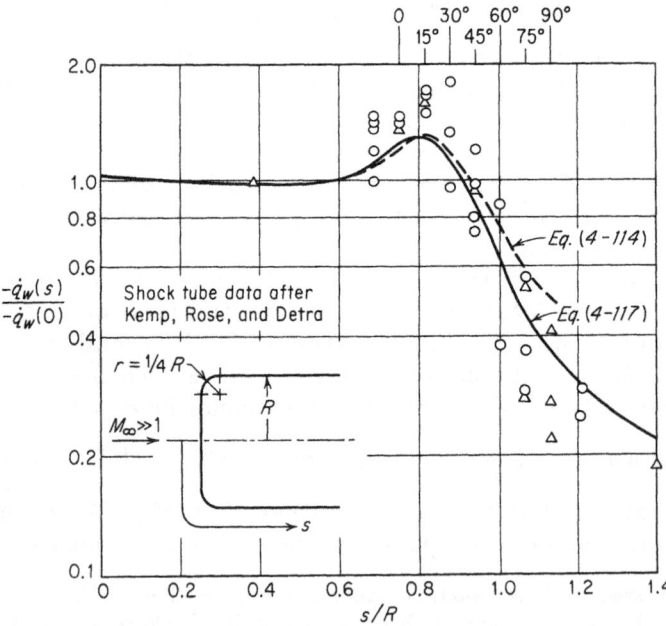

FIG. 4-7. Heat-transfer distribution on a flat-nosed body.

from exact solutions to the equations of motion at the stagnation point whereas Eq. (4-79) was derived under the restrictive assumption that $C = 1$ and $\beta = 0$. Their results can be summarized with the correlation equation

$$\left[\frac{g'(0)}{1 - g(0)}\right]_s \left[\frac{1 - g(0)}{g'(0)}\right]_0 = \frac{1 + 0.096[\beta(s)]^{1/2}}{1.068} \tag{4-115}$$

where
$$\beta = \frac{2\bar{s}}{u_e}\frac{du_e}{d\bar{s}} \tag{4-105}$$

and the subscripts s and 0 denote the position at s along the body and at the stagnation point, respectively. If the wall temperature is constant, then $[g(0)]_s = [g(0)]_0$ and Eq. (4-115) becomes

$$\frac{[g'(0)]_s}{[g'(0)]_0} = \frac{1 + 0.096[\beta(s)]^{1/2}}{1.068} \tag{4-116}$$

It can be shown that these results create a multiplying factor for Eq. (4-114) such that

$$\frac{-\dot{q}_w(s)}{-\dot{q}_w(0)} = \frac{F(s)}{2^{k/2}[(1/u_\infty)(du_e/ds)_0]^{1/2}}\left\{\frac{1 + 0.096[\beta(s)]^{1/2}}{1.068}\right\} \tag{4-117}$$

providing that $-(\dot{q}_w)_0$ is taken as Eq. (4-89) or (4-91) and not Eq. (4-79) and the assumption is made that $C(0,0) = C(s,0)$. If this latter assumption is too restrictive, an equation analogous to Eq. (4-117) develops which differs owing to the use of $\rho_w\mu_w$ rather than $\rho_e\mu_e$ in transformation equations (4-49) and (4-50). Since in some cases the variation $C(0)$ around a body is slight, this difference has been obviated here by simply assuming that $C(0,0) = C(s,0)$ with the understanding that Eq. (4-117) must be modified when this assumption is too restrictive. Some indication of the importance of the multiplying factor in Eq. (4-117) which depends upon β is that β can vary between the limits $0.174 \leq \beta \leq 3.27$ and the factor in the braces will vary within ± 10 per cent. Since for most bodies of practical interest β is well within this range, it can be anticipated that Eqs. (4-117) and (4-114) will give close to the same results.

Measurements of heat-transfer rate on a hemisphere and on a flat-nosed body of revolution which were taken in a shock tube were reported by Kemp, Rose, and Detra and were compared with results of calculations made using Eqs. (4-114) and (4-117). Figure 4-6 shows the results for the hemisphere body, and Fig. 4-7 shows results for the flat-nosed body. The agreement between theory and experiment is good to fair, and both theoretical equations give results within the expected tolerance. Where β is large or small (that is,

$\beta \to 3$ or $\beta \to 0$), the two theories differ by an amount to be expected according to the discussion previously advanced.

Zakkay[1] reports some measurements of heat-transfer rates over a blunted 20° semivertex angle cone which compare quite well with Eq. (4-114) combined with Eq. (4-91). These measurements, taken in a conventional Mach 6 wind tunnel, are quite significant because they indicate that Eq. (4-114) is accurate even when the value of $g(0)$ is not zero but relatively high, about 0.43 in this case, and shows that the concepts leading to Eq. (4-114) are applicable even when $g(0)$ is not precisely zero.

The discussion in this section has described a method of approximate solution for the heat transfer from a dissociating laminar layer to a cool surface based upon the concept of "local similarity." The validity of this method rests with favorable comparisons with measurements and not with any mathematically rigorous examination of the assumptions involved. To accomplish the latter, solutions to the full equations for appropriate boundary conditions are required, the obtaining of which is, unfortunately, a tedious undertaking even with the use of modern computing machines.

4-9. Conclusions. This chapter was concerned with one of the simplest cases of a reacting boundary layer, that of a dissociating laminar boundary layer. The effects of dissociation upon heat transfer to or from a body immersed within a laminar boundary were illustrated with simple and complex solutions to the laminar-boundary-layer equations. Some conclusions can be drawn based upon the material presented in this chapter.

1. While experimental data are sparse, those data which do exist substantiate the validity of the theory within the accuracy of the measurements.

2. It is apparent that the accuracy of engineering applications of the heat-transfer equations derived in this chapter is determined mostly by the accuracy with which the transport properties of high-temperature gas mixtures are known.

3. The effects of dissociation in the laminar boundary layer in the absence of mass transfer or chemical reactions other than dissociation are not appreciable providing the surface of the cool object is catalytic to atom recombination there and providing the driving enthalpy potential is properly defined. The results shown in Figs. 4-2 and 4-5 support this conclusion, since they indicate that, if the surface is catalytic to recombination, the heat transfer to a cool body is insensitive to gas-phase reaction rates. If the gas-phase reaction rates

[1] Victor Zakkay, *J. Aerospace Sci.*, vol. 25, no. 12, pp. 794–795, December, 1958.

are high enough ($C_1 > 10^2$ for the conditions applicable to Fig. 4-2), then the heat transfer to a cool body does not depend upon surface recombination rates at all.   Most cases of practical significance come within the qualifications of this conclusion.

4. Inevitably transition from laminar-boundary-layer flow to turbulent-boundary-layer flow will occur at some position on the body downstream from the nose or leading edge if the body is long enough.   The laminar-boundary-layer analyses presented here are all satisfying to the gas dynamicist but in many instances of importance to designers will be of little use in the turbulent-boundary-layer regime.   Chapter 7 will be concerned with the effects of dissociation in the turbulent boundary layer.

5

Mass Transfer and Chemical Reactions in the Laminar Boundary Layer

5-1. Introduction. This chapter is concerned with the effects of mass transfer and chemical reactions upon laminar-boundary-layer characteristics. The equations which are developed in this chapter can be used to calculate the effects of any number of simultaneously occurring chemical reactions upon the compressible laminar-boundary-layer heat transfer. An example calculation is provided to illustrate how the equations can be used.

As was indicated in Chap. 3, mass transfer may come about in a number of ways, including the direct introduction of a gas into the boundary layer through a porous wall, the gasification or sublimation of a solid wall, or the vaporization of a liquid layer which is either melted undersurface material or a liquid introduced into the boundary layer through upstream openings. In general, we are concerned here with the effect of the laterally introduced gas upon laminar-boundary-layer characteristics, disregarding how the gas arrived at the interface between the gas and liquid or solid state. Figure 5-1 illustrates a number of these mass-transfer circumstances. In each case the technique for mathematical solution of the boundary-layer equations for the gas layer is the same. The boundary conditions are the same in each case as far as the gas layer is concerned.

The effects of chemical reaction are handled in a manner similar to that outlined in Sec. 3-3. However, the treatment presented in this chapter removes several simplifying assumptions involved in the analysis of Chap. 3. In particular, the transport parameters L, P, and S need not be equal to one and the effects of the gas mixture resulting from the chemical reactions upon the important product $\rho\mu$ are taken into account. Thus, the variation of C_H with P and composition and $-\dot{q}_w$ with L will be accounted for in this chapter along with the effects of mass transfer upon $-\dot{q}_w$ which were also neglected in the simplified treatment of Sec. 3-3.

As might be anticipated when we deal with chemical reactions, we must use several concepts from thermochemistry. Rather than present these concepts fully in this chapter, a separate chapter,

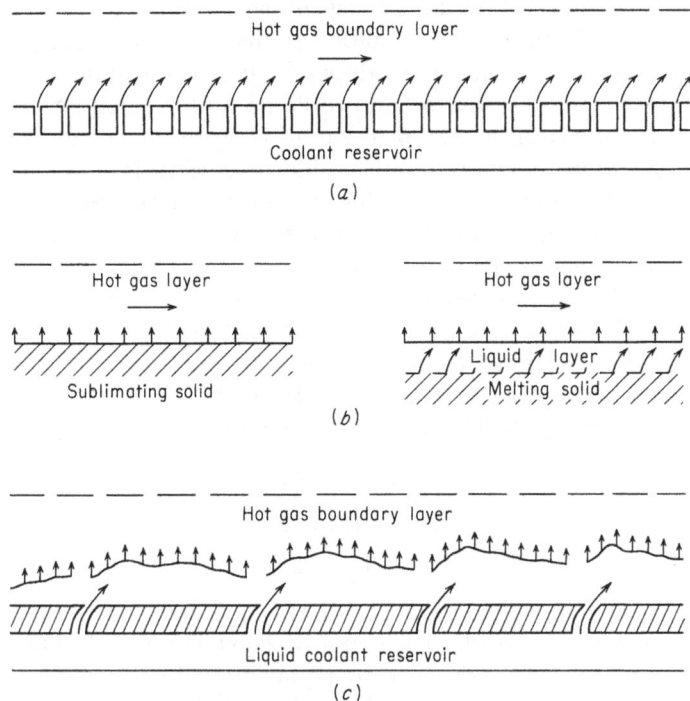

Fig. 5-1. Various conditions resulting in mass-transfer effects upon the gas boundary layer: (a) Gas introduced through a porous wall (transpiration cooling); (b) ablation; (c) liquid-film cooling.

Chap. 9, was prepared which is entirely devoted to thermochemistry and thermodynamic functions. Chapter 9 will be referred to from time to time in reference to certain equations used in this chapter. Chapter 9 is written in such a way as to stand as a separate and complete summary of the derivation of the thermochemistry concepts used in this book.

5-2. Applicability of the Boundary-layer Equations. The objective of this chapter is to treat the combined effects of mass transfer and chemical reaction upon heat transfer to or from a surface from or to a reacting gas stream. The equations to be developed can account for the Lewis number and Prandtl number being other than one for most gas mixtures, for mass transfer, and for simultaneously occurring chemical reactions, including burning of the species being introduced into the gas stream by mass transfer and dissociation of the hot gas stream species. With the use of the results developed in Sec. 4-8, the results of this chapter can be made applicable to blunt-nosed bodies at the stagnation point and away from the stagnation point, and with the use of the results of Sec. 5-11, the results of this chapter can be made applicable to the flat-plate case. All the above results, however, depend upon the applicability of the boundary-layer equations to the problems under consideration.

Consider, for example, the problem of a representative reacting gas stream. Suppose that a combustible material is vaporizing and entering a hot gas stream. Under these circumstances, it is entirely reasonable to assume that the combustible material will react with the hot gas species within the gas stream. For example, graphite (carbon) and oxygen might be the reactants. Presumably the boundary-layer equations presented in Sec. 2-4 apply in this problem. However, when we proceed to make use of these equations, we begin to run into difficulties. For example, even with the simplest of reactions involving O_2, O, C, and CO, which might include

$$C + O \rightleftharpoons CO$$

$$2C + O_2 \rightleftharpoons 2CO$$

$$O_2 \rightleftharpoons 2O$$

we are immediately confronted with the necessity of accounting for four separate species and with using at least three different reaction rates in our species-concentration equation. Furthermore, our observation of flame structure and the knowledge that reaction rates are finite lead us to the conclusion that the chemical reactions occur in a thin but finite reaction zone or flame front. Gradients of concentration and temperature are large in the reaction zone, and hence, terms neglected in deriving the boundary-layer equations of Sec. 2-4 may be significant within the reaction zone. It becomes apparent that there are several conceivable situations involving chemical reactions in a gas stream which cannot be treated using the boundary-layer theory.

Dooley,[1] using a comparatively simple model of a reacting mixture (a thermally decomposing polymer) flowing over a heated flat plate, was able to demonstrate the limitations of the boundary-layer equations in dealing with reacting gas mixtures. For his special case he

[1] Donald A. Dooley, *Proc. 1957 Heat Transfer and Fluid Mech. Inst.*, pp. 321–342, Stanford University Press, Stanford, Calif., 1957.

was able to show that the boundary-layer equations applied up to the point that the flame front or reaction zone became attached to the flat plate.   As soon as this occurred, gradients in the s direction which are neglected in deriving the boundary-layer equations become of significant order of magnitude and the boundary-layer theory no longer applies.

We shall imagine a model for our reacting gas mixture which will preclude the appearance of a reaction zone or flame front attaching itself to the surface normal to the wall.   First we shall assume that the reactions take place at the wall or immediately adjacent to it and that the rest of the hot gas mixture is frozen.   The theory for a frozen laminar boundary layer flowing over a reacting surface will then be developed.   Following this, the method of accounting for gas-phase reactions will be described for use whenever the frozen-gas-layer approximation is believed to be inadequate.

5-3. Frozen-boundary-layer Approximation.   It can be demonstrated that the heat transfer from a boundary layer composed of a mixture of dissociating gases is independent of the location of the reaction zone within the boundary layer to a first approximation, since for most gases the Lewis number is close to unity.   This approximation includes the assumption that the transport properties are independent of the composition of the boundary-layer gas mixture or at least that their variation with composition is of secondary magnitude to their variation with temperature.   We shall make use of this convenient finding here by assuming that the chemical reactions we are concerned with occur at the surface and that gas-phase reactions are frozen.   Thus, under this assumption, the composition of the gas mixture throughout the boundary layer will be determined by the convection and diffusion of products and reactants through the boundary layer, the gas species present in the external stream, and the gas species at the surface, including those which may not be participating in the surface chemical reactions.   Let us first examine the frozen-boundary-layer approximation.

In order to proceed, we define a partial enthalpy I_f such that

$$I_f = \sum_i C_i \left(\int_0^T C_{p_i} \, dT \right) + \frac{u^2}{2} \tag{5-1}$$

that is

$$I = I_f + \sum_i C_i h_i^0 \tag{5-2}$$

Then, it follows that

$$\frac{\partial I_f}{\partial y} = C_{p_f} \frac{\partial T}{\partial y} + u \frac{\partial u}{\partial y} + \sum_i \left(\int_0^T C_{p_i} \, dT \right) \frac{\partial C_i}{\partial y} \tag{5-3a}$$

where
$$C_{p_f} = \sum_i C_i C_{p_i} \qquad (5\text{-}3b)$$

Now, for a mixture of monatomic, diatomic, and triatomic molecules, let us make the convenient assumption that

$$C_{p_i} \simeq C_p = C_{p_f} = \text{const}$$

neglecting the difference among species of the contributions by vibrational, rotational, and electronic modes to a first approximation. Therefore,

$$\sum_i \int_0^T C_{p_i} \, dT \, \frac{\partial C_i}{\partial y} \simeq C_{p_f} T \sum_i \frac{\partial C_i}{\partial y} = 0 \qquad (5\text{-}4)$$

since, by conservation of mass,

$$\sum_i C_i = 1 \qquad (5\text{-}5)$$

thus, to a reasonable approximation,

$$\frac{\partial I_f}{\partial y} \simeq C_{p_f} \frac{\partial T}{\partial y} + u \frac{\partial u}{\partial y} \qquad (5\text{-}6)$$

and
$$\frac{\partial I}{\partial y} = \frac{\partial I_f}{\partial y} + \sum_i h_i^0 \frac{\partial C_i}{\partial y} \qquad (5\text{-}7)$$

Now, providing that chemical reactions are taking place at the wall, heat is transported at the surface by two methods: conduction and diffusion; that is,

$$-\dot{q}_w = \left(k \frac{\partial T}{\partial y} + \rho D_{12} \sum_i h_i \frac{\partial C_i}{\partial y} \right)_w \qquad (5\text{-}8)$$

hence, using Eqs. (5-6) and (5-8),

$$-\dot{q}_w = \left(\frac{k}{C_{p_f}} \frac{\partial I_f}{\partial y} + \rho D_{12} \sum_i h_i \frac{\partial C_i}{\partial y} \right)_w \qquad (5\text{-}9)$$

since $u = 0$ at the surface. Also using Eq. (5-7) with Eq. (5-9), we can write, in view of Eq. (5-4),

$$-\dot{q}_w = \left[\frac{k}{C_{p_f}} \frac{\partial I}{\partial y} + (L - 1) \sum_i h_i \frac{\partial C_i}{\partial y} \right]_w \qquad (5\text{-}10)$$

Hence, if $L = C_{p_f} \rho D_{12}/k = 1$, heat transfer to the surface $-\dot{q}_w$ is independent of the location of the chemical reactions *providing* k/C_{p_f} is independent of composition and *providing* the solution to the energy equation for $I(\eta)$ can be uncoupled from the solution to the conservation-of-species equation. Since L is close to one and, in addition, transport properties are relatively insensitive to composition for many gas mixtures and the parameter $C = \rho\mu/\rho_e\mu_e$ and hence $f(\eta)$ are insensitive to composition, our statement that heat transfer is insensitive to the location of the chemical-reaction zone in the boundary layer becomes plausible and the frozen-boundary-layer approximation becomes tenable.

5-4. Boundary-layer Heat Transfer. *Frozen-boundary-layer Equations.* In order to determine heat transfer to the surface using Eq. (5-9), we must solve for variations of I_f and C_i through the boundary layer. In order to do this, we need a differential equation for I_f. We obtain the required equation by using the boundary-layer equations developed in Sec. 2-4. It will be recalled that, if

$$u = u_e f'(\eta) \tag{5-11}$$

$$C_i = (C_i)_e z_i(\eta) \tag{5-12}$$

and
$$I = I_e g(\eta) \tag{5-13}$$

then it was found that under transformations

$$\bar{s} = \int_0^s \rho_e\mu_e r_0^{2k} u_e \, ds \tag{5-14}$$

and
$$\eta = \frac{\rho_e u_e r_0^k}{(2\bar{s})^{1/2}} \int_0^y \frac{\rho}{\rho_e} \, dy \tag{5-15}$$

our boundary-layer equations become the transformed conservation-of-species equation

$$\left(\frac{C}{S} z_i'\right)' + f z_i' = \frac{2\bar{s} f' z_i}{(C_i)_e} \frac{d(C_i)_e}{d\bar{s}} - \frac{2\bar{s}\dot{w}_i}{\rho\rho_e u_e^2 \mu_e r_0^{2k}(C_i)_e} \tag{5-16}$$

The energy equation becomes

$$\left(\frac{C}{P} g'\right)' + f g' = \frac{2\bar{s} f' g}{I_e} \frac{dI_e}{d\bar{s}} + \left[\frac{C}{S}\left(\frac{1}{L} - 1\right) \sum_i \frac{h_i(C_i)_e}{I_e} z_i'\right]'$$
$$+ \frac{u_e^2}{I_e}\left[\left(\frac{1}{P} - 1\right) C f' f''\right]' \tag{5-17}$$

and the momentum equation was shown to be

$$(C f'')' + f f'' = \frac{2\bar{s}}{u_e} \frac{du_e}{d\bar{s}}\left[(f')^2 - \frac{\rho_e}{\rho}\right] \tag{5-18}$$

Boundary conditions for these equations are

$$At\ y = 0,\ \eta = 0 \qquad\qquad As\ y \to \infty,\ \eta \to \infty$$

$$z_i(0) = (z_i)_w = \frac{(c_i)_w}{(c_i)_e} \qquad\qquad z_i \to 1$$

$$f'(0) = 0 \qquad f(0) = f_w \qquad\qquad f' \to 1 \qquad f \to 0$$

$$g'(0) = g'_w \qquad\qquad\qquad g' \to 0$$

$$g(0) = g_w \qquad\qquad\qquad g \to 1$$

Now, to proceed in the present case, define

$$g_f = \frac{I_f}{(I_f)_e} \tag{5-19}$$

Then, from Eqs. (5-2), (5-12), (5-13), and (5-19), we obtain

$$g = \frac{(I_f)_e}{I_e} g_f + \sum_i \frac{(C_i)_e}{I_e} h_i^0 z_i \tag{5-20}$$

Substitute Eq. (5-20) into Eq. (5-17), and combine the result with Eq. (5-16). There results

$$\left(\frac{C}{P} g_f'\right)' + f g_f' = \frac{u_e^2}{(I_f)_e}\left[\left(\frac{1}{P} - 1\right)Cf'f''\right]' - \frac{1}{(I_f)_e}\left\{\sum_i (C_i)_e h_i^0\left[\left(\frac{C}{S} z_i'\right)' + f z_i'\right]\right\} \tag{5-21}$$

where in Eq. (5-17) we have set $dI_e/d\bar{s} = 0$ (isoenergetic flow) and in Eq. (5-16) we have put $d(C_i)_e/d\bar{s} = 0$ (frozen external stream composition) and $\dot{w}_i = 0$ (frozen gas phase). Now Eq. (5-16) with the forementioned assumption becomes

$$\left(\frac{C}{S} z_i'\right)' + f z_i' = 0 \tag{5-22}$$

Hence Eq. (5-21) becomes

$$\left(\frac{C}{P} g_f'\right)' + f g_f' = \frac{u_e^2}{(I_f)_e}\left[\left(\frac{1}{P} - 1\right)Cf'f''\right]' \tag{5-23}$$

Let us simplify matters by assuming that $P \simeq 1$ but not always exactly unity, that $u_e^2/(I_f)_e = (\gamma_e - 1)M_e^2 \simeq 0$, and that either $du_e/d\bar{s} = 0$ (flat plate and cone flow) or

$$\frac{du_e}{d\bar{s}} \to 0 \quad \text{or} \quad (f')^2 - \frac{\rho_e}{\rho} \simeq 0$$

Then the terms on the right-hand side of Eqs. (5-18) and (5-23) are negligible relative to those on the left-hand side of these equations, and Eqs. (5-18), (5-22), and (5-23) all become of the form

$$\left(\frac{C}{Z} \lambda'\right)' + f\lambda' = 0 \tag{5-24}$$

where $Z = 1, S,$ or P, depending upon whether Eq. (5-24) represents Eq. (5-18), (5-22), or (5-23) and where the boundary conditions are

$$\begin{array}{cc} At\ \eta = 0 & As\ \eta \to \infty \\ \lambda = \lambda(0) & \lambda \to 1 \end{array}$$

and $\lambda'(0) = \text{const}$

Effect of Mass Transfer. The solution to Eq. (5-24) is, by quadrature,

$$\lambda = [1 - \lambda(0)]\frac{G(\eta;Z)}{G(\infty;Z)} - \lambda(0) \tag{5-25}$$

where

$$G(\eta;Z) = \int_0^\eta \frac{Z}{C}\exp\left(-\int_0^{\eta'}\frac{Z}{C}f\,d\eta''\right)d\eta' \tag{5-26}$$

and where

$$\lambda'(0) = \left(\frac{Z}{C}\right)_w \frac{1 - \lambda(0)}{G(\infty;Z)} \tag{5-27}$$

Letting $\lambda' = z_i'$, g_f', and f'', respectively, and using the boundary conditions to Eqs. (5-18), (5-22), and (5-23) with Eq. (5-27) result in

$$(z_i')_w = \left(\frac{S}{C}\right)_w \frac{1 - (z_i)_w}{G(\infty;S)} \tag{5-28}$$

$$(g_f')_w = \left(\frac{P}{C}\right)_w \frac{1 - (g_f)_w}{G(\infty;P)} \tag{5-29}$$

$$(f'')_w = \left(\frac{1}{C}\right)_w \frac{1}{G(\infty;1)} \tag{5-30}$$

where $G(\infty;Z)$ is given by Eq. (5-26) evaluated at $\eta = \infty$. In order to evaluate Eq. (5-26) for any value of Z, we must have an expression for $f(\eta)$ including the value $f(0)$. The relation between $f(0)$ and the mass transfer at the wall can be obtained using Eqs. (5-14) and (5-15) to find

$$\rho v = -\rho_e u_e \mu_e r_0^k \left[2(2\bar{s})^{1/2} f' \frac{\partial \eta}{\partial \bar{s}} + \frac{f}{(2\bar{s})^{1/2}} \right]$$

since v is given by

$$\frac{v}{\bar{R}T} = -\frac{\partial \psi}{\partial s} - \frac{\psi}{p_e r_0^k}\frac{d(p_e r_0^k)}{ds}$$

and

$$\psi = \frac{(2\bar{s})^{1/2}}{p_e r_0^k} f(\eta)$$

Then, since $f'(0) = 0$,

$$(\rho v)_w = -\rho_e u_e \mu_e r_0^k \frac{f(0)}{(2\bar{s})^{1/2}}$$

or

$$f(0) = \frac{-(\rho v)_w (2\bar{s})^{1/2}}{\rho_e u_e \mu_e r_0^k} \tag{5-31}$$

The quantity $Z(2/C)^{1/2}[G(\infty;Z)]^{-1} = 2^{1/2}\lambda'(0)$ was obtained for various values of mass-transfer parameter $-(2/C)^{1/2}f(0)$ using the tabulated

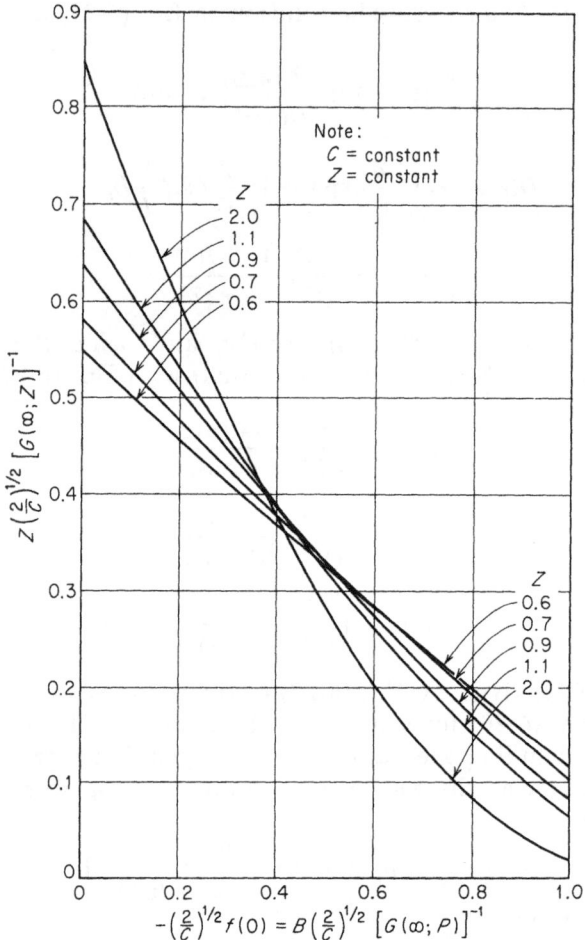

FIG. 5-2. The effect of mass transfer upon the skin-friction and heat-transfer parameter $G(\infty;Z)$.

results for a related solution to the constant-property boundary-layer equations reported by Mickley et al.[1] and which are plotted on Fig. 5-2. The parameter $C = \rho\mu/\rho_e\mu_e$ was assumed to be constant through the boundary layer, since Young and Janssen[2] have shown that, if $C = C(T')$ where, for $M_e < 5.6$,

$$\frac{T'}{T_e} = 0.42 + 0.032M_e^2 + 0.58\frac{T_w}{T_e} \tag{5-32}$$

[1] H. S. Mickley, R. C. Ross, A. L. Squyers, and W. E. Stewart, *NACA TN* 3208, 1954.
[2] G. B. W. Young and E. Janssen, *J. Aeronaut. Sci.*, vol. 19, no. 4, pp. 229–236, 238, April, 1952.

the resulting solutions for skin friction and heat transfer will equal those obtained by letting C vary through the boundary layer. In this way we preserve the effects of variable $\rho\mu$ through the boundary layer upon our solutions and greatly simplify the presentation of results, since Fig. 5-2 and Eqs. (5-28) through (5-30) represent the desired solutions to the compressible laminar-boundary-layer equations with mass transfer present providing

$$C = \frac{\rho(T')\mu(T')}{\rho_e\mu_e} \tag{5-33}$$

where T'/T_e is given by Eq. (5-32). Our chief assumption in presenting the desired results in this way is that the parameters S and P are constant throughout the boundary layer (but not necessarily equal to one). We shall return in Sec. 5-10 to examine the assumptions involved in using a reference temperature in this manner.

Now, using Eq. (5-15), it follows that

$$\frac{\partial}{\partial y} = \frac{\rho u_e r_0^k}{(2\bar{s})^{1/2}} \frac{\partial}{\partial \eta} \tag{5-34}$$

whence, using Eqs. (5-12), (5-19), and (5-34) in Eq. (5-9) results in

$$-\dot{q}_w = \frac{C\rho_e\mu_e u_e r_0^k}{P(2\bar{s})^{1/2}} \left[(I_f)_e g'_f + L\sum_i h_i(C_i)_e z'_i \right]_w \tag{5-35}$$

where

$$\rho_w\mu_w = \frac{\rho_w\mu_w}{\rho_e\mu_e} \rho_e\mu_e = C\rho_e\mu_e$$

Combining Eqs. (5-28) and (5-29) with Eq. (5-35) results in

$$-\dot{q}_w = \frac{\rho_e u_e \mu_e r_0^k}{(2\bar{s})^{1/2}} \frac{(I_f)_e}{G(\infty;P)}$$

$$\times \left\{ 1 - (g_f)_w + \frac{1}{(I_f)_e} \frac{G(\infty;P)}{G(\infty;S)} \sum_i (h_i)_w (C_i)_e [1 - (z_i)_w] \right\} \tag{5-36}$$

where Lewis number $L = P/S$. From Eqs. (5-2), (5-12), and (5-19) and the boundary conditions that $(g_f)_e = 1$ and $(z_i)_e = 1$, it follows that

$$I_e - I_w = (I_f)_e [1 - (g_f)_w] + \sum_i (C_i)_e h_i^0 [1 - (z_i)_w] \tag{5-37}$$

Hence, combining Eqs. (5-36) and (5-37) results in

$$-\dot{q}_w = C_H \rho_e u_e (I_e - I_w) \left\{ 1 + \left[\frac{G(\infty;P)}{G(\infty;S)} - 1 \right] \frac{\sum_i (C_i)_e h_i^0 [1 - (z_i)_w]}{I_e - I_w} \right.$$

$$\left. + \frac{G(\infty;P)}{G(\infty;S)} \frac{\sum_i (C_i)_e \left(\int_0^T C_{p_i} dT \right)_w [1 - (z_i)_w]}{I_e - I_w} \right\} \tag{5-38}$$

where we define

$$C_H \rho_e u_e = \frac{\rho_e u_e \mu_e r_0^k}{(2\bar{s})^{1/2} G(\infty;P)} \qquad (5\text{-}39)$$

According to the frozen-boundary-layer approximation, Eq. (5-38) gives the heat-transfer rate at the surface for a compressible laminar boundary layer of chemically reactive species which incorporates the effects of mass transfer. The effects of mass transfer enter in through the functions $G(\infty;Z)$, which, as Fig. 5-2 shows, are functions of the mass-transfer rate at the surface. This latter quantity, in turn, depends upon the chemical reactions which are assumed to occur at the surface.

When $f(0) = 0$, that is, $(\rho v)_w = 0$, or no heterogeneous surface reactions occur, it has been shown that [see Eq. (4-78)]

$$G(\infty;Z) \simeq 2.11 \frac{Z^{2/3}}{C^{1/2}} \qquad (5\text{-}40)$$

Hence for the case, $(\rho v)_w = 0$ and Eq. (5-38) becomes

$$-\dot{q}_w = C_H \rho_e u_e (I_e - I_w) \left\{ 1 + (L^{2/3} - 1) \frac{\sum_i (C_i)_e h_i^0 [1 - (z_i)_w]}{I_e - I_w} \right\} \qquad (5\text{-}41)$$

since

$$\sum_i (C_i)_e \left(\int_0^T C_{p_i} dT \right)_w [1 - (z_i)_w] \simeq (C_{p_f} T_w) \sum_i [(C_i)_e - (C_i)_w] = 0 \qquad (5\text{-}42)$$

in accordance with Eqs. (5-4) and (5-5) and the assumptions stated before in Eq. (5-4).

Equation (5-41) is a generalization of the expression for heat transfer from a dissociating compressible laminar boundary layer, since it can account for any number of simultaneous gas-phase reactions. See, for example, Eqs. (4-89) and (4-91).

In the next section we shall use Eq. (5-38) in deriving an expression for heat transfer into the interior of a surface which is subliming or otherwise undergoing a heterogeneous chemical reaction. That is, we deal with processes which involve the wall material passing from the solid to gaseous state.

5-5. Heat Transfer to a Solid Surface. In this section we shall derive an equation for the heat transfer into a solid material which is undergoing a surface reaction at the interface between the solid and the gas.

Consider the situation illustrated in the sketch below:

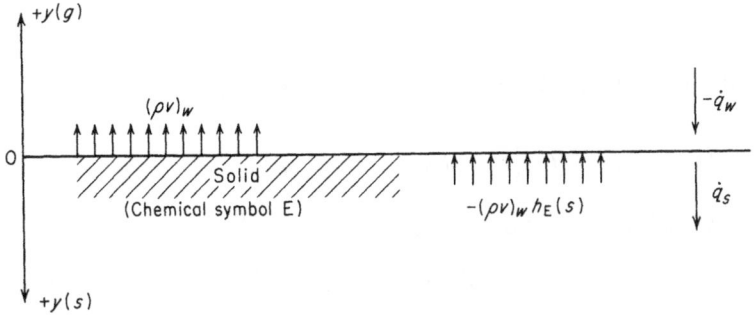

Using the coordinate system defined as shown in the sketch, we can write an equation for the heat flux across the interface $y(g) = 0 = y(s)$. This is, neglecting radiation to and from the boundary-layer gases as being additive factors,

$$(\dot{q}_s)_w - (\rho v)_w [h_E(s)]_w = -\dot{q}_w - (\rho v)_w h_w \tag{5-43}$$

where we use E to denote the chemical symbol of the solid material. Now if the solid is a sublimator, we can write

$$[h_E(s)]_w + L_{v_E} = [h_E(g)]_w \tag{5-44}$$

where L_{v_E} is the heat of vaporization of the material E. Therefore, combining Eqs. (5-43) and (5-44), we obtain

$$(\dot{q}_s)_w = -\dot{q}_w - (\rho v)_w h_w + (\rho v)_w [h_E(g)]_w - (\rho v)_w L_{v_E} \tag{5-45}$$

Equations (5-38) and (5-45) can now be combined to obtain

$$\dot{q}_s = C_H \rho_e u_e \left\{ I_e - I_w - \sum_i h_i^0 [(C_i)_e - (C_i)_w] \right.$$

$$\left. + \frac{G(\infty;P)}{G(\infty;S)} \sum_i (h_i)_w [(C_i)_e - (C_i)_w] - B_3 h_w + B_3 [h_E(g)]_w - B_3 L_{v_E} \right\} \tag{5-46}$$

where we define the convenient mass-transfer parameter B_3 as

$$B_3 = \frac{(\rho v)_w}{\rho_e u_e C_H} \tag{5-47}$$

Now Eq. (5-2) can be used with Eq. (5-46) to obtain

$$\dot{q}_s = C_H \rho_e u_e \left((I_f)_e - (I_f)_w + \frac{G(\infty;P)}{G(\infty;S)} \left\{ \sum_i (h_i)_w [(C_i)_e - (C_i)_w] \right. \right.$$

$$\left. \left. - B_4 h_w + B_4 [h_E(g)]_w \right\} - B_3 L_{v_E} \right) \tag{5-48}$$

where the parameter B_4 is related to B_3 by the equation

$$B_4 = \frac{G(\infty;S)}{G(\infty;P)} B_3 \qquad (5\text{-}49)$$

Now define a term h_c, called the chemical enthalpy potential, such that

$$
\begin{aligned}
h_c &= \sum_i (h_i)_w[(C_i)_e - (C_i)_w] - B_4 h_w + B_4[h_E(g)]_w \\
&= \sum_i (h_i)_w[(C_i)_e - (C_i)_w] - B_4 \sum_i (C_i)_w(h_i)_w + B_4[h_E(g)]_w \\
&= \sum_{i \neq E} (h_i)_w[(C_i)_e - (1 + B_4)(C_i)_w] + [h_E(g)]_w[B_4 - (1 + B_4)(C_E)_w] \quad (5\text{-}50)
\end{aligned}
$$

where $(C_E)_e \equiv 0$, since all material leaving the solid surface is confined to the boundary layer. Then Eq. (5-48) becomes

$$\dot{q}_s = C_H \rho_e u_e \left[(I_f)_e - (I_f)_w + \frac{G(\infty;P)}{G(\infty;S)} h_c - B_3 L_{v_E} \right] \qquad (5\text{-}51)$$

where C_H is defined by Eq. (5-39) and h_c is defined by Eq. (5-50). Defining a driving enthalpy difference as

$$\Delta I = (I_f)_e - (I_f)_w + h_c \qquad (5\text{-}52)$$

Eq. (5-51) becomes

$$\dot{q}_s = C_H \rho_e u_e \Delta I \left\{ 1 + \left[\frac{G(\infty;P)}{G(\infty;S)} - 1 \right] \frac{h_c}{\Delta I} - \frac{B_3 L_{v_E}}{\Delta I} \right\} \qquad (5\text{-}53)$$

when $B_3 = 0$, $B_4 = 0$, and $C_E = 0$ and Eq. (5-53) then becomes Eq. (5-41) when Eq. (5-42) is taken into account, since, for $B_3 = 0$, $\Delta I = I_e - I_w$.

Equation (5-53) gives the heat transfer to the solid material in the presence of mass transfer which results owing to a heterogeneous reaction at the surface and including a term h_c which represents the heat released or absorbed because of chemical reactions among gas species near the surface. Before our problem is completely determined, we need to know the mass fractions of the various species at the surface and at the edge of the boundary layer, which, in turn, are used with Eq. (5-50) to determine the chemical enthalpy potential h_c. We also need a method to calculate the mass-transfer rate $(\rho v)_w$ as a function of surface chemistry. These latter two aspects of our problem will be dealt with next.

5-6. Diffusion-controlled Surface Reactions. There are two possible controlling factors upon the rate of mass loss from a reacting surface over which a gas is flowing. One factor is the chemical kinetics of the heterogeneous surface reaction, and the other is the rate at which reactants can be convected and diffused through the boundary layer to the reacting surface and the rate at which the products of reaction can be diffused and convected away from the surface. We shall assume here that the diffusion and convection process is the limiting one and return to examine this assumption in a later section.

Making the assumption that diffusion and convection determine the distribution of species in the gas layer and that the chemical reaction goes to equilibrium at the surface will determine the mass-loss rate. The material lost from the surface is one of the reactants and will be consumed, under our assumptions, at a rate sufficient to supply a concentration in chemical equilibrium with the maximum concentration of reactants which can be diffused and convected through the boundary layer from the external stream to the surface.   It should be possible, then, to relate the mass-loss parameter B_4 to the concentration of species at the surface and in the external stream using the equation for conservation of species.   This will be done in that which follows.

Consider our transformed equation for conservation of species for a frozen-boundary-layer flow, Eq. (5-22),

$$\left(\frac{C}{S}z_i'\right)' + fz_i' = 0 \tag{5-22}$$

or, since $z_i(C_i)_e = C_i$ by Eq. (5-12) and replacing the index i by k,

$$\left(\frac{C}{S}C_k'\right)' + fC_k' = 0 \tag{5-54}$$

Now multiply Eq. (5-54) by $r_{i,k}$ and sum over all k.   There results, in view of the relation that $\sum_k r_{i,k}C_k = \bar{C}_i$,

$$\left(\frac{C}{S}\bar{C}_i'\right)' + f\bar{C}_i' = 0 \tag{5-55}$$

where $\bar{C}_i$ is the mass fraction of the *element* i present in our gaseous mixture.

Integrating Eq. (5-55) with respect to η gives

$$\frac{C}{S}\bar{C}_i' + \int f\bar{C}_i' \, d\eta = c$$

or integrating the second term by parts, we obtain

$$\frac{C}{S}\bar{C}_i' + f\bar{C}_i - \int \bar{C}_i f' \, d\eta = c$$

Evaluating this expression at $\eta = 0$ gives

$$\left(\frac{C}{S}\bar{C}_i'\right)_w + f(0)(\bar{C}_i)_w = c$$

Now $[(C/S)\bar{C}_i']_w = 0$ when $f(0) = 0$, since $(\bar{C}_i)_w = \bar{C}_i = \text{const}$ in that case. Hence, we can evaluate the constant of integration c as being zero and we have, for all species i not equal to E,

$$\left(\frac{C}{S}\bar{C}_i'\right)_w + f(0)(\bar{C}_i)_w = 0 \qquad i \neq \text{E} \tag{5-56}$$

Now, from Eqs. (3-1), (5-12), and (5-28),

$$(\bar{C}_i')_w = \left(\frac{S}{C}\right)_w \frac{(\bar{C}_i)_e - (\bar{C}_i)_w}{G(\infty;S)} \tag{5-57}$$

Hence, solving Eqs. (5-56) and (5-57) for $(\bar{C}_i)_w$ results in

$$(\bar{C}_i)_w = \frac{(\bar{C}_i)_e}{1 + B_4} \qquad i \neq \text{E} \tag{5-58}$$

where Eqs. (5-31), (5-39), and (5-49) were used to define B_4 and $(\bar{C}_i)_e$ will be given for any problem. B_4 determines the value of $f(0)$ as follows:

$$B_4 = \frac{G(\infty;S)}{G(\infty;P)} B_3 = \frac{(\rho v)_w(2\bar{s})^{1/2}G(\infty;S)}{\rho_e u_e \mu_e r_0^k} = -f(0)G(\infty;S)$$

Furthermore, for $i = \text{E}$, the solid-material specie which participates as a reactant, we can write Eq. (5-55) as

$$\left(\frac{C}{S}\bar{C}_\text{E}'\right)' + f\bar{C}_\text{E}' = 0 \tag{5-59}$$

whence, integrating with respect to η and evaluating the integral at the surface $\eta = 0$ in the manner that led to Eq. (5-56), we obtain

$$\left(\frac{C}{S}\bar{C}_\text{E}'\right)_w + f(0)(\bar{C}_\text{E})_w = f(0) \tag{5-60}$$

since

$$\left(\frac{C}{S}\bar{C}_\text{E}'\right)_w = 0 \qquad \text{when } \bar{C}_\text{E} = 1$$

Now, from Eqs. (3-1), (5-12), and (5-28) we can derive

$$\left(\bar{C}_\text{E}'\right)_w = \left(\frac{S}{C}\right)_w \frac{(\bar{C}_\text{E})_e - (\bar{C}_\text{E})_w}{G(\infty;S)} \tag{5-61}$$

Hence, solving Eqs. (5-60) and (5-61) for $(\bar{C}_E)_w$ results in

$$(\bar{C}_E)_w = \frac{B_4 + (\bar{C}_E)_e}{1 + B_4} \tag{5-62}$$

where B_4 is defined by Eq. (5-49) as before.

Equations (5-58) and (5-62) relate the mass-transfer parameter B_4 to the *element* species-concentration terms at the surface and in the free stream. The equilibrium constants for the surface chemical reaction along with Dalton's law of partial pressures and the fact that pressure is constant through the boundary layer provide enough information to determine the $(\bar{C}_i)_w$. Hence our solution to the problem is now completely specified.

While the specification of all relevant equations is complete at this point, it may not be apparent how these equations can be applied. To illustrate the application of these equations, an example problem was worked out which will be given in the next section.

5-7. An Example Calculation. Consider an experiment wherein a cool supersonic stream composed of a mixture of CO_2 with N_2 is blown over a heated graphite right circular cone.[1] Provided that the surface temperature is sufficiently high, the CO_2 in the boundary layer will react with the graphite surface to form CO in significant amounts. This reaction will, in turn, have some effect upon the heat transfer from the cone to the stream and upon the mass loss from the cone. The experiment is schematically represented in the sketch below along with some assumed test parameters used in our example calculation.

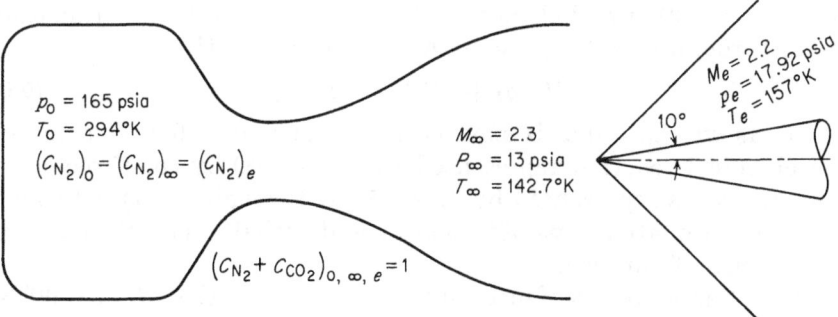

It is assumed that local heat-transfer rates from the cone to the stream could be measured under varying wall temperatures and

[1] This experiment was suggested to the author by W. S. Bradfield and J. J. Sheppard of the Convair Scientific Research Laboratory.

proportions of CO_2 to N_2. The theory of the present chapter will now be applied to this experiment by way of illustration.

In the assumed temperature range of interest to us here ($1000°K \leq T \leq 2500°K$), experiments by Thorne and Winslow[1] have shown that the principal components of gaseous carbon which sublimate from a graphite surface are C, C_2, and C_3 and that C_3 is the predominant species present. To simplify matters, we shall assume that C_3 is the only graphite species which breaks from the graphite solid matrix and enters into chemical reaction with the gaseous CO_2. Because the partial pressures of C, C_2, and C_3 are all small relative to the assumed total pressure of our experiment, it turns out that the results to be obtained are independent of whether we assume that C, C_2, or C_3 or all of them together are the gaseous carbon species participating in a chemical reaction with CO_2 providing, only, that the concentrations of C, C_2, and C_3 are present in equilibrium. Proceeding, then, we assume that our reaction is

$$\tfrac{1}{3}C_3(s) + CO_2(g) \rightarrow 2CO(g) \qquad (5\text{-}63)$$

We can replace this one reaction with four equivalent reactions, viz.,

$$\tfrac{1}{3}C_3(s) \rightarrow \tfrac{1}{3}C_3(g) \qquad (5\text{-}64)$$

$$\tfrac{1}{3}C_3(g) \rightarrow C(g) \qquad (5\text{-}65)$$

$$2C(g) + 2O(g) \rightarrow 2CO(g) \qquad (5\text{-}66)$$

$$CO_2(g) \rightarrow C(g) + 2O(g) \qquad (5\text{-}67)$$

Equation (5-64) is the heterogeneous reaction and is endothermic. The heat absorbed is the latent heat of vaporization of C_3. Since the sublimation reaction is treated as a separate term in our heat-transfer equation (5-53), we shall not deal with it further here. Equations (5-65) through (5-67) represent the gas-phase reaction

$$\tfrac{1}{3}C_3(g) + CO_2(g) \rightarrow 2CO(g) \qquad (5\text{-}68)$$

which, as we shall find, is exothermic. Our reason for replacing the one chemical reaction represented by Eq. (5-68) by the three equivalent reactions represented by Eqs. (5-65) through (5-67) is to make use of the tabulated heats of formation and enthalpies for the reactions and species of concern.

At the temperatures of interest to us here ($1000°K \leq T \leq 2500°K$), there is little indication that compounds involving C and N will form, and hence no account will be taken of such reactions as

$$C_2(g) + N_2(g) \rightarrow C_2N_2(g)$$

[1] R. J. Thorne and G. H. Winslow, *J. Chem. Phys.*, vol. 26, no. 1, pp. 186–197, 1957.

The Chemical Enthalpy Potential h_c. Let us simplify Eq. (5-50), the expression for h_c. Consider a reaction

$$\sum_i^m a_i A_i \to \sum_i^n b_i B_i$$

where a_i, b_i are the stoichiometric coefficients and A_i, B_i are the chemical symbols of reactants and products. Then, we can write

$$\Delta H = \sum_i^n b_i H_{B_i} - \sum_i^m a_i H_{A_i} \tag{5-69}$$

where H_{A_i}, H_{B_i} = molar enthalpy of A_i and B_i.
Then define the specific enthalpies

$$h_{A_i} = \frac{H_{A_i}}{M_{A_i}} \qquad \text{enthalpy/unit mass of } A_i$$

$$h_{B_i} = \frac{H_{B_i}}{M_{B_i}} \qquad \text{enthalpy/unit mass of } B_i$$

where
$$M_{A_i},\, M_{B_i} = \frac{\text{mass}}{\text{mole}} = \text{mol. wt}$$

Then Eq. (5-69) can be written

$$\Delta Q_{B_1} = \frac{-\Delta H}{M_{B_1}} = \sum_{i=1}^m \frac{a_i H_{Ai}}{M_{B_1}} - \sum_{i=2}^n \frac{b_i H_{Bi}}{M_{B_1}} - b_1 h_{B_1} \tag{5-70}$$

which gives the specific-heat release per unit mass of *product* B_1. Note that Eq. (5-70) differs from Eq. (3-8) which gives the specific-heat release per unit mass of reactant A_1. Equation (5-70) is better suited for use in determining h_c according to Eq. (5-50) and its derivatives.

For example, for the reverse reaction to that given by Eq. (5-67)

$$C(g) + 2O(g) \to CO_2(g) \tag{5-67a}$$

Eq. (5-70) gives

$$\Delta Q_{CO_2} = \frac{2M_O}{M_{CO_2}} h_O + \frac{M_C}{M_{CO_2}} h_C - h_{CO_2} \tag{5-71}$$

$$|\Delta Q_{CO_2}| = \frac{\text{heat}}{\text{unit mass of } CO_2}$$

In a similar way, for the reaction of Eq. (5-66)

$$\Delta Q_{CO} = \frac{M_C}{M_{CO}} h_C + \frac{M_O}{M_{CO}} h_O - h_{CO} \tag{5-72}$$

$$|\Delta Q_{CO}| = \frac{\text{heat}}{\text{unit mass of } CO}$$

and for the inverse of the reaction of Eq. (5-65), $3C \rightarrow C_3$,

$$\Delta Q_{C_3} = h_C - h_{C_3} \tag{5-73}$$

$$|\Delta Q_{C_3}| = \frac{\text{heat}}{\text{unit mass of } C_3}$$

Using the tables of thermodynamic properties[1] to determine ΔQ_{C_3}, ΔQ_{CO}, and ΔQ_{CO_2} we find that

$$\Delta Q_{C_3} = 14{,}950 \text{ Btu/lb } C_3 \tag{5-74}$$

$$\Delta Q_{CO} = 16{,}430 \text{ Btu/lb } CO \tag{5-75}$$

$$\Delta Q_{CO_2} = 15{,}430 \text{ Btu/lb } CO_2 \tag{5-76}$$

where the ΔQ's are essentially the negative of the specific heats of formation at zero degrees neglecting the slight difference in species specific heats. We now introduce the ΔQ's into our equation for h_c. Write out Eq. (5-50) for h_c for our example as ($E = C_3$)

$$
\begin{aligned}
h_c =&(h_{N_2})_w[(C_{N_2})_e - (C_{N_2})_w(1 + B_4)] + (h_{CO})_w[(C_{CO})_e - (C_{CO})_w(1 + B_4)] \\
&+ (h_{CO_2})_w[(C_{CO_2})_e - (C_{CO_2})_w(1 + B_4)] + (h_{C_3})_w[B_4 - (1 + B_4)(C_{C_3})_w] \\
&+ (h_O)_w[(C_O)_e - (C_O)_w(1 + B_4)] + (h_C)_w[(C_C)_e - (1 + B_4)(C_C)_w]
\end{aligned} \tag{5-77}
$$

In order to proceed, express the element mass fractions $\bar{C}_C$, $\bar{C}_O$, and $\bar{C}_N$ in terms of the species mass fractions. That is,

$$\bar{C}_O = C_O + \frac{M_O}{M_{CO}} C_{CO} + 2 \frac{M_O}{M_{CO_2}} C_{CO_2} \tag{5-78}$$

$$\bar{C}_C = C_C + C_{C_3} + \frac{M_C}{M_{CO}} C_{CO} + \frac{M_C}{M_{CO_2}} C_{CO_2} \tag{5-79}$$

and
$$\bar{C}_N = C_{N_2} \tag{5-80}$$

where, if we have a molecule $C_n O_m$,

$$r_{C, C_n O_m} = \frac{n M_C}{M_{C_n O_m}} \tag{5-81}$$

With the help of Eqs. (5-58) and (5-80) we can write

$$(C_{N_2})_e - (1 + B_4)(C_{N_2})_w = 0 \tag{5-82}$$

and hence the first term in Eq. (5-77) is identically zero as might be

[1] See, for example, J. S. Gordon, *WADC Tech. Rept.* 57-33, January, 1957; and L. V. Feigenbutz, G. L. Stiehl, and G. L. Katz, *Convair (San Diego Div.) Rept.* ZR-600-001, 1958.

anticipated, since N_2 enters into no gas-phase reactions in our example. Furthermore, using Eqs. (5-58) and (5-78), we can write

$$-[(C_O)_e - (1 + B_4)(C_O)_w] = \frac{M_O}{M_{CO}} [(C_{CO})_e - (1 + B_4)(C_{CO})_w]$$

$$+ 2 \frac{M_O}{M_{CO_2}} [(C_{CO_2})_e - (1 + B_4)(C_{CO_2})_w] \quad (5\text{-}83)$$

and using Eqs. (5-62) and (5-79), we can write, where E = C in Eq. (5-62),

$$-[(C_C)_e - (1 + B_4)(C_C)_w] = (C_{C_3})_e + B_4 - (1 + B_4)(C_{C_3})_w$$

$$+ \frac{M_C}{M_{CO}} [(C_{CO})_e - (1 + B_4)(C_{CO})_w] + \frac{M_C}{M_{CO_2}} [(C_{CO_2})_e - (1 + B_4)(C_{CO_2})_w]$$

$$(5\text{-}84)$$

Inserting Eqs. (5-82), (5-83), and (5-84) into Eq. (5-77) we obtain

$$h_c = \left[(h_{CO})_w - \frac{M_O}{M_{CO}} (h_O)_w - \frac{M_C}{M_{CO}} (h_C)_w \right][(C_{CO})_e - (1 + B_4)(C_{CO})_w]$$

$$+ \left[(h_{CO_2})_w - 2\frac{M_O}{M_{CO}} (h_O)_w - \frac{M_C}{M_{CO_2}} (h_C)_w \right][(C_{CO_2})_e - (1 + B_4)(C_{CO_2})_w]$$

$$+ [(h_{C_3})_w - (h_C)_w][B_4 - (1 + B_4)(C_{C_3})_w] - (C_{C_3})_e(h_C)_w \quad (5\text{-}85)$$

Now we can introduce our ΔQ's. Referring to Eqs. (5-71), (5-72), and (5-73), we can write Eq. (5-85), remembering that $(C_{C_3})_e = 0$, as

$$h_c = -\Delta Q_{CO}[(C_{CO})_e - (1 + B_4)(C_{CO})_w] - \Delta Q_{CO_2}[(C_{CO_2})_e$$

$$- (1 + B_4)(C_{CO_2})_w] - \Delta Q_{C_3}[B_4 - (1 + B_4)(C_{C_3})_w] \quad (5\text{-}86)$$

The first term in Eq. (5-86) is the heat released owing to the formation of CO, the second term the heat sink due to the reduction of CO_2, and the last term a heat sink due to dissociation of C_3. It remains to determine $(C_{C_3})_w$, $(C_{CO_2})_w$, and $(C_{CO})_w$. It is helpful to note that $(C_{C_3})_e = (C_{CO})_e = 0$ and $(C_{CO_2})_e + (C_{N_2})_e = 1$. The numerical magnitudes of ΔQ_{C_3}, ΔQ_{CO}, and ΔQ_{CO_2} are given by Eqs. (5-74) through (5-76).

The Species Mass Fractions. Consider the reaction

$$\tfrac{1}{3}C_3(g) + CO_2(g) \rightarrow 2CO(g)$$

We have available to us the following equations, which can be used to solve for the various species mass fractions at the surface $(C_i)_w$. Dalton's law gives, since $p = p_e = p_w$ in the boundary layer,

$$p_e = (p_{C_3})_w + (p_{CO})_w + (p_{CO_2})_w + (p_{N_2})_w \quad (5\text{-}87)$$

The equilibrium constant for the reaction gives[1]

$$K_1(T) = \frac{p_{C_3}^{1/3} p_{CO_2}}{p_{CO}^2} \tag{5-88}$$

where $K_1(T)$ was determined from the equilibrium constants for the equivalent set of reactions given by Eqs. (5-65) through (5-67) using the equilibrium constants given in thermodynamic tables.

Furthermore, the data of Thorne and Winslow[2] give

$$K_2(T) = p_{C_3} \tag{5-89}$$

The relation between partial pressure and species mass fraction is given by

$$p_i = \frac{C_i}{M_i \chi} \tag{5-90}$$

where

$$\frac{1}{\chi} = \sum_i p_i M_i \tag{5-91}$$

and M_i is the molecular weight of species i. Equation (5-90) combined with Eq. (5-82) gives

$$(p_{N_2})_w = \frac{\chi_e}{\chi_w} \frac{(p_{N_2})_e}{1 + B_4} \tag{5-92}$$

B_4 can be eliminated from Eq. (5-92) by using Eqs. (5-83) and Eq. (5-90). We obtain, since $(p_O)_e = (p_O)_w = p_O \equiv 0$,

$$\frac{\chi_e}{\chi_w} \frac{1}{1 + B_4} = \frac{(p_{CO})_w + 2(p_{CO_2})_w}{2(p_{CO_2})_e} \tag{5-93}$$

Hence, combining Eqs. (5-92) and (5-93), we obtain

$$(p_{N_2})_w = \frac{(p_{N_2})_e}{2(p_{CO_2})_e} [(p_{CO})_w + 2(p_{CO_2})_w] \tag{5-94}$$

and using Eqs. (5-88) and (5-89) with (5-94),

$$(p_{N_2})_w = \frac{1}{2} \frac{(p_{N_2})_e}{(p_{CO_2})_e} \left(p_{CO} + 2 \frac{K_1 p_{CO}^2}{K_2^{1/3}} \right)_w \tag{5-95}$$

Thus, using Eqs. (5-88) and (5-89) in Eq. (5-94) to replace $(p_{CO})_w$, we obtain

$$(p_{N_2})_w = \frac{1}{2} \frac{(p_{N_2})_e}{(p_{CO_2})_e} \left[\frac{(p_{CO_2})^{1/2} K_2^{1/6}}{K_1^{1/2}} + 2(p_{CO_2}) \right]_w \tag{5-96}$$

[1] See Eq. (9-93).
[2] Thorne and Winslow, op. cit.

Solving Eqs. (5-95) and (5-96) for p_{CO} gives

$$(p_{CO})_w = \frac{a}{4}\left(\left\{1 + \frac{16}{a}\left[p_{CO_2} + \frac{1}{2}(ap_{CO_2})^{1/2}\right]\right\}_w^{1/2} - 1\right) \quad (5\text{-}97)$$

where

$$a = \frac{K_2^{1/3}}{K_1} = \frac{p_{CO}^2}{p_{CO_2}} \quad (5\text{-}98)$$

We can now express $(p_{CO})_w$ and $(p_{N_2})_w$ in terms of $(p_{CO_2})_w$ and $(p_{C_3})_w$ in terms of T in Eq. (5-87). There results

$$K_2 + \frac{1}{2}\frac{(p_{N_2})_e}{(p_{CO_2})_e}[2p_{CO_2} + (ap_{CO_2})^{1/2}] + \frac{a}{4}\left(\left\{1 + \frac{8}{a}[2p_{CO_2} + (ap_{CO_2})^{1/2}]\right\}^{1/2} - 1\right)$$
$$+ p_{CO_2} = p_e \quad (5\text{-}99)$$

It is understood that in Eq. (5-99) K_2 and a are evaluated at T_w and p_{CO_2} is the value of p_{CO_2} for temperature T_w and pressure p_e.

Once Eq. (5-99) is solved for $(p_{CO_2})_w$, Eq. (5-96) gives $(p_{N_2})_w$ and Eq. (5-97) gives $(p_{CO})_w$. Equation (5-89) gives $(p_{C_3})_w$. After all the partial pressures are obtained, Eq. (5-90) can be used to obtain the $(C_i)_w$ and Eq. (5-92) can be used to obtain B_4.

The Equilibrium Constants. The necessary equilibrium constants can be found using experimental and tabulated data. Thorne and Winslow[1] give

$$\log_{10}(K_2) = \log_{10}(p_{C_3}) = 9.811 - \frac{40296}{T} \quad (5\text{-}100)$$

where p_{C_3} is in atmospheres and T is in degrees Kelvin. Equation (5-100) was used in our calculations.

An equilibrium constant was not available for the reaction

$$\tfrac{1}{3}C_3(g) + CO(g) \rightleftharpoons 2\,CO(g) \qquad K_1$$

so we obtained the desired equilibrium constant from those constants available for the reactions

$$C(g) + O(g) \rightleftharpoons CO(g) \qquad K_3$$
$$C(g) + 2O(g) \rightleftharpoons CO_2(g) \qquad K_4$$
$$C_3(g) \rightleftharpoons 3C(g) \qquad K_5$$

which are available in the literature.[2] It is an easy matter to show that

$$K_1 = \frac{p_{C_3}^{1/3}p_{CO_2}}{p_{CO}^2} = \frac{K_5^{1/3}K_3^2}{K_4}$$

[1] Thorne and Winslow, *op. cit.*

[2] Gordon, *op. cit.*; and Feigenbutz, Stiehl, and Katz, *op. cit.*

and a fit to a plot of K_1 so obtained in the temperature region of interest to us is

$$\log_{10} K_1 = -5.03 - \frac{7 \times 10^3}{T} \qquad (5\text{-}101)$$

where K_1 is in atmospheres and T in degrees Kelvin.

The important parameter a used in Eq. (5-99) is found to be

$$a = \frac{p_{CO}^2}{p_{CO_2}} = \frac{K_2^{1/3}}{K_1}$$

or, using Eqs. (5-100) and (5-101),

$$\log_{10} a = 8.30 - \frac{6432}{T} \qquad (5\text{-}102)$$

where a is in atmospheres and T in degrees Kelvin.

Results of Calculations. The surface concentrations were determined as a function of T_w and $(p_{N_2})_e/(p_{CO_2})_e$ using Eqs. (5-89), (5-90), and (5-97). Equation (5-92) was then used to find B_4. Next Eqs. (5-74) to (5-76) and (5-86) were used to find h_c. Equations (5-1), (5-39), (5-52), and (5-53) were then used to determine the heat transfer to the solid $\dot{q}_s$. Figure 5-2 was used to determine $G(\infty;P)$. For the gas mixture of concern to us, it can be calculated that

$$P \simeq P_{N_2} = 0.70, \qquad S \simeq S_{N_2} = 0.965, \qquad L = P/S \simeq L_{N_2} = 0.725$$

using the methods for calculating transport properties outlined in Chap. 10. The values of ρ_e, u_e, and μ_e were calculated making use of M_e, T_e, and the tabulated values of supersonic cone flow for a $10°$ semivertex angle cone. $\mu_e \simeq (\mu_{N_2})_e$ and is essentially independent of external stream composition according to the methods outlined in Sec. 10-3. ρ_e is dependent upon the ratio of $(C_{N_2})_e$ to $(C_{CO_2})_e$. Once the heat-transfer rate $\dot{q}_s$ was found, Eq. (5-53) and the definition of B_4, Eq. (5-49), were used to find the mass-loss rate; viz.,

$$(s)^{1/2}(\rho v)_w = (s)^{1/2}\rho_e u_e C_H B_4 \qquad (5\text{-}103)$$

The results of these calculations are shown in Figs. 5-3 to 5-5. Figures 5-3 and 5-4 show the partial pressure of CO_2 in equilibrium at the surface as a function of the ratio $(p_{N_2})_e/(p_{CO_2})_e$. The expected results are obtained; namely, as T_w increases, $(p_{CO_2})_w$ decreases, and as $(p_{CO_2})_e$ decreases, $(p_{CO_2})_w$ decreases. That is, the amount of CO_2 present at the surface is decreased by increasing the surface temperature or by decreasing the partial pressure of CO_2 available for reduction by decreasing the pressure of CO_2 in the free stream.

Figure 5-5 presents the heat-transfer rate to the surface and the mass-loss rate as a function of surface temperature and the ratio $(p_{N_2})_e/(p_{CO_2})_e$. The following conclusions can be drawn:

1. Heat is transferred from the cone toward the stream primarily

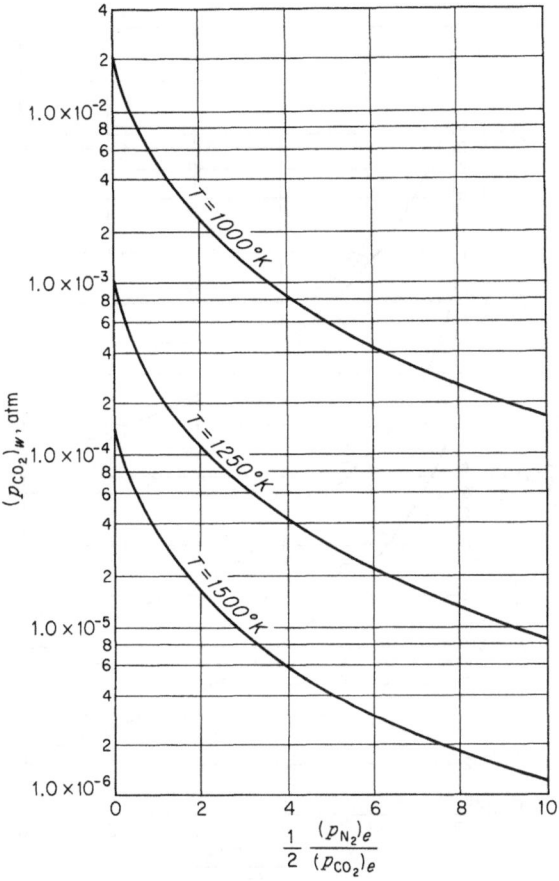

FIG. 5-3. Effect of surface temperature and partial pressure of CO_2 at the boundary-layer edge upon partial pressure of CO_2 at a heated graphite surface for $p_e = 1.224$ atm.

because the surface temperature is much higher than the local free-stream temperature and the reaction $C_3(s) \rightarrow C_3(g)$ is endothermic.

2. Enriching the N_2 stream with increasing amounts of CO_2 increased the heat transfer from the cone, keeping the surface temperature constant because the reaction

$$\tfrac{1}{3}C_3(s) + CO_2(g) \rightarrow 2CO(g)$$

is endothermic and heat must be supplied to maintain the surface temperature constant.

3. Enriching the N_2 stream with increasing amounts of CO_2 increased the mass loss as might be expected.

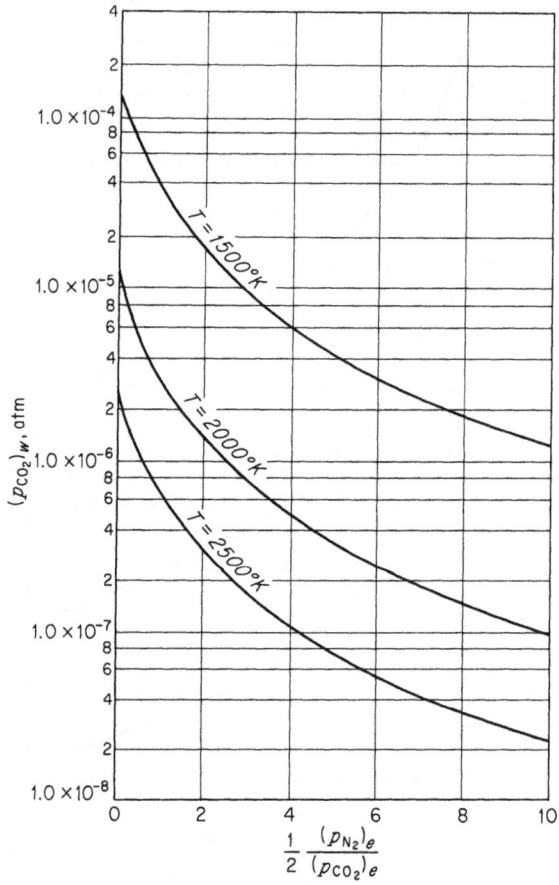

Fig. 5-4. Effect of surface temperature and partial pressure of CO_2 at the boundary-layer edge upon partial pressure of CO_2 at a heated graphite surface for $p_e = 1.224$ atm.

4. As surface temperature increases, heat-transfer rate increases but mass-loss rate decreases for constant values of the enrichment ratio $(p_{N_2})_e/(p_{CO_2})_e$. This latter behavior occurs because the mass-transfer parameter B_4 is fixed by the diffusion or conservation-of-species equations and is constant with increasing surface temperature, whereas $\rho_e u_e C_H$ is decreased through the decrease in $(C)^{1/2}$ as

temperature increases because the average density in the boundary layer decreases as temperature increases.

5. There should be significant and measurable differences in heat-transfer and mass-loss rates with variation of the concentration of

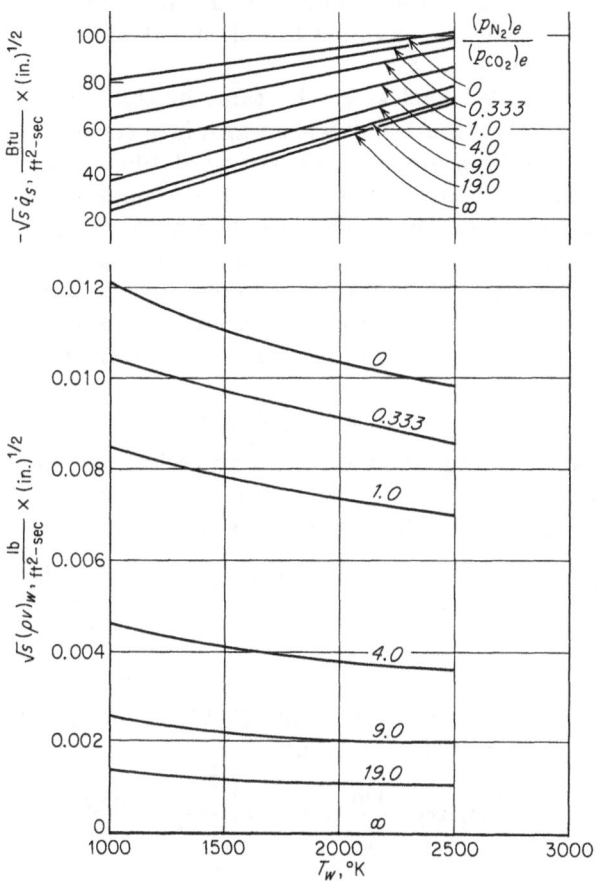

FIG. 5-5. Rates of surface heat transfer and mass loss for the case of a mixture of N_2 and CO_2 flowing over a heated 10° semivertex angle graphite cone. $M_e = 2.2$, $T_e = 157°K$, and $p_e = 1.224$ atm.

CO_2 at any surface temperature if the surface mass-loss rate is diffusion limited.

We shall discuss the validity of the assumption of diffusion-controlled reaction rates in the next section.

5-8. Kinetically Controlled Surface Reactions. Under the frozen-boundary-layer approximation, the surface-mass-loss rate will be determined by the rate at which products and reactants can be diffused and convected through the boundary layer to the reacting surface where the products and reactions are present in chemical equilibrium. It is entirely possible that the reaction rates necessary under this assumption require surface-mass-loss rates exceeding those allowable according to the kinetics of surface reactions. To demonstrate this point, let us examine the chemical kinetics of the surface reaction discussed in Sec. 5-7.

One of the uncertainties which stand in the way of any critical examination of the kinetics of carbon-graphite surface reactions is the many different solid-state structures of available commercial graphite. Graphite comes in many forms having many possible crystal orientations, depending upon the process used to prepare it. Of course, the solid-state structure has a direct bearing upon the rate of reaction at the surface with gas reactants. Keeping this uncertainty in mind, we shall use some reaction rates measured and reported by Blyholder and Eyring[1] to calculate mass-loss rates assuming the kinetics of the surface reaction

$$\tfrac{1}{3}C_3(s) + CO_2(g) \rightarrow 2\,CO(g)$$

as the rate-controlling process. An expression for the surface-mass-loss rate can be written based upon the experimental results of Blyholder and Eyring. This expression is

$$(\rho v)_w = 4.9 \times 10^{-10} \frac{T_w}{1000} \exp\left(-\frac{1006}{T_w}\right)\left[\frac{(C_{CO_2})_w}{(C_{CO_2})_w + 1.57(C_{N_2} + C_{CO})_w}\right]^{1/2}$$

(5-104)

where T_w is in degrees Kelvin and $(\rho v)_w$ in lb/ft²-sec. When Eq. (5-104) is used with Eqs. (5-5), (5-39), (5-47), (5-49), (5-82), and (5-83), the following equation for B_4 results:

$$2.04\rho_e u_e C_H \frac{G(\infty;P)}{G(\infty;S)} B_4 \exp\frac{1006}{T_w} \times 10^9 = \left[\frac{(C_{CO_2})_e - 1.66B_4}{1.57 - 0.57(C_{CO_2})_e + 2.52B_4}\right]^{1/2}$$

(5-105)

Even without solving Eq. (5-105) for B_4 using representative values of $(C_{CO_2})_e$ and T_w, it can be seen that B_4 will be of the order of 10^{-8}. When Eq. (5-105) is solved for one typical case of $T_w = 1000°K$ and $(p_{N_2})_e/(p_{CO_2})_e = 0$, a case which according to Fig. 5-5 has the highest mass-loss rate of those calculated, we find that

$$B_4 = 2.99 \times 10^{-9}(s)^{1/2}$$

[1] G. Blyholder and H. Eyring, J. Phys. Chem., vol. 61, no. 5, pp. 682–688, 1957; and G. Blyholder and H. Eyring, J. Phys. Chem., vol. 63, no. 6, pp. 1004–1008, 1959.

where s is in inches. The diffusion-limited value of B_4 for this case is

$$B_4 = 0.273$$

Furthermore, when the mass loss is kinetically limited and $(s)^{1/2} = 1$,

$$(C_{CO_2})_w = 0.999999992$$

which can be compared with

$$(C_{CO_2})_w = 0.0246$$

when the process is diffusion limited.

We cannot escape the conclusion that the kinetics of the heterogeneous surface reaction may dominate the mass-loss rate, and if they do, *no appreciable effect of chemical reactions will be observed during the assumed experiments.* However, the kinetics are sufficiently in doubt to make the experiments worthwhile. In fact, they may well cast some light upon the situation.

5-9. Kinetics of Gas-phase Reactions. In Sec. 5-3 it was shown that the heat transfer from a reacting boundary layer is relatively insensitive to the location of the reaction zone providing the transport properties are relatively insensitive to gas composition and providing $L \simeq 1$. Under such conditions it is convenient to assume that the reactions take place at the surface only and that no gas-phase reactions occur in the boundary layer. Sections 5-4 through 5-6 present a theory applicable to a reacting laminar boundary layer based on the approximation that it is reasonable to assume that all reactions take place at the surface and no gas-phase reactions occur.

However, it is possible that the assumption of a frozen gas layer may not be satisfactory for some cases where the reacting-boundary-layer equations are still suitable. It then becomes necessary to incorporate into the calculations the kinetics for whatever chemical reactions may be occurring in the gas phase. We shall demonstrate how this can be done by choosing a particular gas-phase reaction, that of the dissociating ideal gas used in Sec. 4-6, where results of calculations of heat transfer to the stagnation point of a blunt body were presented for various values of gas-phase reaction-rate parameters which were presented but not derived there. In particular, we are concerned with the dissociation reaction

$$A_2(g) + P(g) \underset{k_R}{\overset{k_D}{\rightleftharpoons}} A(g) + A(g) + P(g) \tag{5-106}$$

where P is either A or A_2. The relevant species-concentration equation is given by Eq. (5-16) with $i = A$ and $(C_A)_e = $ const and is

$$\left(\frac{C}{S}z'_A\right)' + fz'_A = -\frac{2\bar{s}\dot{w}_A}{\rho\rho_e\mu_e u_e^2 r_0^{2k}\alpha_e} \tag{5-107}$$

where $\alpha_e = (\rho_A/\rho)_e$. Furthermore, for stagnation-point flow it was shown in Sec. 4-6 that

$$\frac{2\bar{s}\dot{w}_A}{\rho\rho_e\mu_e u_e^2 r_0^{2k}\alpha_e} = \frac{\dot{z}_A}{(k+1)(du_e/ds)_0} \tag{5-108}$$

where, at the stagnation point,

$$r_0 = s$$

$$u_e = s\left(\frac{du_e}{ds}\right)_0$$

$$\frac{2\bar{s}}{u_e}\frac{du_e}{d\bar{s}} = \frac{1}{k+1}$$

and $\dot{w}_A$ is related to $\dot{z}_A$ by the equation

$$\dot{w}_A = \rho\dot{\alpha} = \rho\alpha_e\dot{z}_A$$

Thus, Eq. (5-107) becomes

$$\left(\frac{C}{S}z'_A\right)' + fz'_A + \frac{\dot{z}_A}{(k+1)(du_e/ds)_0} = 0 \tag{5-109}$$

where $z_A = \alpha/\alpha_e$, and we require an equation for $\dot{z}_A = \dot{\alpha}/\alpha_e$ appropriate to the reaction given by Eq. (5-106). We can write for the production of atoms according to the reaction of Eq. (5-106)

$$\frac{1}{N_0}\left(\frac{dn_A}{dt}\right)_D = k_D\frac{n_P n_{A_2}}{N_0^2}$$

a second-order reaction. For the production of molecules we can write

$$\frac{1}{N_0}\left(\frac{dn_{A_2}}{dt}\right)_R = k'_R\frac{n_P n_A^2}{N_0^3}$$

a third-order reaction. But we can also write for any process changing the number atoms and molecules

$$\frac{dn_A}{dt} = -2\frac{dn_{A_2}}{dt}$$

Hence

$$\frac{1}{N_0}\left(\frac{dn_A}{dt}\right)_R = -2k'_R\frac{n_P n_A^2}{N_0^3} = -k_R\frac{n_P n_A^2}{N_0^3}$$

and the *net* rate of change of the number density of atoms is, then,

$$\frac{dn_A}{dt} = \left(\frac{dn_A}{dt}\right)_D + \left(\frac{dn_A}{dt}\right)_R$$

$$= k_D \frac{n_P n_{A_2}}{N_0} - k_R \frac{n_P n_A^2}{N_0^2}$$

$$= -k_R \frac{n_P n_A^2}{N_0^2}\left[1 - \frac{N_0 k_D}{k_R}\frac{n_{A_2}}{(n_A)^2}\right] \tag{5-110}$$

Now

$$\alpha = \frac{\rho_A}{\rho} = \frac{\rho_A}{\rho_A + \rho_{A_2}}$$

and

$$\rho_A = n_A m_A \tag{5-111}$$

$$\rho_{A_2} = 2 n_{A_2} m_A \tag{5-112}$$

whence

$$\alpha = \frac{n_A}{n_A + 2 n_{A_2}} \tag{5-113}$$

Furthermore, for conservation of mass,

$$\frac{\rho_A}{\rho} + \frac{\rho_{A_2}}{\rho} = 1 \tag{5-114}$$

Hence, using Eqs. (5-111) through (5-114), it can be shown that

$$n_A = \frac{\rho \alpha}{m_A} \tag{5-115}$$

and

$$n_{A_2} = \frac{\rho}{2 m_A}(1 - \alpha) \tag{5-116}$$

In addition, when $dn_A/dt = 0$, the system is in chemical equilibrium and, from Eq. (5-110) with $dn_A/dt = 0$,

$$N_0 k_D = k_R \frac{(n_A^*)^2}{n_{A_2}^*} \tag{5-117}$$

where n_A^* and $n_{A_2}^*$ denote equilibrium values of n_A and n_{A_2}, respectively. Combination of Eq. (5-110) with Eq. (5-117) gives

$$\frac{dn_A}{dt} = -k_R \frac{(n_A + n_{A_2}) n_A^2}{N_0^2}\left[1 - \left(\frac{n_A^*}{n_A}\right)^2 \frac{n_{A_2}}{n_{A_2}^*}\right] \tag{5-118}$$

where

$$n_P = n_A + n_{A_2}$$

Use of Eqs. (5-115) and (5-116) in Eq. (5-118) results in

$$\frac{dn_A}{dt} = -k_R \frac{\rho^3(1 + \alpha)}{2 N_0^2 m_A^3} \cdot \frac{\alpha^2(1 - \alpha^*) - \dfrac{\rho^*}{\rho}(1 - \alpha)(\alpha^*)^2}{1 - \alpha^*} \tag{5-119}$$

Now
$$p = p^*$$

and, since we are evaluating the rate of approach to chemical equilibrium for the properties evaluated at local equilibrium temperature

$$T = T^*$$

we find that, from the equation of state,

$$\frac{\rho^*}{\rho} = \frac{1 + \alpha}{1 + \alpha^*} \tag{5-120}$$

Using Eq. (5-120) in Eq. (5-119) results in

$$m_A \frac{dn_A}{dt} = -k_R \frac{\rho^3(1 + \alpha)}{2N_0^2 m_A^2} \frac{\alpha^2 - (\alpha^*)^2}{1 - (\alpha^*)^2} \tag{5-121}$$

but from Eq. (5-115),

$$\dot{w}_A = \rho\dot{\alpha} = m_A \frac{dn_A}{dt} = \rho\alpha_e \dot{z}_A$$

Hence, from Eq. (5-121),

$$\dot{z}_A = -\frac{k_R \rho^2(1 + \alpha)}{2\alpha_e N_0^2 m_A^2} \frac{\alpha^2 - (\alpha^*)^2}{1 - (\alpha^*)^2}$$

or, since $p_0 = p = \rho(1 + \alpha)RT$ and $N_0 k = R$,

$$\dot{z}_A = -\frac{2k_R p_0^2 T^{-2}\alpha_e}{R^2[1 - (\alpha^*)^2]} \frac{z_A^2 - (z_A^*)^2}{1 + \alpha_e z_A} \tag{5-122}$$

Hence, if, as was assumed in Sec. 4-6,

$$k_R \propto T^{-1.5}$$

and if we define

$$\theta(\eta) = \frac{T(\eta)}{T_e} = \frac{T(\eta)}{T_0}$$

then Eq. (5-109) becomes, for $k + 1 = 2$,

$$\left(\frac{C}{S} z_A'\right)' + f z_A' + C_1 \theta^{-3.5}\alpha_e \frac{z_A^2 - (z_A^*)^2}{1 + \alpha_e z_A} = 0 \tag{5-123}$$

where
$$C_1 = K_1 p_0^2 T_0^{-3.5}\left[\left(\frac{du_e}{ds}\right)_0\right]^{-1} \tag{5-124}$$

$$K_1 = \frac{k_R T^{-1.5}}{R^2[1 - (\alpha^*)]^2} \qquad \alpha^* \simeq \text{const} \tag{5-125}$$

and, according to our previous assumption,

$$k_R T^{-1.5} = \text{const}$$

Equation (5-123) is Eq. (4-82) used in Sec. 4-6, and Eq. (5-124) is Eq. (4-81) given in Sec. 4-6 without derivation. The value of α^* is given by the equilibrium constant for the reaction given by Eq. (5-106). That is,

$$K_p = \frac{(p_A^*)^2}{p_{A_2}^*} = \frac{\rho_A^*(k/m_A)^2 T^2}{\rho_{A_2}^*(k/2m_A)T} \tag{5-126}$$

but

$$p = p_A^* + p_{A_2}^* = (\rho_A^* + \rho_{A_2}^*)R^*T \tag{5-127}$$

where

$$R^* = \sum_i C_i^* R_i = (1 + \alpha^*)\frac{k}{2m_A} \tag{5-128}$$

whence, using Eqs. (5-127) and (5-128) with Eq. (5-126), we obtain

$$\frac{(\alpha^*)^2}{1 - (\alpha^*)^2} = \frac{K_p(T)}{4p_0} \tag{5-129}$$

where $K_p(T)$ can be determined by the methods outlined in Sec. 9-8. Our system of equations for treating the chemically nonequilibrium boundary-layer flow of a simple dissociating gas mixture is now complete. Equations (5-125) and (5-127) to (5-129) would be used with Eq. (4-82) to (4-85) to obtain α, T, ρ, u, and v throughout the dissociating boundary layer.

It should be stated that any gas-phase reaction can be treated in this manner providing the chemical kinetics can be described as was done above for the chemical reaction given by Eq. (5-106). The problem is somewhat more complicated when more than one gas-phase reaction is occurring, but the number of equations available will always equal the number of unknowns to be solved for, and the problem is similar in principle to that problem described here. However, similarity of the solutions breaks down for all gas-phase reactions except those occurring on the stagnation-point streamline. That is, the term on the right side of Eq. (5-107) will not be a function of η alone in regions away from the stagnation point.

It should be mentioned that the chemical kinetics used here are those describing the simplest situation imaginable. For example, the particle P which participates in the reaction equation (5-106) was assumed to be either A or A_2 without regard to whether A or A_2 was in an excited state or not. If, when it collides with another particle, A_2 has a fully excited vibrational mode, it is more likely to dissociate than if it possesses an unexcited vibrational mode. Both types of collisions are apt to occur. Furthermore, both A and A_2 have different collision efficiencies, depending upon whether or not they possess excited electronic modes. In general, the more excited electronic modes possessed by a molecule, the larger the molecules and the more frequent the collisions during any interval of time. Also an excited molecule possesses more internal energy available for transfer between modes upon collision. Realistically, a number of rate constants k_R might apply, depending upon the molecular species participating in the reaction. However, we take the position that the simple kinetic theory as used here adequately describes the chemical reaction providing the rate constants k_R and k_D are determined experimentally under conditions similar to those present in a reacting boundary layer.

Dorrance[1] has described the behavior of a simple reacting flow system, the flow of a pure diatomic gas through a strong normal shock wave, under conditions where the approach to chemical equilibrium is occurring simultaneously with the approach to equilibrium of the vibrational modes of molecules downstream of the shock wave. It was found that these competing processes create some interesting density, composition, and temperature variations depending upon the relative time constants of the two processes.

5-10. The Reference Temperature. In Sec. 5-4 the solutions to the reacting-boundary-layer equations were obtained under the assumptions that

1.
$$C = \frac{\rho\mu}{\rho_e\mu_e} = \text{const}$$

2.
$$C = \frac{\rho\mu}{\rho_e\mu_e} = \frac{\rho(T')\mu(T')}{\rho_e\mu_e}$$

where T' is a reference temperature given by Eq. (5-32); viz.,

$$\frac{T'}{T_e} = 0.42 + 0.032M_e^2 + 0.58\frac{T_w}{T_e} \qquad (5\text{-}32)$$

We shall demonstrate here that both assumptions 1 and 2 above are related to solutions to the boundary-layer equations obtained making no assumption relative to $C(\eta)$.

Relation between C_{av} and $C(T')$. The so-called "reference-temperature" or "reference-enthalpy" method was originally advanced as a semiempirical method of correlating calculations for skin friction and heat transfer for a compressible laminar boundary layer flowing over a flat plate. In essence, the method is predicated on the assumption that there is some reference temperature (or reference enthalpy) which, when used to evaluate the properties inserted into the flat-plate constant-property equations for skin friction and heat transfer,

[1] W. H. Dorrance, *J. Aerospace Sci.*, vol. 28, no. 1, pp. 43–50, January, 1961.

will yield values of skin friction and heat transfer for the compressible laminar boundary layer. Obviously, this reference temperature must depend upon both Mach number and the ratio of surface temperature to free-stream temperature (or viscous dissipation and heat conduction). That is, there is a $T'(T_w/T_e, M_e)$ such that

$$\tau_w\left(M_e, \frac{T_w}{T_e}\right) = C_{f_{M_e=0}}(T')\rho(T')\frac{u_e^2}{2} \tag{5-130}$$

where $C_{f_{M_e=0}}(T')$ refers to C_f evaluated for $\rho = \text{const} = \rho(T')$ and $\mu = \mu(T')$. Several empirically determined functional relationships for $T'(M_e, T_w/T_e)$ have been advanced which differ very little from one another and which give satisfactory results for engineering calculations. The method has been regarded as a rather fortuitous empirical method and cautiously accepted as such.

In this section we shall show that the reference-temperature method is somewhat more than a fortuitous empiricism and, in fact, has a basis in the numerical solutions to the boundary-layer equations. First, let us associate the reference temperature (or reference enthalpy when appropriate) with our boundary-layer constant $C = \rho\mu/\rho_e\mu_e$ arising in our solutions presented in Sec. 5-4.

Using Eq. (5-10), it follows that

$$\tau_w = \mu_w\left(\frac{\partial u}{\partial y}\right)_w = \mu_w u_e f''(0)\left(\frac{d\eta}{dy}\right)_w$$

or, from Eq. (5-15),

$$\left(\frac{d\eta}{dy}\right)_w = \frac{\rho_w u_e}{(2\rho_e u_e \mu_e s)^{1/2}}$$

where, from Eq. (5-14), for a flat-plate boundary layer,

$$\bar{s} = \rho_e u_e \mu_e s$$

Thus, since $\rho_w\mu_w/\rho_e\mu_e = C_w$,

$$\tau_w = C_w \frac{\rho_e u_e^2 f''(0)}{(2R_e)^{1/2}} \tag{5-131}$$

where

$$R_e = \frac{\rho_e u_e s}{\mu_e}$$

and, from the result given in Sec. 4-5 for $(\rho v)_w = 0$,

$$f''(0) = \frac{0.47}{(C_w)^{1/2}} \tag{5-132}$$

Now from Eq. (5-131) and the definition of C_f,

$$C_f = \frac{2\tau_w}{\rho_e u_e^2} = \frac{(2)^{1/2}f''(0)C_w}{(R_e)^{1/2}}$$

or, in view of Eq. (5-132),

$$C_f = \frac{0.666 C_w^{1/2}}{(R_e)^{1/2}} \tag{5-133}$$

Thus, for $M_e = 0$, $T_w/T_e = 1$, and $C_w = 1$, we obtain

$$C_{f_{M_e=0}} = \frac{0.666}{(R_e)^{1/2}} \tag{5-134}$$

Now according to Eqs. (5-130), (5-131), (5-133), and (5-134),

$$\tau_w = \frac{\rho_e u_e^2}{2} (0.666) \frac{(C_w)^{1/2}}{(R_e)^{1/2}} = \frac{\rho_e u_e^2}{2} \frac{\rho(T')}{\rho_e} \frac{0.666}{[\rho(T')u_e s/\mu(T')]^{1/2}}$$

Hence, from the above,

$$C_w = \frac{\rho(T')\mu(T')}{\rho_e \mu_e}$$

and this is the relation used in Sec. 5-4. Since the solutions developed in Sec. 5-4 assume that C is constant through the boundary layer, then

$$C = C_w = \frac{\rho(T')\mu(T')}{\rho_e \mu_e} \tag{5-135}$$

We could then use any of the recommended functional forms for $T'(M_e, T_w/T_e)$ in our solution. However, we have yet to show that the use of a reference temperature in this manner bears any relation to a rigorous solution to the boundary-layer equations. We do that in the following.

Significance of the Reference Temperature. We shall now relate the reference temperature to an average temperature procured from the numerical solutions to the boundary-layer equations obtained assuming that C is a variable given by $C(\eta)$. In order to proceed, let us make some simplifying assumptions, which do not, however, invalidate that which follows. We assume that

$$L = P = 1 \qquad C_p = \text{const}$$

and a flat-plate solution. Then as was shown in Chap. 2, the momentum and energy equations become uncoupled and our equation to be solved is

$$(Cf'')' + ff'' = 0 \tag{5-136}$$

and the solution to the energy equation is Crocco's integral given by Eq. (2-103). This is

$$g = g_w + f'(1 - g_w) \tag{5-137}$$

where

$$g = \frac{I}{I_e}$$

Since $C_p = \text{const}$ and $I = C_p T + u^2/2$, Eq. (5-137) becomes

$$\frac{T}{T_e}(u) = a_1 + b_1 \frac{u}{u_e} + c_1 \left(\frac{u}{u_e}\right)^2 \qquad (5\text{-}138)$$

where, for $\gamma = C_p/C_v = 1.4$,

$$a_1 = \frac{T_w}{T_e}$$

$$b_1 = 1 + 0.2 M_e^2 - \frac{T_w}{T_e}$$

$$c_1 = -0.2 M_e^2$$

The solution to Eq. (5-136) is, by quadrature,

$$C_w f''(0) = \frac{1}{G(\infty;1)} \qquad (5\text{-}139)$$

where

$$G(\infty;1) = \int_0^\infty \frac{1}{C} \exp\left(-\int_0^\eta \frac{1}{C} f \, d\eta'\right) d\eta \qquad (5\text{-}140)$$

and where

$$f''(\eta) = \frac{1}{C(\eta)G(\infty;1)} \exp\left(-\int_0^\eta \frac{f}{C} \, d\eta\right) \qquad (5\text{-}141)$$

and where we have used the boundary conditions

$$f'(0) = 0 \text{ and } f'(\infty) = 1$$

Now we seek a constant value of C, say C_{av}, such that

$$(\tau_w)_{C_{av}} = (\tau_w)_{C(\eta)}$$

or using Eqs. (5-131) and (5-132), we seek a C_{av} such that

$$C_{av} f''(0;C_{av}) = C_w f''[0;C(\eta)]$$

or where

$$\frac{1}{C_{av}} = \frac{f''(0;C_{av})}{C_w f''[0;C(\eta)]} \qquad (5\text{-}142)$$

Using Eqs. (5-139) and (5-140) in Eq. (5-142), we have

$$\frac{1}{C_{av}} = \frac{\displaystyle\int_0^\infty \frac{1}{C} \exp\left(-\int_0^\eta \frac{1}{C} f \, d\eta'\right) d\eta}{\displaystyle\int_0^\infty \exp\left(-\frac{1}{C_{av}} \int_0^\eta f \, d\eta'\right) d\eta}$$

or, taking note of Eqs. (5-140) and (5-141),

$$\frac{1}{C_{av}} \simeq \int_0^\infty \frac{1}{C} f''(\eta) \, d\eta$$

but

$$f''(\eta) = \frac{d}{d\eta}\frac{u}{u_e} = \frac{1}{u_e}\frac{du}{d\eta}$$

whence

$$\frac{1}{C_{av}} = \int_0^1 \frac{1}{C}\frac{du}{u_e} \tag{5-143}$$

Now,

$$C = \frac{\rho\mu}{\rho_e\mu_e} = \left(\frac{T_e}{T}\right)^{1-n} \tag{5-144}$$

if we assume $\mu \propto T^n$. Hence we find that C_{av} must be evaluated at a temperature T_{av} such that

$$\left(\frac{T_{av}}{T_e}\right)^{1-n} = \int_0^1 \left[\frac{T(u)}{T_e}\right]^{1-n}\frac{du}{u_e} \tag{5-145}$$

If we have a constant-viscosity boundary layer (but *not* constant density), then $n = 0$ and, using Eq. (5-138) in Eq. (5-145) with $n = 0$, we find that

$$\frac{T_{av}}{T_e} = 0.5 + 0.033M_e^2 + 0.5\frac{T_w}{T_e} \tag{5-146}$$

If we have $\mu \propto T^{1/2}$, then using Eq. (5-145) with $n = \frac{1}{2}$ gives

$$\frac{T_{av}}{T_e} = \left\{\int_0^1 \left[\frac{T(u)}{T_e}\right]^{1/2}\frac{du}{u_e}\right\}^2 \tag{5-147}$$

and if we assume that $(T_w/T_e) \gg 1$ or $0.2M_e^2$, as it is in our example of Sec. 5-7, then it can be shown that

$$\left[\frac{T(u)}{T_e}\right]^{1/2} = \left(\frac{T_w}{T_e}\right)^{1/2}\left[1 + \frac{1}{2}\frac{b_1}{a_1}\frac{u}{u_e} + \frac{1}{2}\frac{c_1}{a_1}\left(\frac{u}{u_e}\right)^2 + \cdots\right]$$

and Eq. (5-147) becomes

$$\frac{T_{av}}{T_e} \simeq 0.5 + 0.033M_e^2 + 0.5\frac{T_w}{T_e} + \cdots + 0\left[\frac{1}{72}\left(\frac{b_1}{a_1}\right)^2; \frac{1}{60}\left(\frac{c_1}{a_1}\right)^2\right]$$

which can be compared with Eq. (5-146).

Eckert[1] gives a reference temperature

$$\frac{T'}{T_e} = 0.5 + 0.038M_e^2 + 0.50\frac{T_w}{T_e}$$

and Young and Janssen[2] give a reference temperature

$$\frac{T'}{T_e} = 0.42 + 0.032M_e^2 + 0.58\frac{T_w}{T_e}$$

[1] E. R. G. Eckert, *J. Aeronaut. Sci.*, vol. 22, no. 8, pp. 585–587, August, 1955.

[2] G. B. W. Young and E. Janssen, *J. Aeronaut. Sci.*, vol. 19, no. 4, pp. 229–236, April, 1952.

and both of these equations are very close to Eq. (5-146) derived above, which is *exact* for a compressible *laminar boundary layer when* $L = P = 1$, $\mu = $ const, and $C_p = $ const and close to the correct value for $L = 1$, $P = 1, C_p = $ const, $\mu \propto T^{1/2}$, or $\mu \propto T^n$ for *any* value of n providing that $T_w/T_e \gg 1$ or $0.2 M_e^2$. When we relax the restrictions that $L = P = 1$ and $C_p = $ const, it is apparent that the analyses presented here will be somewhat more complex. However, in principle, a T_{av}/T_e could be found for any conditions postulated for the compressible laminar boundary layer. The indirect method of correlating numerical calculations according to Eq. (5-130) is equivalent to performing an analysis such as that described here which relates use of a reference temperature (the "average" temperature found in this analysis) to numerical solutions of the boundary-layer equations and, as such, removes some of the empiricism involved.

Romig[1] and Cohen[2] have shown that $C(\eta)$ can be correlated as a function of $h(\eta)$ for an equilibrium air mixture including dissociated air. This being so, the momentum equation is uncoupled from the species-conservation equation, providing the equation for $h(\eta)$ can be uncoupled from the species equation as it is when $L = P = 1$ when Crocco's integral, Eq. (5-137), applies. For air, this requirement is almost met and Eq. (5-137) might reasonably be used. In this case we obtain

$$\left(\frac{h_{av}}{h}\right)^{1-n} = \int_0^1 \left[\frac{h(u)}{h_e}\right]^{1-n} \frac{du}{u_e}$$

where
$$\frac{\rho(h_{av})\mu(h_{av})}{\rho_e\mu_e} = \left(\frac{h_{av}}{h}\right)^{1-n}$$

It is easy to show that for $n = 0$, for example,

$$\frac{h_{av}}{h_e} = \frac{1}{2}(h_e + h_w) + 0.0833u_e^2$$

a value remarkably close to Eckert's[3] reference enthalpy equation for $P = 0.70$

$$\frac{h'}{h_e} = \frac{1}{2}(h_e + h_w) + 0.092u_e^2$$

a result not to be unexpected in view of our analysis of the situation. Thus we see that the use of a reference temperature or reference enthalpy bears a close relation to numerical solutions of the boundary-layer equations for a variable $C(\eta)$.

[1] Mary F. Romig, *Jet Propulsion*, vol. 26, no. 12, pp. 1098–1101, 1956.
[2] Nathaniel B. Cohen, *NASA TN* D-194, 1960.
[3] Eckert, *op. cit.*

The reference-temperature concept has proved valuable in establishing heat-transfer relations useful for engineering calculations. Romig,[1] for example, was able to develop a useful expression for heat transfer of a laminar-boundary-layer stagnation point which represents the experimental measurements shown on Fig. 4-4 as well as or better than the results of numerical integration of the boundary-layer equations.

5-11. The Recovery Factor. The equations for heat transfer to or from a reacting laminar boundary layer developed in Secs. 4-5 and 5-4 are based upon several assumptions including the one that the viscous dissipation term in the transformed energy equation is negligible or equal to zero. That is, the energy equation, Eq. (5-23), is

$$\left(\frac{C}{P}g_f'\right)' + fg_f' = \left[\frac{u_e^2}{(I_f)_e}\left(\frac{1}{P} - 1\right)Cf'f''\right]' \qquad (5\text{-}23)$$

and the solutions developed in Secs. 4-5 and 5-5 contained the assumption that the term on the right-hand side of Eq. (5-23) was equal to zero. Of course, this is strictly true if $P = 1$ or $u_e = 0$. The latter condition is correct for the stagnation-point solution, for example. In both of these cases the recovery factor r is equal to one, where the recovery factor is defined by the equation

$$(I_f)_r = (I_f)_e + (r - 1)\frac{u_e^2}{2} = I_e - \left[\sum (C_i)_e h_i^0 + \frac{u_e^2}{2}\right] + r\frac{u_e^2}{2}$$

or
$$r = \frac{2[(I_f)_r - (I_f)_e]}{u_e^2} + 1 \qquad (5\text{-}148)$$

where $(I_f)_r$ is the partial enthalpy at the surface when heat transfer is zero. If $u_e = 0$ or $P = 1$, then $(I_f)_r = (I_f)_e$ and $r = 1$. Let us estimate the magnitude of the recovery factor when $P \neq 1$ and $u_e \neq 0$ as, for example, in the flat-plate solution.

Let both P and C be constant in Eq. (5-23). Then we obtain

$$g_f'' + \frac{P}{C}fg_f' = \frac{u_e^2}{(I_f)_e}(1 - P)[f'f''' + (f'')^2] \qquad (5\text{-}149)$$

The solution to Eq. (5-149) is[2]

$$g_f' = \frac{dg_f}{d\eta} = \exp\left(-\frac{P}{C}\int_0^\eta f\,d\eta'\right)$$

$$\times \left\{\int_0^\eta \exp\left(\frac{P}{C}\int_0^{\eta'} f\,d\eta''\right)\frac{u_e^2}{(I_f)_e}(1 - P)[f'f''' + (f'')^2]\,d\eta' + a\right\} \qquad (5\text{-}150)$$

[1] Romig, *op. cit.*, with Addendum, vol. 27, no. 12, p. 1255, 1957.
[2] See, for example, Eq. (8-9) and its solution.

and
$$g_f = \int_0^\eta g_f'(\eta') \, d\eta' + b \tag{5-151}$$

where $g_f'(\eta')$ is given by Eq. (5-150) and a and b are constants of integration. Since $g_f' = g_f'(0)$ when $\eta = 0$ and $g_f = g_f(0)$ when $\eta = 0$, then

$$a = g_f'(0) \tag{5-152}$$

and
$$b = g_f(0) = (g_f)_w \tag{5-153}$$

Now we assumed previously that

$$f \simeq f''(0) \frac{\eta^2}{2} + \cdots$$

since $f(0) = f'(0) = 0$. Then it follows that

$$(f'')^2 + ff''' \simeq [f''(0)]^2 \tag{5-154}$$

If we recall that

$$G(\eta;Z) = \frac{Z}{C} \int_0^\eta \exp\left(-\frac{Z}{C} \int_0^{\eta'} f \, d\eta''\right) d\eta' \tag{5-26}$$

so that
$$\frac{dG}{d\eta} = \frac{Z}{C} \exp\left(-\frac{Z}{C} \int_0^\eta f \, d\eta'\right) \tag{5-155}$$

and if we define a function $J(\eta;P)$ such that

$$J(\eta;P) = \int_0^\eta \frac{dG}{d\eta'}\left[\int_0^{\eta'} \frac{(f'')^2 + f'f'''}{dG/d\eta''} \, d\eta''\right] d\eta' \tag{5-156}$$

then when Eqs. (5-26) and (5-152) to (5-156) are substituted into Eqs. (5-150) and (5-151), there results

$$g_f = (g_f)_w + g_f'(0) \frac{C}{P} G(\eta;P) + \frac{u_e^2}{2(I_f)_e} (1 - P) \frac{[f''(0)]^2}{2} 4 \frac{C}{P} J(\eta;P) \tag{5-157}$$

Now
$$-(\dot{q}_w)_c = \rho_e u_e C_{H_e}[(I_f)_r - (I_f)_w] = k_w\left(\frac{\partial T}{\partial y}\right)_w$$

or
$$-(q_w)_c = \frac{(I_f)_e}{P} \mu_w g_f'(0)\left(\frac{\partial \eta}{\partial y}\right)_w$$

but for a flat-plate boundary layer, $r_0^k = 1$ and

$$\left(\frac{\partial \eta}{\partial y}\right)_w = \frac{\rho_w u_e}{\mu_e(2R_e)^{1/2}}$$

where $R_e = \rho_e u_e s/\mu_e$. Since $\rho_w \mu_w/\rho_e \mu_e = C$, we have

$$-(\dot{q}_w)_c = \rho_e u_e C_{H_e}[(I_f)_r - (I_f)_w] = \frac{\rho_e u_e(I_f)_e}{(2R_e)^{1/2}} \frac{g_f'(0)C}{P} \tag{5-158}$$

We can evaluate Eq. (5-157) at $\eta = \infty$ to obtain, since $g_f(\infty) = 1$,

$$g_f'(0)\frac{C}{P} = G(\infty;P)^{-1}\left\{1 - (g_f)_w\right.$$

$$\left. - \frac{u_e^2}{2(I_f)_e}(1 - P)\frac{[f''(0)]^2}{2}4\frac{C}{P}[J(\infty;P)]\right\} \quad (5\text{-}159)$$

thus, combining Eq. (5-158) with Eq. (5-159), we find that

$$\rho_e u_e C_{H_e} = \frac{\rho_e u_e}{(2R_e)^{1/2}G(\infty;P)} \quad (5\text{-}160)$$

as is given by Eq. (5-39) developed previously when applied to a flat-plate boundary layer, and, in view of Eq. (5-148),

$$r = 1 + (P - 1)2\frac{C}{P}J(\infty;P) \quad (5\text{-}161)$$

Now we have shown previously that

$$f''(0) \simeq \frac{0.47}{C^{1/2}}$$

Hence, from Eqs. (5-154) and (5-156),

$$J(\infty;P) = \frac{(0.47)^2}{C}\int_0^\infty \frac{dG}{d\eta}\left[\int_0^\eta \left(\frac{dG}{d\eta'}\right)^{-1}d\eta'\right]d\eta \quad (5\text{-}162)$$

and since the integral involved in Eq. (5-162) can be shown to be of magnitude 1 by numerical integration,[1] then

$$r \simeq 1 + (P - 1)\frac{0.882}{2P} \simeq [(P)^{1/2}]^{0.882P^{-1}} + \cdots \quad (5\text{-}163)$$

For $P \simeq 1$, as it is for most gas mixtures,

$$r \simeq P^{1/2} \quad (5\text{-}164)$$

and this is the value for r customarily used for flat-plate solutions.[2] If $P = 0.72$, the value of r given by Eq. (5-164) differs from that given by Eq. (5-163) by less than 3 per cent, for example. Thus the origin of the recovery-factor relation for the flat-plate laminar boundary layer given by Eq. (5-164) is explained and related to numerical solutions of the boundary-layer equations.

It is now apparent how the equations developed in Secs. 4-4 and 5-4 can be applied to the flat-plate laminar boundary layer under

[1] The value of the integral is 0.97 when $P = 1$ and $C = 1$, for example.

[2] See, for example, K. Pohlhausen, *Z. angew, Math. Mech.*, vol. 1, pp. 120–121, 1921.

conditions of zero mass transfer. Equation (5-160) is used to define C_{H_c}, where the driving enthalpy difference becomes $(I_f)_r - (I_f)_w$ instead of $(I_f)_e - (I_f)_w$ and $r = (P)^{1/2}$ instead of $r = 1$ in Eq. (5-148). Otherwise, the results of Secs. 4-4 and 4-5 remain unchanged. When mass transfer is present, the developments of this section do not apply.

5-12. Conclusions. In this chapter we have dealt with the reacting laminar boundary layer. Section 5-2 established the nature of the problems for which the boundary-layer equations are applicable. Section 5-3 then dealt with a justification for using equations for frozen gas-flow boundary layers under certain conditions for which the chemical reactions are assumed to occur at or very near the surface.

The method for treating the effects of chemical reactions upon the laminar boundary layer under the frozen-boundary-layer approximation was dealt with in Secs. 5-5 and 5-6. An example calculation was presented in Sec. 5-7.

The role of chemical kinetics was dealt with in Secs. 5-8 and 5-9. In Sec. 5-8 it was shown that the chemical kinetics of surface reactions may well determine the mass-loss rate and surface-heat-transfer rate under the frozen-boundary-layer approximation just as they determined the heat-transfer rate in the treatment of the dissociating laminar boundary layer in Sec. 4-7. In Sec. 5-9 we developed equations suitable for use when the frozen-boundary-layer approximation is not used and gas-phase dissociation is assumed to occur.

Section 5-10 was concerned with an examination of the origins of the so-called reference temperature or reference enthalpy. It was found that the reference temperature or enthalpy is directly related to an average temperature or enthalpy obtained from numerical solutions to the boundary-layer equations.

Section 5-11 gave the derivation of the expression customarily used for recovery factor in the calculation of heat transfer from a compressible laminar boundary layer flowing over a flat plate with zero mass transfer.

6

Leading-edge Bluntness, Shock-wave
Interaction, and Vorticity Effects

6-1. Introduction. It was mentioned in the introduction to
Chap. 4 that there are essentially two classes of problems concerned
with hypersonic boundary-layer flow: those primarily associated with
blunt-leading-edge bodies and those associated with sharp-leading-
edge bodies. The former class of problems was dealt with in Chap. 4.
In this chapter we shall be concerned with problems largely associated
with relatively sharp-leading-edge two-dimensional bodies in motion
at hypersonic speeds.

The first problem we shall describe is that of the interaction of the
leading-edge shock wave and the boundary layer. Equations will be
derived which will account for the effects of the interaction of
leading-edge shock wave and the boundary layer upon surface skin
friction and heat transfer. The flow regime where such effects are
significant will be defined.

Another problem we shall describe is concerned with the effects of
leading-edge blunting upon boundary-layer behavior comparatively
far downstream from the leading edge. We mean comparatively
far downstream in the sense that the effects we shall be concerned
with in this chapter are significant in regions downstream from the
vicinity of the stagnation-point region where the problems and
methods of Chap. 4 apply. In particular we shall deal with the
effects upon boundary-layer behavior of introduction of vorticity
into the flow external to the boundary layer. The vorticity is
intimately connected with the curvature of the leading-edge shock
wave, which is relatively significant when the leading edge is blunted
a significant amount. The flow regions where such effects are
apparent will be bounded in terms of gas-dynamic parameters.

144

6-2. Interactions of the Leading-edge Shock Wave and the Boundary Layer. Since Prandtl originally advanced his concept of the thin boundary layer within which the effects of viscosity are largely confined in the flow of a gas over a body, it has been accepted that the boundary layer has a displacement effect in the sense that the effective shape of the body is altered or enlarged because of the reduced mass flux within the boundary layer. For most cases of interest to the aerodynamicist, this displacement effect is negligible in so far as the "displacement thickness" δ^*, defined in Eq. (6-1) below, is usually a negligible fraction of the lateral dimension of a body, say, the local body radius of a body of revolution, and, as such, cannot have appreciable effect upon the external "inviscid flow." This displacement distance can be described as the distance δ^* which the flow external to the boundary layer is displaced owing to the diminution of mass flux within the boundary layer. That is,

$$\rho_e u_e (\delta - \delta^*) = \int_0^\delta \rho u \, dy \qquad \delta \to \infty$$

or, rearranging,

$$\delta^* = \int_0^\infty \left(1 - \frac{\rho u}{\rho_e u_e} \right) dy \qquad (6\text{-}1)$$

Now, within a hypersonic boundary layer, the temperature is apt to be high, which means that the density will be low and, as Eq. (6-1) serves to indicate, that the displacement thickness will be large, in fact, will almost be as large as the boundary-layer thickness itself if $T_w \gg T_\infty$. Another way of describing this phenomenon is to state that the mass flux per unit area normal to the flow in a hot hypersonic layer is almost zero and certainly much less than the mass flux per unit area in the undisturbed stream. Ordinarily, this would be of no particular consequence except close to the leading edge of a sharp-nosed object. Near the sharp leading edge, where the displacement thickness is increasing from zero somewhere close to the leading edge, the turning of the flow external to the boundary layer causes an initial compression accompanied by a shock wave which becomes quite severe at high Mach numbers. This compression with the following flow expansion, in turn, affects the build-up of the boundary layer, and thus we have the ingredients for a problem of interaction between the external stream and the boundary layer. It is just this problem, which is accentuated at hypersonic Mach numbers, with which this section is concerned.

Consider Fig. 6-1 where is sketched the hypersonic flow over a sharp-edged flat plate along with the coordinate system we shall use in that which follows. Because of the turning of the gas stream in the region between the edge of the boundary layer and the shock wave, the pressure and Mach number at the edge of the boundary layer will vary with s just as they would if we replaced the flat plate by a shape defined by the boundary-layer edge (actually the locus of the displacement thickness) in an inviscid hypersonic gas stream. Let us acquire some familiarity with the problem by determining the pressure at the edge of the boundary layer in an approximate manner.

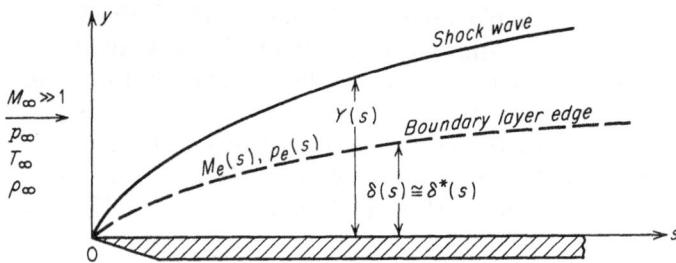

FIG. 6-1. The problem of interaction between the leading-edge shock wave and the boundary-layer flow over a flat plate.

It can be shown for two-dimensional objects (similar results can be developed for bodies of revolution) that, if $M_\infty \gg 1$ and $M_\infty(d\delta^*/ds) \ll 1$, then[1]

$$\frac{p_e}{p_\infty} = 1 + \gamma M_\infty \frac{d\delta^*}{ds} + \frac{\gamma(\gamma+1)}{4}\left(M_\infty \frac{d\delta^*}{ds}\right)^2 + \cdots \qquad (6\text{-}2)$$

Also, if $M_\infty \gg 1$ and $M_\infty(d\delta^*/ds) \gg 1$, then it can be shown, using oblique shock relations, that[2]

$$\frac{p_e}{p_\infty} \simeq \frac{\gamma+1}{2}\gamma\left(M_\infty \frac{d\delta^*}{ds}\right)^2 \qquad (6\text{-}3)$$

but δ^* itself is affected by p_e/p_∞ as we shall see, and this, in effect, represents the crux of the problem. Now from laminar-boundary-layer theory for compressible flows in the absence of interaction of the leading-edge shock wave and the boundary layer, it is known that [see, for example, Eq. (2-59)]

$$(\delta^*)^2 \propto \frac{\bar{\mu}s}{\bar{\rho}u_e} \qquad (6\text{-}4)$$

where $\bar{\mu}$ and $\bar{\rho}$ are some average values of μ and ρ within our heated boundary layer. Now, $u_e \simeq u_\infty$ for $M_\infty \gg 1$ and $\bar{p} = p_e$

[1] See, for example, W. H. Dorrance, *J. Aeronaut. Sci.*, vol. 19, no. 9, pp. 593–600, 1952, Eq. (26) with $C_p = 2/\gamma M_\infty^2[(p_e/p_\infty) - 1]$, $\delta = d\delta^*/ds$, and $f = 1$.
[2] *Ibid.*, Eq. (25) with $C_p = (2/\gamma M_\infty^2)[(p_e/p_\infty) - 1]$, $\omega = d\delta^*/ds$, and $M_\infty \to \infty$.

according to the boundary-layer concept, which Shen[1] has shown to be valid for the present problem.   Hence, Eq. (6-4) can be written

$$(\delta^*)^2 \propto C M_\infty^4 \frac{p_\infty}{p_e} \frac{s^2}{R_\infty} \tag{6-5}$$

and thus

$$\frac{d\delta^*}{ds} \propto \frac{\delta^*}{s} \tag{6-6}$$

where we have used

$$\frac{\bar{\mu}}{\mu_\infty} = C \frac{\bar{T}}{T_\infty}$$

$$\frac{p_e \rho_\infty}{p_\infty \bar{\rho}} = \frac{\bar{T}}{T_\infty} \propto M_\infty^2 \qquad \text{for } M_\infty \gg 1$$

and

$$R_\infty = \frac{\rho_\infty u_\infty s}{\mu_\infty}$$

Now it is apparent from Eqs. (6-2) and (6-3) that we can examine two extremes, the weak-interaction case and the strong-interaction case, where $M_\infty \gg 1$ for both cases but $M_\infty(d\delta^*/ds) \ll 1$ for the former case and $M_\infty(d\delta^*/ds) \gg 1$ in the latter case.   For weak interactions and $p_e/p_\infty \simeq 1$ according to Eq. (6-2), combination of Eqs. (6-2) and (6-6) gives

$$\frac{p_e}{p_\infty} - 1 \propto \bar{\chi} \qquad \bar{\chi} < 1 \qquad M_\infty \gg 1 \tag{6-7}$$

where $\bar{\chi}$ is defined as

$$\bar{\chi} = \frac{C^{1/2} M_\infty^3}{(R_\infty)^{1/2}} \tag{6-8}$$

For strong interactions p_e/p_∞ is proportional to $M_\infty^2(d\delta^*/ds)^2$ according to Eq. (6-3), and combination of Eqs. (6-3) and (6-6) gives

$$\frac{p_e}{p_\infty} \propto \bar{\chi} \qquad \bar{\chi} \gg 1 \qquad M_\infty \gg 1 \tag{6-9}$$

and these asymptotic behaviors are borne out by the more exact treatments of the problem.   Relations (6-7) and (6-9) yield at least three valuable conclusions relative to this problem:

1. $\bar{\chi} = C^{1/2} M_\infty^3/(R_\infty)^{1/2}$ is the proper similarity parameter for the

---

[1] S. F. Shen, *J. Aeronaut. Sci.*, vol. 19, no. 7, pp. 500–501, 1952.

theory of interaction of a hypersonic leading-edge shock wave and a laminar boundary layer.

2. For strong interactions, combination of relations (6-5) and (6-9) reveals that

$$\delta^* \propto s^{3/4}$$

3. For strong interactions, relation (6-9) reveals that

$$\frac{p_e}{p_\infty} \propto s^{-1/2}$$

These conclusions, which are of considerable importance to developing a rigorous strong-interaction theory, were independently arrived at by Li and Nagamatsu[1] and Lees[2] upon developing strong-interaction theories. Stewartson[3] later showed that not only do these similarity considerations hold within the viscous layer in the strong-interaction region but apparently the inviscid region between the shock wave and the boundary layer yields to similarity considerations such that $Y(s) \propto s^{3/4}$ and $Y(s) \propto \delta^*(s)$ wherever $M_\infty \gg 1$ and $\bar{\chi} \gg 1$. $Y(s)$ is shown in Fig. 6-1. Thus the treatment of the entire flow field between the shock wave and the plate is similar to the approximations of these theories within the strong-interaction region, that is, where $\bar{\chi} \gg 1$.

Strong-interaction-theory Equations. Having thus gained some insight into the problem, let us proceed to develop a theory for the interaction of a leading-edge shock wave and a boundary layer for the laminar-boundary layer. We shall restrict ourselves here to the strong-interaction problem for the following reasons: (1) The weak-interaction theory is well described elsewhere,[4] (2) the results of the weak-interaction theory show that heat transfer is relatively insensitive to weak interactions, and (3) an accurate (in the limit as $M_\infty \to \infty$) theory for strong interactions near the leading edge for hypersonic flow over a flat plate can be developed.

We shall develop the strong-interaction theory here, using the nomenclature and equations in the form consistent within this chapter. Our assumptions made in order to proceed are given in Table 6-1.

[1] T. Y. Li and H. T. Nagamatsu, *J. Aeronaut. Sci.*, vol. 20, no. 5, pp. 345–355, 1953.

[2] L. Lees, *J. Aeronaut. Sci.*, vol. 20, no. 2, p. 143, 1953.

[3] K. Stewartson, *J. Aeronaut. Sci.*, vol. 22, no. 5, pp. 303–309, 1955.

[4] Wallace D. Hayes and Ronald F. Probstein, "Hypersonic Flow Theory," pp. 341–353, Academic Press, Inc., New York, 1959.

TABLE 6-1. ASSUMPTIONS MADE IN DEVELOPING STRONG-INTERACTION THEORY

Assumption	Justification
$k = 0$ in the continuity equation	We deal with a flat plate
$C_p = $ const, $C_v = $ const	Ignore vibrational relaxation and dissociation as second-order effects in this region
$P = L = S = 1$	$P \simeq 1$, $L \simeq 1$, $S \simeq 1$ in real gas, and consequences of their not being unity are of second order of importance in present problem
$u_e = u_\infty$	Exact as $M_\infty \to \infty$ with $\bar{\chi}$ finite. Neglects vorticity due to shock curvature. See Sec. 6-4
$T_e = T_\infty$	Exact as $M_\infty \to \infty$ with $\bar{\chi}$ finite. Neglects vorticity due to shock curvature. See Sec. 6-4
$\mu_e = \mu_\infty$	Follows from $T_e = T_\infty$
$C = 1$	Simplifies calculation; otherwise poor
M_e^2; $M_\infty^2 \gg \dfrac{2}{\gamma - 1}$	Will simplify expression for $p_e(M_e)$ and is consistent with $M_\infty \gg 1$
$\dfrac{h_e}{I_e} \ll 1$	Consistent with $M_e^2 \gg \dfrac{2}{\gamma - 1}$; $M_\infty \gg 1$
$\alpha = $ const	Neglect effects of dissociation as being of second order in interaction region

Our boundary-layer equations are given by Eqs. (2-94) and (2-102) with $I_e = $ const and $P = L = 1$ and are

$$f''' + ff'' + \frac{2\bar{s}}{u_e} \frac{du_e}{d\bar{s}} \left[\frac{p_e}{\rho} - (f')^2 \right] = 0 \qquad (6\text{-}10)$$

and

$$g'' + fg' = 0 \qquad (6\text{-}11)$$

with boundary conditions

$$
\begin{aligned}
f(0) = f'(0) = 0 \qquad & f'(\infty) = 1 \\
g(0) = g_w \qquad\qquad & g(\infty) = 1
\end{aligned}
\qquad (6\text{-}12)
$$

Further simplification of Eq. (6-10) results for $I_e = I_\infty = $ const; then it can be shown that

$$\frac{p_e}{\rho} - (f')^2 = g \frac{h}{I} \frac{I_e}{h_e} - (f')^2$$

where

$$g = \frac{I}{I_e} \qquad \text{and} \qquad f' = \frac{u}{u_e}$$

Now

$$I = h + \frac{u^2}{2}$$

and

$$I_e = h_e + \frac{u_e^2}{2}$$

and thus

$$\frac{p_e}{\rho} - (f')^2 = \frac{I_e}{h_e}[g - (f')^2] + g\frac{I_e}{h_e}\frac{h - I}{I_e} + (f')^2\frac{I_e - h_e}{I_e}$$

but

$$g\frac{I_e}{h_e}\frac{h - I}{h_e} = -g\frac{I_e}{h_e}\frac{u^2}{2I} = -\frac{u^2}{2h_e}$$

and

$$(f')^2\frac{I_e - h_e}{I_e} = \frac{u^2}{2h_e}$$

hence we obtain

$$\frac{p_e}{\rho} - (f')^2 = \frac{I_e}{h_e}[g - (f')^2] \tag{6-13}$$

whence Eq. (6-10) becomes

$$f''' + ff'' + \beta[g - (f')^2] = 0 \tag{6-14}$$

and the coupling of the energy equation (6-11) and the momentum equation (6-14) is explicit. In Eq. (6-14)

$$\beta = \frac{2\bar{s}}{u_e}\frac{du_e}{d\bar{s}}\frac{I_e}{h_e} = \frac{2\bar{s}}{M_e}\frac{dM_e}{d\bar{s}} \tag{6-15}$$

since $I_e = I_\infty$, $C_p = $ const, and $P = 1$. Now M_e is related to the pressure at the edge of the boundary layer by the isentropic flow expression

$$\frac{p_e}{p_\infty} = \left[\frac{[1 + (\gamma - 1)/2]M_\infty^2}{[1 + (\gamma - 1)/2]M_e^2}\right]^{\gamma/(\gamma-1)} \simeq \left(\frac{M_\infty}{M_e}\right)^{2\gamma/(\gamma-1)} \tag{6-16}$$

and according to our analysis conducted earlier, it is assumed that in the strong-interaction region

$$\frac{p_e}{p_\infty} \propto s^{-1/2} \tag{6-17}$$

also, by definition,

$$\bar{s} = \int_0^s \rho_e\mu_e u_e\, ds = \frac{u_\infty\mu_\infty}{RT_e}\int_0^s p_e\, ds \tag{6-18}$$

since, by our assumptions, $T_\infty = T_e$, $u_\infty = u_e$, and $\mu_\infty = \mu_e$. Combine Eqs. (6-15), (6-16), and (6-18) and relation (6-17) in the following manner to obtain a relation for β. From Eq. (6-15),

$$\beta = \frac{2\bar{s}}{M_e}\frac{dM_e}{d\bar{s}} = 2\bar{s}\frac{ds}{d\bar{s}}\frac{d(\log M_e)}{ds}$$

but from Eq. (6-16)

$$\frac{d\log M_e}{ds} = -\frac{\gamma - 1}{2\gamma}\frac{d[\log(p_e/p_\infty)]}{ds}$$

and from Eq. (6-17)

$$\frac{d[\log (p_e/p_\infty)]}{ds} = -\frac{1}{2s}$$

and letting $u_e = u_\infty$, $T_e = T_\infty$, $\mu_e = \mu_\infty$, and $\rho_e/\rho_\infty = (p_e/p_\infty) \propto s^{-1/2}$ in Eq. (6-18), we obtain

$$\bar{s}\frac{ds}{d\bar{s}} = 2s$$

and so, using the above relations, it follows that

$$\beta = \frac{\gamma - 1}{\gamma} \tag{6-19}$$

where the important relation is the relation between p_e and s in Eq. (6-17) revealed by our simplified analysis presented earlier. Equations (6-14) and (6-11) have been solved for different constant values of β subject to boundary conditions (6-12) by Li and Nagamatsu[1] and Cohen and Reshotko.[2] The results of their calculation are $f(\eta)$, $f'(\eta)$, $f''(\eta)$, $g(\eta)$, and $g'(\eta)$ for various initial values $g(0) = g_w$. Li and Nagamatsu employed an analog computer in their calculations, whereas Cohen and Reshotko used a method of numerical integration suitable to a digital computer, which they employed. Li and Nagamatsu particularized on the values of β pertinent to the strong-interaction problem. They chose $\beta = 0.286$ ($\gamma = 1.4$) and $\beta = 0.400$ ($\gamma = 1.667$) suitable for air and helium, for example.

Strong-interaction-theory Solutions. It is now in order to relate the solutions for $f(\eta)$, $f'(\eta)$, $f''(\eta)$, etc., to the various quantities of interest in calculating heat transfer, skin friction, boundary-layer thickness, and pressure distribution. Begin with the displacement thickness δ^*. From Eq. (6-1) and the definition of the independent variable η, we obtain

$$\delta^* = \frac{(2\bar{s})^{1/2}}{\rho_e u_e} \int_0^\infty \left(\frac{\rho_e}{\rho} - f'\right) d\eta$$

since

$$\frac{d\eta}{dy} = \frac{\rho u_e}{(2\bar{s})^{1/2}}$$

Furthermore, since $p = p_e$, then $\rho_e/\rho = T/T_e$ and we can write

$$\frac{\rho_e}{\rho} - f' = \frac{T}{T_e} - f' = \frac{g - (f')^2(u_e^2/2I_e)}{1 - (u_e^2/2I_e)} - f'$$

[1] T. Y. Li and H. T. Nagamatsu, *J. Aeronaut. Sci.*, vol. 20, no. 9, pp. 653–655, 1953. See also T. Y. Li and H. T. Nagamatsu, *Proc. 1954 Heat Transfer and Fluid Mech. Inst.*, pp. 143–157, California Book Co., Ltd., Berkeley, Calif., 1954.

[2] Clarence B. Cohen and Eli Reshotko, *NACA TN 3325*, February, 1955.

since
$$I_e = h_e + \frac{u_e^2}{2} \qquad \text{and} \qquad g = \frac{I}{I_e}$$

Thus
$$\frac{p_e}{\rho} - f' = \frac{I_e}{h_e}\left\{ g - (f')^2 - \frac{h_e}{I_e}[f' - (f')^2] \right\}$$

Now from our assumptions stated in Table 6-1,

$$\frac{h_e}{I_e} \ll 1 \qquad \text{for } M_\infty \gg 1$$

hence
$$\frac{p_e}{\rho} - f' \simeq \frac{I_e}{h_e}[g - (f')^2]$$

Furthermore, since $u_e = u_\infty$, $T_e = T_\infty$, $\mu_e = \mu_\infty$, and $\rho_e/\rho_\infty = p_e/p_\infty \propto s^{-1/2}$, it follows that

$$\int_0^s \rho_e \mu_e u_e \, ds = 2\frac{\mu_\infty u_\infty}{\bar{R} T_\infty} p_e s$$

and
$$\frac{(2\bar{s})^{1/2}}{\rho_e u_e} = \frac{2s}{(R_\infty)^{1/2}}\left(\frac{p_\infty}{p_e}\right)^{1/2}$$

Using these expressions in the equation for δ^* given above results in

$$\left(\frac{p_e}{p_\infty}\right)^{1/2} \frac{\delta^*}{\delta} = (\gamma - 1)\frac{I_0\bar{\chi}}{M_\infty} \qquad (6\text{-}20)$$

where $\bar{\chi}$ is defined by Eq. (6-8) with $C = 1$ and where

$$I_0 = \int_0^\infty [g - (f')^2] \, d\eta \qquad (6\text{-}21)$$

$$\eta = \frac{\rho_e u_e}{(2\bar{s})^{1/2}} \int_0^y \frac{\rho}{\rho_e} \, dy \qquad (6\text{-}22)$$

and
$$f' = \frac{u}{u_e} \qquad (6\text{-}23)$$

and, according to our assumptions for $M_\infty \gg 1$, $h_e/I_e \simeq 0$, which is used in obtaining the integrand of Eq. (6-21). Now since we assume that $p \propto s^{1/2}$, we can write

$$\frac{p_e}{p_\infty} = A\bar{\chi} \qquad (6\text{-}24)$$

Also, since $\delta^* \propto s^{3/4}$, we can write

$$\frac{\delta^*}{s} = \frac{B(\bar{\chi})^{1/2}}{M_\infty} \simeq \frac{\delta}{s} \qquad (6\text{-}25)$$

Combination of Eqs. (6-20), (6-24), and (6-25) gives

$$(A)^{1/2}B = (\gamma - 1)I_0 \tag{6-26}$$

Furthermore, combination of Eqs. (6-3), (6-24), and (6-25) gives

$$A = \tfrac{9}{32}\gamma(\gamma + 1)B^2 \tag{6-27}$$

and combining Eq. (6-26) with Eq. (6-27) gives

$$A = \frac{3}{4}(\gamma - 1)\left[\frac{\gamma(\gamma + 1)}{2}\right]^{1/2}I_0 \tag{6-28}$$

and, from Eqs. (6-27) and (6-28),

$$B = 2[2\gamma(\gamma + 1)]^{-1/4}[\tfrac{2}{3}(\gamma - 1)I_0]^{1/2} \tag{6-29}$$

where I_0 is defined by Eq. (6-21). Equations (6-24) and (6-28) give p_e/p_∞ in terms of $\bar{\chi}$ and the integral I_0; Eqs. (6-25) and (6-29) give the displacement thickness δ^* in terms of $\bar{\chi}$ and the integral I_0. It remains to relate $\bar{\chi}$ and I_0 to the skin-friction coefficient and heat transfer. Now

$$\tau_w = \left(\mu\frac{\partial u}{\partial y}\right)_w = \frac{\rho_w\mu_w u_e^2 f''(0)}{2\rho_\infty u_\infty s}\left(R_\infty\frac{p_\infty}{p_e}\right)^{1/2} \tag{6-30}$$

where Eqs. (6-22) and (6-18) have been used and where

$$R_\infty = \frac{\rho_\infty u_\infty s}{\mu_\infty} \tag{6-31}$$

Since $C = \rho_w\mu_w/\rho_e\mu_e = 1$, $\mu_e = \mu_\infty$, and $p_e/p_\infty = \rho_e/\rho_\infty$, Eq. (6-30) becomes

$$C_f(R_\infty)^{1/2} = C_2(\bar{\chi})^{1/2} \tag{6-32}$$

where

$$C_2 = (A)^{1/2}f''(0) \tag{6-33}$$

and

$$C_f = \frac{2\tau_w}{\rho_\infty u_\infty^2} \tag{6-34}$$

By definition

$$C_H = \frac{-\dot{q}_w}{\rho_\infty u_\infty I_e[1 - g(0)]} \tag{6-35}$$

where, if $P = L = C = 1$,

$$-\dot{q}_w = \left(k\frac{\partial T}{\partial y}\right)_w = \frac{\rho_w\mu_w u_e I_e g'(0)}{2\rho_\infty u_\infty s}\left(\frac{p_\infty}{p_e}R_\infty\right)^{1/2} \tag{6-36}$$

where Eqs. (6-22) and (6-18) have been used in deriving Eq. (6-36). Combination of Eqs. (6-24), (6-35), and (6-36) gives

$$C_H(R_\infty)^{1/2} = D(\bar{\chi})^{1/2} \tag{6-37}$$

where

$$D = \frac{g'(0)}{2[1 - g(0)]}(A)^{1/2} \tag{6-38}$$

and advantage has been taken of the fact that $C = 1$, $p_e/p_\infty = \rho_e/\rho_\infty$, and $\mu_e = \mu_\infty$. Furthermore, making use of Eqs. (6-35), (6-37), and (6-38), it can be shown that

$$\frac{-\dot{q}_w}{(-\dot{q}_w)_{p_e/p_\infty = 1}} = (A\bar{\chi})^{1/2} \qquad \bar{\chi} \gg 1 \qquad (6\text{-}39)$$

and for $\bar{\chi} = 10$ and $A = 0.514$, this ratio is 2.25 for $T_w/T_0 = 1$, and it is apparent from the results of calculations shown in Table 6-2 that

TABLE 6-2. VALUES OF A, B, C_2, D, AND I FOR VARIOUS VALUES OF $g(0) = \dfrac{T_w}{T_0}$

$\dfrac{T_w}{T_0}$	A	B	C_2	D†	I_0
$\beta = 0.286$ ($\gamma = 1.4$)					
0	0.148	0.397	0.208	0.095	0.382
0.20	0.232	0.495	0.279	...	0.596
0.60	0.377	0.632	0.412	0.155	0.970
1.00	0.514	0.738	0.549	0.187	1.322
$\beta = 0.400$ ($\gamma = 1.667$)					
0	0.261	0.457	0.287	0.125	0.350
0.20	0.403	0.568	0.395	...	0.540
0.60	0.660	0.726	0.602	0.210	0.885
1.00	0.921	0.858	0.820	0.252	1.235
2.00	1.506	1.097	1.361	0.342	2.020

† Li's and Nagamatsu's tabulated values of D are actually $2D$.

the shock-wave–boundary-layer interaction has an appreciable effect on heat transfer for this case ($A = 0.514$) and a lesser effect for $T_w/T_0 < 1$, since $A < 0.514$ when $T_w/T_0 < 1$.

Combination of Eqs. (6-32) and (6-37) yields the useful relationship

$$\frac{2C_H}{C_f} = \frac{2D}{C_2} \qquad (6\text{-}40)$$

and this completes the description of the method of calculation.

Example Calculations. Li and Nagamatsu[1] tabulate the functions A, B, C_2, $2D$, and I for $\beta = 0.286$ and 0.40 and various values of $g(0) = T_w/T_0$ ($T_0 = $ stagnation temperature). These tabulations are presented above because of their usefulness.

[1] Ting Yi Li and H. T. Nagamatsu, *Proc. Fourth Midwestern Conf. Fluid Mech.*, pp. 273–287, Purdue University, 1955.

With the use of the tabulated values for D, Fig. 6-2 was prepared which gives the variation of heat-transfer coefficient with wall temperature. It can be seen that reducing the wall temperature markedly reduces the interaction effect as might be anticipated upon reflecting that reducing the surface temperature increases the surface density and hence reduces the displacement thickness, leading to the trend exhibited in Fig. 6-2.

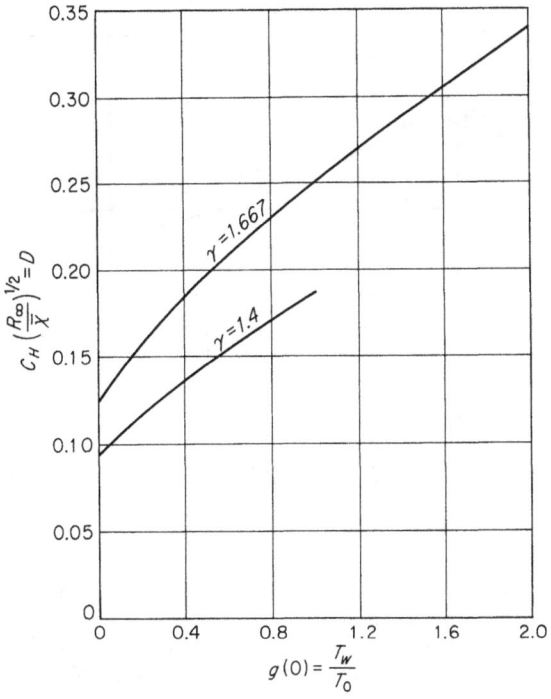

Fig. 6-2. The variation of heat-transfer coefficient with surface temperature in the strong-interaction region. (*After T. Y. Li and H. T. Nagamatsu, J. Aeronaut. Sci., vol. 20, no. 5, pp. 345–355, 1953.*)

Figure 6-3 was prepared with the use of Eq. (6-40) and the tabulated values of C_2 and D. Figure 6-3 shows the variation $2C_H/C_f$ with wall temperature, whereupon it appears that this ratio increases with reduction in wall temperature. Also shown in Fig. 6-3 are results obtained by Pai and Shen,[1] who present a theory applicable for all values of $\bar{\chi}$ which is based upon the von Kármán momentum integral technique and use of a plausible three-term series relation between p_e/p_∞ and $M_\infty(d\delta^*/ds)$. The curve shown in Fig. 6-3 results from the

[1] S. I. Pai and S. F. Shen, *Proc. Fourth Midwestern Conf. on Fluid Mech.,* pp. 259–272, Purdue University, 1955.

leading term of a series expansion inherent in Pai's and Shen's method of solution and corresponds to the strong-interaction case treated here. As can be seen, reasonable agreement exists between the results of the two calculations.

Many others have obtained solutions applicable in the strong-interaction region which can be compared with the treatment presented here for an insulated plate. Taking them in chronological

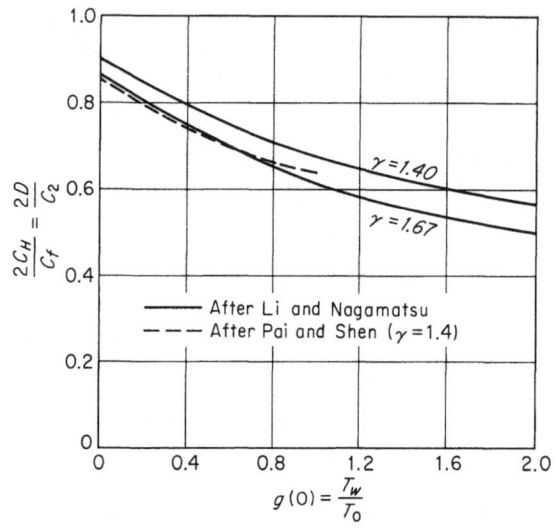

FIG. 6-3. The effect of surface temperature on the ratio $2C_H/C_f$ in the strong-interaction region. (*After T. Y. Li and H. T. Nagamatsu, J. Aeronaut. Sci., vol. 20, no. 5, pp. 345–355, 1953; and S. L. Pai and S. F. Shen, Proc. Fourth Midwestern Conf. on Fluid Mech., pp. 259–272, Purdue University, 1955.*)

order of appearance, Bertram[1] presented an approximate method which, however, does not yield explicit relations in the interaction parameter $\bar{\chi}$ for large $\bar{\chi}$. Li and Nagamatsu[2] presented an approach which made use of the von Kármán momentum integral applied to the entire region between the shock wave and the flat plate and which incorporated an approximation of the velocity distribution in this region by a quartic in y. No distinction was made between the edge of the boundary layer and the location of the shock wave. Lees[3] presented an analysis which made a distinction between the

[1] M. H. Bertram, *NACA TN* 2773, 1952.
[2] T. Y. Li and H. T. Nagamatsu, *J. Aeronaut. Sci.*, vol. 20, no. 5, pp. 345–355, 1953.
[3] Lester Lees, *J. Aeronaut. Sci.*, vol. 20, no. 2, p. 143, 1953.

edge of the boundary layer and the location of the shock wave and is essentially equivalent to the theory presented herein.   Stewartson[1] presented a theory which showed that not only does the boundary layer yield to similarity concepts in the strong-interaction region but also the flow between the shock wave and the boundary layer obeys similarity laws and yields a similar solution.   Stewartson's solution is somewhat more exact than the present solution for $\bar{\chi}$ large but finite, since Eq. (6-3), which is exact only when $\bar{\chi} = \infty$, is not used in Stewartson's theory.   The results of all these theories (except Bertram's) are compared in Table 6-3.

TABLE 6-3. A Comparison of the Results of Various Strong-interaction Theories for the Case of No Heat Transfer

$$\left(\text{Let } \frac{p_e}{p_\infty} = A\bar{\chi}; \quad M_\infty \frac{\delta^*}{s} \simeq M_\infty \frac{\delta}{s} = B\bar{\chi}^{1/2} \right)$$

Source	A	B
Li and Nagamatsu	0.573	
Lees	0.52	
Stewartson	0.555	0.704
Li and Nagamatsu (this book)	0.514	0.738

Assuming that Stewartson's solution represents the most exact treatment of this problem, a comparison with his results shows that the theory given here is accurate to within -7.4 per cent in the parameter A and within $+4.7$ per cent in the parameter B.   However, the theory given here (which is essentially equivalent to that of Lees) is the only theory of those compared in Table 6-3 which accounts for the effects of wall-temperature variation, and therein lies its unique value.

Before this strong-interaction theory is accepted, it is reasonable to inquire as to the accuracy of its comparison with suitable experiment and the accuracy of the fundamental assumptions upon which the theory is based.   One of these assumptions is that $p \propto s^{-1/2}$, which can be checked directly by measurements of surface pressure on a sharp-edged flat plate providing M_∞ is very high and the measurements are taken close to the leading edge (that is, $\bar{\chi} \gg 1$).   Nagamatsu and Sheer[2] report appropriate measurements taken on a flat plate

[1] K. Stewartson, *J. Aeronaut. Sci.*, vol. 22, no. 5, pp. 303–309, 1955.

[2] H. T. Nagamatsu and R. E. Sheer, Jr., *ARS J.*, vol. 30, no. 5, pp. 454–462, 1960.

immersed within a hypersonic stream generated by a shock tunnel. Figure 6-4 shows their results, where the expected linearity of p with $s^{-1/2}$ is demonstrated over a portion of the plate.

Figure 6-5 presents a direct comparison of Eq. (6-24) with Nagamatsu's and Sheer's measurements. It can be seen that the agreement is not noticeably good, although the sought-after linearity of p with $\bar{\chi}$ is obtained outside the leading-edge region ($\bar{\chi} \to \infty$).

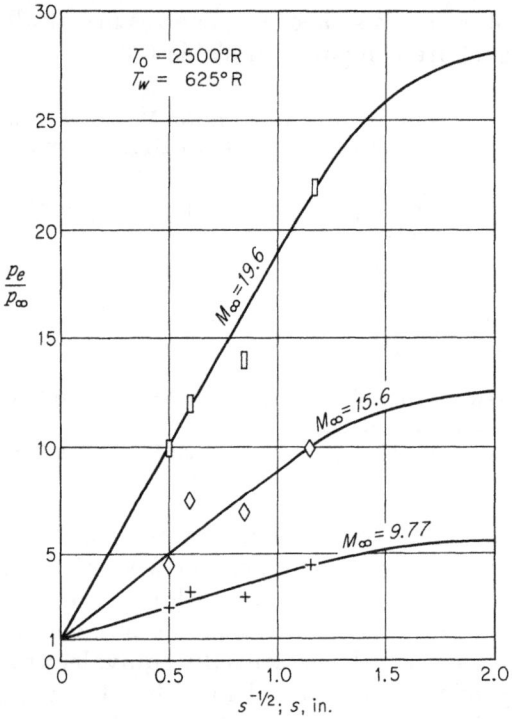

FIG. 6-4. Measured variations of flat surface pressure with distance from the leading edge. (*After H. T. Nagamatsu and R. E. Sheer, Jr., ARS J., vol. 30, no. 5, pp. 454–462, 1960.*)

Figure 6-5 obviously indicates that the strong-interaction theory is completely inadequate near the leading edge ($\bar{\chi} \to \infty$), where the curves shown on Fig. 6-5 depart from a linear variation with $\bar{\chi}$. As the leading edge is approached, it seems reasonable to visualize that the inviscid outer region and viscous boundary layer become indistinguishable and, in fact, merge. Farther forward of this region, a region is reached where the mean free path is of the order of the distance from the leading edge and the continuum-flow theory breaks down completely. The merged layer region forward of the strong-interaction region of this section has been treated by Oguchi,[1] who finds that a similarity prevails in this region such that $Y \propto s^{3/4}$ as in the strong-interaction region but that $p = \text{const}$ ($p \propto s^0$) in contrast to the strong-interaction variation of $p \propto s^{-1/2}$.

It can be concluded that the strong-interaction theory presented here has fair success in predicting strong-interaction effects where it

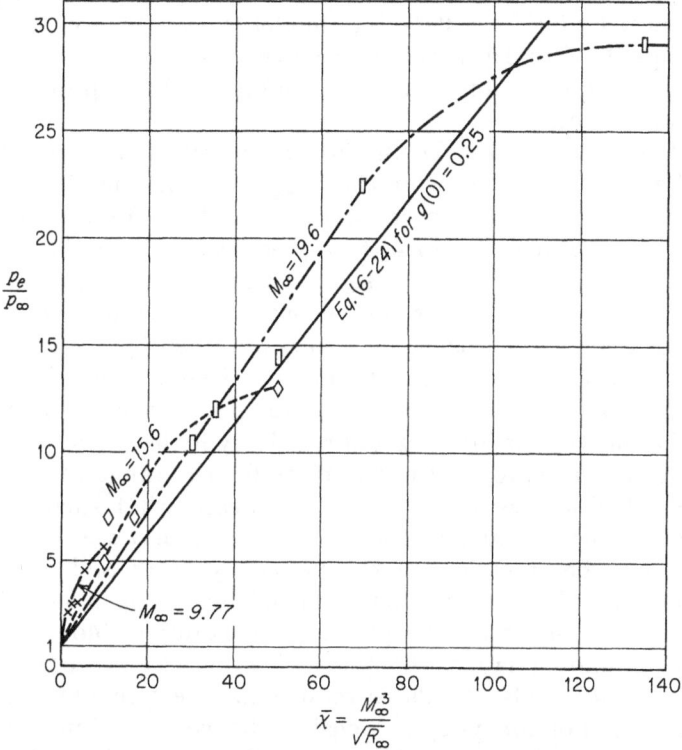

Fig. 6-5. Comparison between theory and measurements of flat-plate surface pressure. (*Measurements after H. T. Nagamatsu and R. E. Sheer, Jr., ARS J., vol. 30, no. 5, pp. 454–462, 1960.*)

is applicable $[1 < \bar{\chi} < O(20)]$. Its lack of wholehearted success in this region when compared with experiment can be attributed to a number of factors which include the following:

1. The assumption of a linear viscosity law with $C = 1$.
2. The omission of the effects of dissociation.

---

[1] Hakuro Oguchi, "The Sharp Leading Edge Problem in Hypersonic Flow," a paper delivered at the Second International Symposium on Rarefied Gas Dynamics, Berkeley, Calif., Aug. 3–6, 1960.

3. The use of Eq. (6-3) relating p_e to $d\delta^*/ds$ which is precise only when $\bar{\chi} = \infty$.

4. The assumption that $T_e = T_\infty$ and $u_e = u_\infty$ at the boundary-layer edge. This neglects the effects of vorticity in the external stream between the boundary layer and the shock wave. This vorticity is due to the curvature in the leading-edge shock wave.

5. The practical uncertainties in attempting to obtain a perfectly sharp leading edge for a flat-plate model or, conversely, the omission of leading-edge-bluntness effects from the theory.

The last two omissions will be the subject taken up in the sections following this one.

Effects of Mass Transfer on Strong-interaction Theory. The possible effects of mass transfer (blowing and sucking) in the strong-interaction region have been studied by Li.[1] Li determined the effect of mass transfer upon surface pressure and boundary-layer and shock-layer thickness for an insulated flat plate set impulsively into uniform motion at a hypersonic velocity. Li's results with no mass transfer are identical with some results obtained by Stewartson[2] with which useful comparisons can be made. Keeping in mind that Li's results are for the case of zero heat transfer to the flat plate, it was found that suction or blowing had relatively little effect (± 8 per cent) upon surface pressures of the flat plate for the range of blowing and sucking velocities assumed by Li. This somewhat surprising result for the zero heat-transfer case comes about because mass transfer affects both the thickness of the boundary layer and the recovery temperature of the plate in a manner resulting in opposing effects upon the induced pressure on the plate. For example, blowing tends to increase boundary-layer thickness, leading to a higher surface pressure, while it also tends to reduce surface temperature, leading to increased boundary-layer density, hence, a thinner boundary layer and reduced surface pressure. These opposing effects result in a lessened effect of blowing or sucking on the surface pressure in the strong-interaction region when there is no heat transfer at the surface. If the surface is maintained at a constant temperature lower than the recovery temperature, it is found that blowing increases the surface pressure in the strong-interaction region in a monatomically increasing manner with increasing blowing parameter.[3]

[1] Ting Yi Li, paper presented to the American Rocket Society Semiannual Meeting, Los Angeles, May 10, 1960.

[2] K. Stewartson, *Proc. Cambridge Phil. Soc.*, vol. 51, part 1, p. 202, 1955.

[3] T. Y. Li and J. F. Gross, *Proc. 1961 Heat Transfer and Fluid Mech. Inst.*, pp. 146–160, Stanford University Press, Stanford, Calif., 1961.

6-3. Effects of Leading-edge Bluntness. It was seen that the agreement between the strong-interaction theory presented in Sec. 6-2 and experiment as shown on Fig. 6-5 was not so good as should be expected if the theory is to be relied upon. Can it be that certain assumptions implicit in the theory and effects overlooked in the theory lead to significant differences between experiment and theory in practical cases? As will be brought out in this section and the next one, there are several discernible factors which contribute to the inaccuracy of the theory.

Two principal effects are omitted in the strong-interaction theory of Sec. 6-2. They are in the order that we shall take them up:

1. The effect of leading-edge bluntness upon the inviscid flow field. This effect would be present even in the absence of a boundary layer. It is this effect which we shall discuss in this section.

2. The effects of curvature in the boundary layer induced leading-edge shock wave taken up in the section following this one. This curvature results in entropy gradients in the stream external to the boundary layer between the boundary layer and the shock wave. The entropy gradients directly lead to the existence of vorticity at the boundary-layer edge. This vorticity was not incorporated into the solution of the boundary equations as presented in Sec. 6-2 and indirectly creates a variation of velocity and temperature at the boundary-layer edge with distance from the leading edge. This latter variation was neglected by the boundary-layer solutions for the strong-interaction region presented in Sec. 6-2, where it was assumed that $u_e = u_\infty$ and $T_e = T_\infty$.

It has long been known that expansion of the flow from the stagnation-point region of a blunt two-dimensional body around a corner to a direction parallel to the free-stream velocity does not result in the pressure dropping immediately to free-stream pressure when the free-stream Mach number is sufficiently supersonic. A series of expansion waves will fill the flow field between the leading-edge shock wave and the surface of the blunted object which is parallel to the free-stream velocity through which the flow will accelerate until the surface pressure drops to free-stream static pressure. Bertram and Henderson[1] report the results of calculations of pressure distribution along the surface of a blunted flat plate aligned parallel to the incident velocity vector which were made using the method of characteristics for rotational flow given by Ferri.[2] Calculations were made for several flat plates having a wedge leading edge where the wedge angle was chosen at each hypersonic Mach number such that sonic flow

---

[1] Mitchel H. Bertram and Arthur Henderson, Jr., *NACA TN* 4301, July, 1958.
[2] Antonio Ferri, *NACA Rept.* 841, 1956.

existed on the wedge surface.   The flow is allowed to expand around
the wedge-plate corner through a Prandtl-Meyer expansion to the
plate surface parallel to the free-stream velocity vector.   The
method of characteristics for rotational flow is then used to calculate
the variation of surface pressure downstream from the wedge-plate
corner.   The Prandtl-Meyer expansion waves are reflected from the

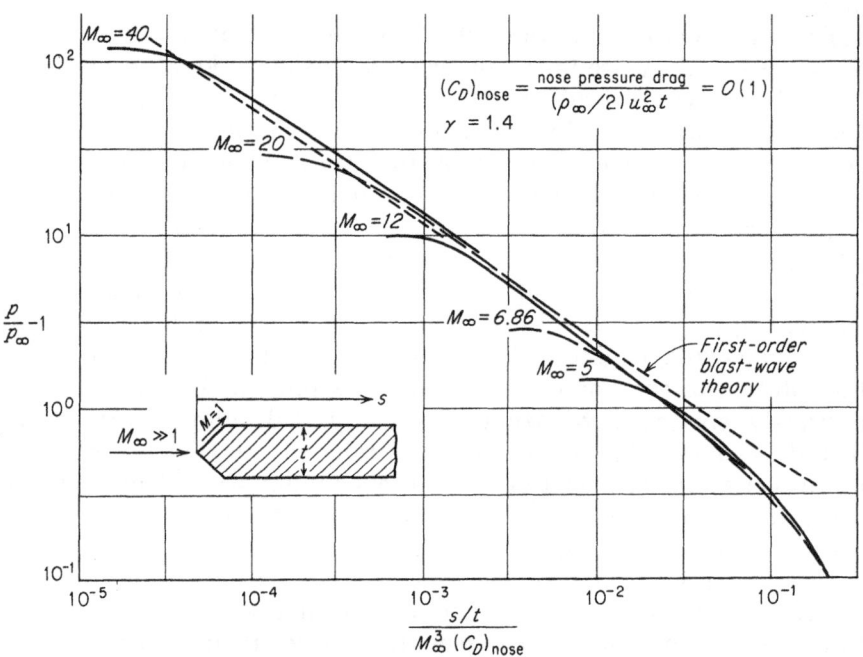

FIG. 6-6. The effect of leading-edge bluntness upon pressure distribution of
inviscid flow over a flat plate.   (*After Mitchel H. Bertram and Arthur Henderson,
Jr., NACA TN 4301, July, 1958.*)

leading-edge shock wave as expansion waves (thereby weakening the
shock wave) and from the surface as expansion waves, thereby
causing an acceleration of the flow at the surface and diminution of
the surface pressure.   The results of these calculations are plotted in
Fig. 6-6.   It can be seen, for fixed M_∞ and nose thickness t, that
p/p_∞ decreases with distance s from the leading edge.   Furthermore,
p/p_∞ is quite high in the leading-edge region.
    The parameter $(s/t)M_\infty^{-3}C_D^{-1}$ used to correlate the calculations shown
in Fig. 6-6 was revealed as the proper parameter by blast wave

theory advanced by Cheng and Pallone[1] and Lees and Kubota.[2] A curve calculated using the blast wave theory is shown in Fig. 6-6 for comparison with the exact (in the absence of viscous effects) calculations made using the method of characteristics. Good agreement between the two is shown.

It must be remembered that all real models will have a finite bluntness, related to t here, no matter what case is taken in their construction and that the effect illustrated in Fig. 6-6 will therefore result. However, it is equally apparent that, if the effective bluntness parameter t is kept small enough, the overpressure shown on Fig. 6-6 will be limited to a region which is also small, being proportional to t.

Calculations for blunt-nosed cylinders, similar in nature to those reported by Bertram and Henderson for blunt-nosed plates, have been reported by Feldman.[3] Feldman's calculations of surface pressure on a blunted cylinder, made using the method of characteristics, were compared with results for surface pressure given by second-order blast wave theory, and good agreement between the results of the two theories is shown. It can be concluded for blunted two- and three-dimensional bodies that there will be a region where the bluntness effects upon pressure distribution compete with and overshadow the effects of shock-wave–boundary-layer interaction. These effects will be limited to regions close to the nose or leading edge and will be proportional in length to the leading-edge thickness or nose radius. This region can be bounded as follows: Concerning ourselves always with hypersonic flows, that is, flows where $M_\infty \gg 1$, Sec. 6-2 reveals that strong-interaction effects are important for two-dimensional bodies where

$$\bar{\chi} > 1 \qquad (6\text{-}41)$$

Figure 6-6 shows that for two-dimensional bodies bluntness-induced effects are important where

$$\frac{s}{tM_\infty^3} \ll 1$$

or where

$$\frac{M_\infty^{3/2}}{(R_t)^{1/2}} \ll \bar{\chi} \qquad (6\text{-}42)$$

where

$$R_t = \frac{\rho_\infty u_\infty t}{\mu_\infty}$$

[1] H. K. Cheng and A. J. Pallone, *J. Aeronaut. Sci.*, vol. 23, no. 7, pp. 700–702, 1956.

[2] Lester Lees and Toshi Kubota, *J. Aeronaut. Sci.*, vol. 24, no. 3, pp. 195–202, 1957.

[3] Saul Feldman, *ARS J.*, vol. 30, no. 5, pp. 463–467, 1960.

Thus there is a region for $\bar{\chi}$ for two-dimensional bodies where strong-interaction effects are important. This region is, from inequalities (6-41) and (6-42),

$$1 < \bar{\chi} < \frac{M_\infty^{3/2}}{(R_t)^{1/2}} \tag{6-43}$$

It is apparent from inequalities (6-43) that, if R_t is large enough, inequalities (6-43) cannot be satisfied and the bluntness-induced effects described in this section will prevail over the strong-interaction effects described in Sec. 6-2. The methods of Chap. 4 apply in this region.

6-4. Vorticity Effects on the Laminar Boundary Layer. We turn now to the effects of vorticity in the inviscid stream between

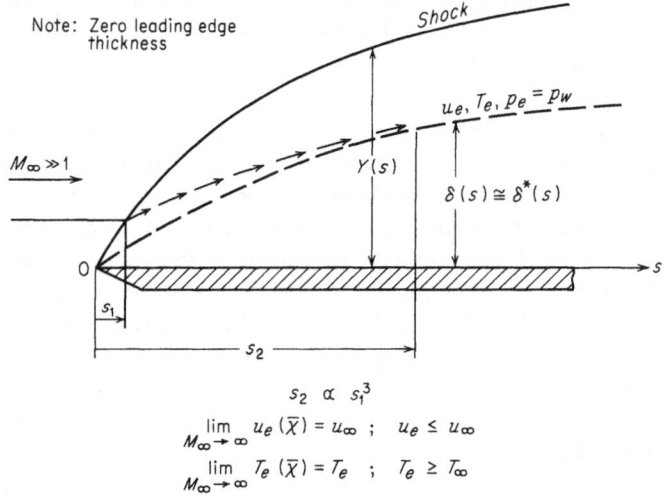

FIG. 6-7. Effect of curvature of the leading-edge shock wave upon fluid properties at the edge of the boundary layer.

the leading-edge shock wave and the boundary-layer edge. Consider the ideally sharp ($R_t = 0$) flat plate depicted in Fig. 6-7. We see that even in the absence of leading-edge bluntness the shock wave is curved as might be expected, since the inviscid flow field is similar to that which would result if an (imaginary) inviscid gas were to flow at hypersonic speeds past a two-dimensional body with a shape represented by the edge of the boundary layer. (The displacement thickness is almost equal to the boundary-layer thickness if $T_w \cong T_0$, since the density of the boundary layer is quite low.) Since the leading-edge shock wave is curved, there are entropy gradients in the inviscid stream between it and the boundary layer, and furthermore, the streamlines entering the boundary layer will have different values of u_e, T_e, and ρ_e at the point of entry into the boundary layer, since each streamline crosses the shock wave at different points where the shock wave has different strengths because of the curvature. The entropy gradient at the point where the streamline enters the boundary layer will equal that at the point where the streamline crossed the leading-edge shock wave. This entropy gradient will be manifested by a finite value of $\partial u/\partial y$ at the edge of the boundary layer, a boundary condition not taken into account in solving the boundary-layer equations in Sec. 6-2.

Table 6-4. Assumptions Made in Determining the Effects of Vorticity Due to Shock-wave Curvature upon Strong-interaction Theory

Assumption	*Justification*
Strong-interaction-theory results of Sec. 6-2 can be used as a first approximation	These results are exact in the limit as $M_\infty \to \infty$ for $\bar{\chi}$ finite as will be shown
$\sin\left(\dfrac{dY}{ds}\right) = \tan\left(\dfrac{dY}{ds}\right) = \dfrac{d\dot{Y}}{ds}$	$\dfrac{dY}{ds} \ll 1$, especially as $M_\infty \to \infty$
$M_\infty^2 \left(\dfrac{dY}{ds}\right)^2 \gg \dfrac{\gamma - 1}{2\gamma}$	True for $\bar{\chi}$ finite as $M_\infty \to \infty$
$T_e > T_\infty$	This will be true for $\bar{\chi}$ finite and M_∞ large but not infinite as will be shown
$Y(s) = G\delta(s)$ $G = \text{const.}$	Results from Stewartson's[†] similarity solution for the strong-interaction region on a flat plate
$C = 1$	Necessary in order to use results of Sec. 6-2; otherwise poor

[†] K. Stewartson, *J. Aeronaut. Sci.*, vol. 22, no. 5, pp. 303–309, 1955.

Some notion of the magnitude of these effects can be obtained by using the results of Sec. 6-2 as a first approximation and by obtaining the perturbation to these results caused by the effects under consideration. This procedure was followed by Lees[1] and will be outlined here. As with most gas-dynamics problems we shall make some assumptions in order to proceed. These assumptions, leading to errors of smaller order than those due to neglecting vorticity, are listed in Table 6-4 above.

Referring to Fig. 6-7, our objective will be to determine the velocity $u_e \le u_\infty$, the temperature $T_e \ge T_\infty$, and the vorticity $(\partial u/\partial y)_e$ at the point where a streamline enters the boundary layer as a function

[1] Lester Lees, *J. Aeronaut. Sci.*, vol. 23, no. 6, pp. 594–612, 1956.

of the position of entry s_2, with R_∞ and M_∞ as parameters. Once these quantities are determined, the effects of $u_e \neq u_\infty$, $T_e \neq T_\infty$, and $(\partial u/\partial y)_e \neq 0$ upon the strong-interaction-theory results of Sec. 6-7 are to be determined, since it was assumed that $u_e \equiv u_\infty$, $T_e \equiv T_\infty$, and $(\partial u/\partial y)_e \equiv 0$ in that theory. A further desirable quantity is the relation between the point of entry of the streamline into the boundary layer s_2 and the point at which the streamline crosses the leading-edge shock wave s_1. In that which follows, the subscript ∞ will denote free-stream values, 1 the value of the quantity on the downstream side of the shock wave at the point where the streamline crosses the shock wave, and 2 or e the value of the quantity at the point where it enters the boundary layer.

Along a streamline it can be written that

$$\frac{T_e}{T_1} = \left(\frac{p_e}{p_1}\right)^{(\gamma-1)/\gamma} \tag{6-44}$$

and from oblique shock relations, consistent with the assumptions of Table 6-4,

$$\frac{p_1}{p_\infty} = \frac{2\gamma M_\infty^2}{\gamma+1}\left(\frac{dY}{ds}\right)_1^2 \tag{6-45}$$

and

$$\frac{T_1}{T_\infty} = \frac{2\gamma(\gamma-1)M_\infty^2(dY/ds)_1^2}{(\gamma+1)^2} \tag{6-46}$$

Therefore, from Eqs. (6-44) to (6-46) combined,

$$\frac{T_e}{T_\infty} = \frac{2\gamma(\gamma-1)}{(\gamma+1)^2}\left(\frac{\gamma+1}{2\gamma}\right)^{(\gamma-1)/\gamma}\left[M_\infty^2\left(\frac{dY}{ds}\right)_1^2\right]^{1/\gamma}\left(\frac{p_e}{p_\infty}\right)^{(\gamma-1)/\gamma} \tag{6-47}$$

Equation (6-47) relates the temperature ratio T_e/T_∞ to the pressure ratio p_e/p_∞ and the shock-wave inclination at the point where the streamline entering the boundary layer originally crossed the leading-edge shock wave. This equation has little value until we locate that point. To proceed, along a streamline,

$$C_p T_e + \frac{u_e^2}{2} = C_p T_\infty + \frac{u_\infty^2}{2} \tag{6-48}$$

or, consistent with the assumptions of Table 6-4,

$$\left(\frac{u_e}{u_\infty}\right)^2 = 1 - \frac{2}{\gamma-1}\frac{1}{M_\infty^2}\frac{T_e}{T_\infty} \tag{6-49}$$

or

$$\frac{u_e}{u_\infty} \simeq 1 - \frac{1}{\gamma-1}\frac{1}{M_\infty^2}\frac{T_e}{T_\infty} \tag{6-50}$$

Equations (6-47) and (6-50) thus relate u_e/u_∞ to p_e/p_∞ and the inclination angle of the leading-edge shock wave at the point where the streamline in question crossed the leading-edge shock wave.   Let us locate this point.   By continuity-of-mass considerations

$$Y(s_1)\rho_\infty u_\infty = \psi(s_2,\delta) \tag{6-51}$$

but according to the assumptions of Table 6-4 and the results presented in Sec. 6-4,

$$Y = G\delta \simeq G\delta^* \tag{6-52}$$

and

$$\delta^* = \frac{B}{M_\infty} \bar{\chi}^{1/2}s \propto s^{3/4} \tag{6-53}$$

Thus, it will be found that

$$\left(\frac{dY}{ds}\right)_1 = \frac{3}{4}\frac{Y_1}{s_1} = \frac{3}{4}\left[\frac{(GB)^4 M_\infty^2}{\rho_\infty u_\infty/\mu_\infty}\right]^{1/3} Y_1^{-1/3} \tag{6-54}$$

since, from Eqs. (6-52) and (6-53),

$$s_1 = \frac{Y_1^{4/3}(\rho_\infty u_\infty/\mu_\infty)^{1/3}}{(GB)^{4/3}M_\infty^{2/3}} \tag{6-55}$$

Now, in Eq. (6-51) it can be shown that

$$\psi(s_2,\delta) = \int_0^\delta \rho u\, dy = (2\bar{s})^{1/2}f_\delta \tag{6-56}$$

where

$$f_\delta = f[\eta(\delta)]$$

Furthermore, to a first approximation, using Eq. (6-18),

$$(2\bar{s})^{1/2} = 2\left[A\frac{M_\infty^3}{(R_\infty)^{1/2}}\rho_\infty\mu_\infty u_\infty s_2\right]^{1/2} \tag{6-57}$$

since, to a first approximation,

$$u_e = u_\infty$$

and

$$\frac{\rho_e}{\rho_\infty} = \frac{p_e}{p_\infty} = A\bar{\chi} \tag{6-24}$$

Equations (6-51), (6-54), (6-56), and (6-57) can now be combined to obtain an expression relating $(dY/ds)_1$ to $\bar{\chi}_2$ and M_∞; viz.,

$$\left(\frac{dY}{ds}\right)_1 = \frac{3}{4}\left[\frac{(GB)^4}{2A^{1/2}f_\delta}\right]^{1/3}\left\{\frac{M_\infty}{[(R_\infty)_2]^{1/2}}\right\}^{1/6} \tag{6-58}$$

or alternately

$$\left(\frac{dY}{ds}\right)_1 = \frac{3}{4}\left[\frac{(GB)^4}{2A^{1/2}f_\delta}\right]^{1/3}\left(\frac{\bar{\chi}_2}{M_\infty^2}\right)^{1/6} \tag{6-59}$$

Equation (6-58) or (6-59) can be combined with Eqs. (6-47) and (6-50) to obtain

$$\frac{u_e}{u_\infty} = 1 - F(\gamma)\left(\frac{\bar{\chi}_2}{M_\infty^2}\right)^{1-(2/3\gamma)} \tag{6-60}$$

and

$$\frac{T_e}{T_\infty} = (\gamma - 1)M_\infty^2 F(\gamma)\left(\frac{\bar{\chi}_2}{M_\infty^2}\right)^{1-(2/3\gamma)} \tag{6-61}$$

where

$$F(\gamma) = \frac{(0.71\gamma)^{1/\gamma} A^{[1-(4/3\gamma)]}(GB)^{8/3\gamma}}{f_\delta^{2/3\gamma}(\gamma+1)^{(\gamma+1)/\gamma}} \tag{6-62}$$

Thus for $\bar{\chi}_2$ finite, $u_e = u_\infty$ in the limit as $M_\infty \to \infty$. Furthermore, from Eqs. (6-60) and (6-48), in the limit as $M_\infty \to \infty$, $T_e = T_\infty$. Equation (6-61) applies for M_∞ large but not infinite and shows that, for M_∞ large and $\bar{\chi}_2$ finite, $T_e > T_\infty$. Thus for $M_\infty \gg 1$, $T_e \geq T_\infty$ for all M_∞ including $M_\infty = \infty$. This shows that, to the order of accuracy of the analysis used, the results of Sec. 6-2 are most accurate in the limit as $M_\infty \to \infty$. Figure 6-5 supports this conclusion. It is seen in Fig. 6-5, except near the leading edge, that the results of Sec. 6-2 are most accurate for the highest test Mach number, as might be anticipated from this analysis.

Equation (6-54) combined with Eq. (6-55) giving Y_1 can be equated to Eq. (6-59) to obtain a relation between $(R_\infty)_1$ and $(R_\infty)_2$. This is

$$[(R_\infty)_2]^{1/3} = \frac{1}{A^{1/2}M_\infty^{5/6}f_\delta}\left[\frac{(GB)^4}{2A^{1/2}f_\delta}\right]^{1/3}(R_\infty)_1 \tag{6-63}$$

or, for given M_∞ and $T_w/T_0 = g(0)$,

$$s_2 \propto (s_1)^3 \tag{6-64}$$

Equation (6-63) reveals that the streamlines entering the boundary layer cross the leading-edge shock wave at points far upstream and closer to the leading edge according to the proportionality relation (6-64). This is consistent with the concept that very little mass flux occurs within the boundary layer, and hence the displacement thickness δ^* is almost equal to the boundary-layer thickness δ.

In developing the strong-interaction theory of Sec. 6-2 the boundary-layer equations were solved for boundary conditions including $(\partial u/\partial y)_e = 0$. It can be shown that $(\partial u/\partial y)_e$ is finite for the case under consideration here and, as such, should be included within the solution. Referring to Fig. 6-7, it is apparent that the vorticity at the edge of the boundary layer is equal to the vorticity on the streamline entering the boundary layer at the point in question. Furthermore, the vorticity is related to the entropy gradient at the point where the streamline crosses the leading-edge shock wave. According to Crocco's theorem[1] this is given by

$$2\mathbf{V} \times \boldsymbol{\omega} = -T\nabla s \tag{6-65}$$

where in cartesian coordinates, for our problem,

$$\mathbf{V} = u\mathbf{i} + v\mathbf{j} \simeq u_e\mathbf{i} \tag{6-66}$$

$$\boldsymbol{\omega} = \frac{1}{2}\left(\frac{\partial v}{\partial s} - \frac{\partial u}{\partial y}\right)\mathbf{k} \simeq -\frac{1}{2}\frac{\partial u}{\partial y}\mathbf{k} \tag{6-67}$$

and, for the gradient of the entropy s,[2]

$$\nabla s \simeq \frac{\partial s}{\partial y}\mathbf{j} \tag{6-68}$$

Combining Eqs. (6-65) to (6-68) gives

$$\left(\frac{\partial u}{\partial y}\right)_e = -\frac{T_e}{u_e}\left(\frac{\partial s}{\partial y}\right)_e = -\frac{p_e}{\bar{R}}\left(\frac{\partial s}{\partial \psi}\right)_e = -\frac{p_e}{\bar{R}}\left(\frac{\partial s}{\partial \psi}\right)_1 \tag{6-69}$$

Now since s is constant along the streamline, then

$$\frac{\partial s}{\partial \psi} = \frac{C_v}{p}\frac{\partial p}{\partial \psi} \tag{6-70}$$

and, by Eq. (6-45),

$$p_1 \propto \left(\frac{dY}{ds}\right)_1^2$$

From Eq. (6-54) we have

$$\left(\frac{dY}{ds}\right)_1 \propto \frac{Y_1}{s_1}$$

from Eq. (6-55) we obtain

$$s_1 \propto Y_1^{4/3}$$

and from Eq. (6-57) we have

$$\psi_\delta(s_2) \propto Y_1$$

Then the intervening relations can be used with Eq. (6-70) to obtain

$$\frac{1}{p_1}\left(\frac{\partial p}{\partial \psi}\right)_1 = -\frac{2}{3\psi_\delta(s_2)} \tag{6-71}$$

Combining Eqs. (6-24), (6-57), and (6-69) to (6-71) gives

$$\left(\frac{du}{dy}\right)_e = \frac{1}{3}\frac{\rho_\infty T_\infty C_v A^{1/2}}{f_\delta \mu_\infty}\left(\frac{\bar{\chi}_2}{M_\infty^2}\right)^{3/2} \propto s^{-3/4} \tag{6-72}$$

---

[1] L. Crocco, *ZAMM*, vol. 17, pp. 1–7, 1937.

[2] The lower-case s in Eqs. (6-65) and (6-68) to (6-70) denotes specific entropy and should not be confused with the independent variable s.

or, for f_e'',

$$f_e'' = \frac{2}{3} \frac{1}{\gamma(\gamma - 1)f_\delta M_\infty^2} \tag{6-73}$$

Thus, the velocity gradient $(\partial u/\partial y)_e$ is finite for M_∞ finite and varies with distance from the leading edge according to Eq. (6-72). However, in terms of the reduced stream function f, the velocity gradient function f_e'' is constant for any Mach number M_∞ and becomes zero in the limit as $M_\infty \to \infty$. It appears that for M_∞ large the boundary conditions used in Sec. 6-2, that is, $u_e = u_\infty$, $T_e = T_\infty$, and $f'' = 0$, are justified and proper in the limit as $M_\infty \to \infty$. When M_∞ is finite, f_e'' is finite and its effect remains to be determined.

Solutions of the constant-density, constant-temperature, two-dimensional boundary-layer equations, including the boundary condition of finite vorticity at the edge of the boundary layer, have been presented by Li[1] and Ting[2] among others. These authors show that positive vorticity increases skin friction, other considerations remaining constant. Li's and Ting's work clearly shows that the proper parameter for vorticity effects is

$$\Omega = -\frac{2\omega\delta}{u_\infty} \simeq \frac{\delta}{u_\infty} \frac{\partial u}{\partial y} \tag{6-74}$$

where, for example, $C_f(R_\infty)^{1/2}$ for constant-density flow over a flat plate increases monatomically as Ω increases. In the present case, using Eqs. (6-53), (6-67), and (6-72) in Eq. (6-74), we obtain, in terms of the reduced stream function,

$$\Omega_e = \frac{1}{3} \frac{(A)^{1/2}B}{f_\delta \gamma(\gamma - 1)} \tag{6-75}$$

Thus, it appears that the vorticity number Ω_e is independent of s in the strong-interaction region. For air, the value of Ω_e is about 0.11 for an insulated plate. According to the results of Ting, this would represent an increase in skin friction for a constant-density boundary layer of about 250 per cent over values given by the theory of a flat-plate constant-density boundary layer with vorticity absent.

6-5. Conclusions. The region where the leading-edge shock wave and the boundary layer strongly interact was investigated in Sec. 6-2. It was found that the proper similarity parameter for this region is $\bar{\chi} = C^{1/2} M_\infty^3/(R_\infty)^{1/2}$ and that the interaction effects are significant when $\bar{\chi} \gg 1$. The interaction effect at any value of $\bar{\chi}$ was reduced by

[1] Ting Yi Li, *J. Aeronaut. Sci.*, vol. 22, no. 9, pp. 651–652, 1955. See also Ting Yi Li, *Rensselaer Polytech. Instit. Rept.* TR AE 5813, 1958.

[2] Lu Ting, *Phys. Fluids*, vol. 3, no. 1, pp. 78–82, 1960.

heat transfer to the surface because heat transfer results in a cooler gas layer and hence higher density in the layer.   The effects of strong interaction on heat transfer vary as $\bar{\chi}^{1/2}$ and thus can be quite significant for $\bar{\chi} \gg 1$, other factors being equal.

Leading-edge blunting and vorticity effects were investigated in Secs. 6-3 and 6-4, respectively.   Some conclusions which can be drawn from the analysis presented in these sections include the following:

1. Leading-edge bluntness creates a region of high pressure aft of the leading edge which overshadows the strong-interaction effect. This region is bounded by the inequality

$$\bar{\chi} \gg \frac{M_\infty^{3/2}}{(R_t)^{1/2}}$$

Thus there exist bounds upon the region where strong-interaction effects predominate as given by the inequalities

$$1 < \bar{\chi} < \frac{M_\infty^{3/2}}{(R_t)^{1/2}}$$

Hence, if R_t is large enough for any free-stream Mach number M_∞, there will be no region where strong-interaction effects prevail. Rather, the bluntness-induced effects will predominate to the exclusion of the strong-interaction effects discussed in Sec. 6-2.

2. Because the leading-edge shock wave is curved, even in the absence of leading-edge bluntness, there will be vorticity in the external stream between the shock wave and boundary layer in the strong-interaction region.   Furthermore, for finite M_∞, $u_e < u_\infty$ and $T_e > T_\infty$ contrary to the assumptions implicit in the strong-interaction theory of Sec. 6-2.   This will result in p_e/p_∞ being greater than is predicted by the strong-interaction theory of Sec. 6-2, and as $M_\infty \to \infty$, the value of p_e/p_∞ will approach p_e/p_∞ given by the strong-interaction theory of Sec. 6-2 from *above*.   This is in keeping with the experimental results shown in Fig. 6-5.   This conclusion can be anticipated using the relation that

$$\left(\frac{p_e}{p_\infty}\right)^{(1)} = A \frac{M_\infty^3}{(p_\infty u_e s/\mu_e)^{1/2}} = \left(\frac{p_e}{p_\infty}\right)^{(0)} \left(\frac{u_e \mu_\infty}{u_\infty \mu_e}\right)^{-1/2}$$

and if we assume that

$$\frac{\mu_\infty}{\mu_e} \propto \frac{T_\infty}{T_e}$$

and use the results of our analyses in Sec. 6-4, which show that

$$u_e \leq u_\infty \qquad T_e \geq T_\infty$$

then, from the above,

$$\frac{p_e^{(1)}}{p_\infty} \geq \frac{p_e^{(0)}}{p_\infty}$$

where the superscript (0) denotes the strong-interaction result of Sec. 6-2 and the superscript (1) denotes a first approximation incorporating corrections for vorticity of Sec. 6-4.

3. The streamlines entering the boundary layer in the strong-interaction region have previously crossed the leading-edge shock wave far upstream, where the shock-wave curvature is significant in producing vorticity in the external stream between the boundary layer and shock wave.

4. All these effects modify the strong-interaction solutions presented in Sec. 6-2 in a direction bringing the theory into closer agreement with experiment. A paper by Oguchi[1] which came to the author's attention after writing this chapter bears out all the above conclusions with a detailed analysis of the problem.

[1] Hakuro Oguchi, First Order Approach to a Strong Interaction Problem in Hypersonic Flow Over an Insulated Flat Plate, *Aeronaut. Research Inst., Univ. Tokyo, Rept.* 330, 1958.

7

The Dissociated Turbulent Boundary Layer

7-1. Introduction. This chapter is concerned with the dissociating compressible turbulent boundary layer. The purpose of this chapter is to present some of the methods used to calculate heat transfer and skin friction for a dissociating compressible turbulent boundary layer flowing over a surface under conditions such as those of hypersonic flight. The boundary-layer equations for a reacting turbulent boundary layer will be derived as a part of the analyses.

It is accepted by most workers in gas dynamics that the details of the structure of the turbulent boundary layer are imperfectly described by existing theories. In fact, such theories, including those presented in this book, invariably include some empiricism before their formulation is brought to the point where skin friction or heat-transfer rates can be calculated. That is, the theory cannot be completed in closed form proceeding from first principles without resorting to experimentally determined constants somewhere along the way. This is not to suggest, however, that useful and accurate semiempirical theories cannot be derived. The approach taken in this chapter presents a semiempirical theory which agrees remarkably well with experiment over a wide range of flow conditions. The weakness of such a theory is, of course, its uncertain accuracy when applied under flow conditions for which prior experimental confirmation is lacking. To a certain extent this criticism might be applied to the laminar-boundary-layer theory as well, except for the fact that no empirical constants appear in the laminar-boundary-layer theory (disregarding the fact that the interaction potentials used in evaluating the collision integrals in the transport coefficients employ empirical constants). The uncertainty in the laminar-boundary-layer theory lies chiefly in questions as to what extent the boundary-layer equations are valid under various flow conditions and not whether the solutions to those equations are valid, since these solutions agree with experiment remarkably well when the laminar-boundary-layer equations are known to be appropriate to the flow system under analysis. The turbulent-boundary-layer solutions presented herein share this latter uncertainty with laminar-boundary-layer theory along with the additional uncertainty that certain constants incorporated into the theory can be determined only by

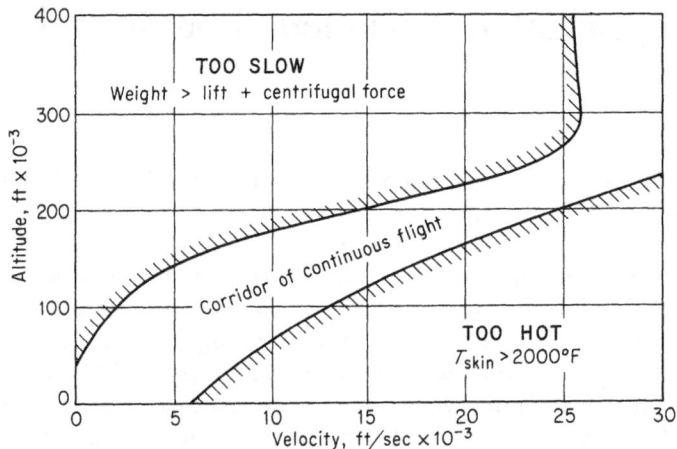

FIG. 7-1. Possible flight regime in the earth's atmosphere. (*After D. J. Masson and C. Gazley, Jr., Aeronaut. Eng. Rev., vol. 15, pp. 46–55, 1956.*)

suitable comparison with experiment. However, since the worker in hypersonic gas dynamics cannot choose to ignore the possibility of the existence of the turbulent boundary layer, there is some merit in presenting a semiempirical turbulent-boundary-layer theory which has enjoyed success in matching measured values of skin friction and heat transfer, keeping always in mind the uncertainties involved.

The importance to the designer of having some method for estimating skin friction and heat transfer of a hypersonic turbulent boundary layer is graphically illustrated using Figs. 7-1 and 7-2. Figure 7-1 was prepared by Masson and Gazley[1] to delineate the region of continuous flight at speeds up to satellite speeds within the earth's atmosphere. The "corridor of continuous flight" shown in Fig. 7-1 refers to those vehicles which depend upon aerodynamic lift and centrifugal force to counterbalance weight in maintaining their altitude during sustained flight within the atmosphere. Purely

[1] D. J. Masson and C. Gazley, Jr., *Aeronaut. Eng. Rev.*, vol. 15, pp. 46–55, 1956.

ballistic objects were not considered in preparing Fig. 7-1. The upper region of Fig. 7-1 is ruled out of consideration for sustained flight by vehicles of reasonable lift coefficient because weight exceeds the sum of lift and centrifugal force there. The region in the lower right-hand corner is ruled out for sustained flight because equilibrium skin temperatures[1] are likely to be unreasonably high there.

Figure 7-2 gives the information of Fig. 7-1 in the alternative fashion of a map of Reynolds number per foot (based upon free-stream

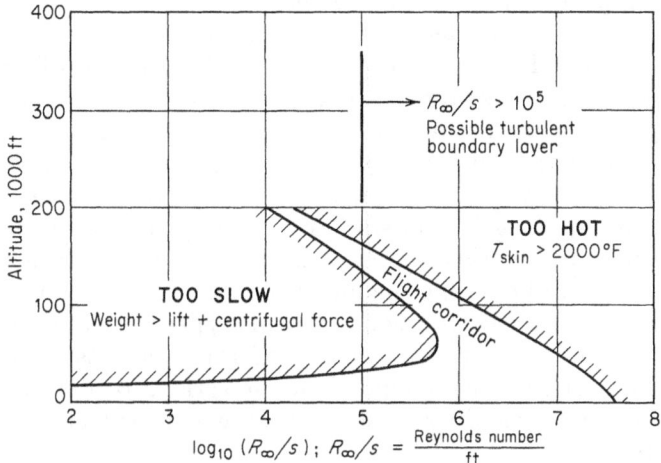

FIG. 7-2. Reynolds numbers per foot in possible flight regime in earth's atmosphere.

conditions) versus altitude. Note that Reynolds numbers per foot exceeding 10^5 exist over an appreciable portion of the flight corridor, and by cross-correlating with Fig. 7-1, these Reynolds numbers per foot apply up to flight Mach numbers of 19. Since turbulent flow might be expected for Reynolds numbers exceeding 10^5, depending upon numerous factors not dealt with in this book, it appears obvious that aerodynamicists must be prepared to contend with a turbulent boundary layer under many circumstances of practical interest for objects of practicable size.

We shall not deal with or predict the occurrence of boundary-layer transition from laminar flow to turbulent flow. Eventually, transition will occur as the boundary-layer gas flows along the hypersonic body providing the Reynolds number is "high enough." The

[1] The equilibrium skin temperature as used here is the temperature at which the convection heat transfer to the surface is equal to the radiation heat transfer from the hot surface.

prediction of when or where transition will occur is not yet an exact science in the author's opinion and is best left out of a book of this kind. Such predictions should be, and usually are, based upon experimental data, and such data are continually being obtained. The author can do no better than to refer the interested reader to the current technical literature on this subject, including the numerous and ever-increasing number of research reports being issued by research agencies throughout the world. The problem of predicting the occurrence of boundary-layer transition remains in the first rank of unsolved problems in gas dynamics.

Early work in turbulent-boundary-layer theory was admirably summed up by Prandtl.[1] Following a description of turbulent-boundary-layer flow as largely a confused and utterly random motion of fluid particles as distinct from the macroscopically orderly fluid motion of laminar-boundary-layer flow, Prandtl goes on to describe O. Reynolds' early work in turbulent flow; the concept of eddy viscosity, mixing length, and similarity theory; and empirical resistance formulas for flows in pipes and over plates. Since the work in compressible and reacting turbulent boundary layers to be described here is an extension of the condition of hypersonic and/or reacting turbulent boundary layers of many of the concepts described by Prandtl, his work is recommended to the serious reader as a helpful preamble to understanding much of that which will be presented in this chapter.

7-2. The Reacting-turbulent-boundary-layer Equations. The equations of change and state for the flow of a compressible, reacting turbulent boundary layer over a flat plate will be derived first. The case of the flat plate was chosen because of its geometrical simplicity, the wealth of experimental data available for comparison, and its probable direct applicability to surfaces over which a turbulent boundary layer is apt to be flowing (wing surfaces and near-cylindrical surfaces). Throughout the development that follows the following assumptions will apply:

1. The geometry being treated is that of two-dimensional flow over a flat plate. Figure 7-3 shows the geometry and coordinates used.

2. The boundary-layer equations of motion apply, and the mean flow is in a steady state.

3. The reacting gas is assumed to be a mixture of perfect gases. For simplicity, air will be approximated as a mixture of one species of atoms and one species of molecules.

---

[1] L. Prandtl, "The Mechanics of Viscous Fluids, Aerodynamic Theory," vol. III, div. G, pp. 119–155, Durand Reprinting Committee, California Institute of Technology, Pasadena, 1943.

4. Interaction effects of leading-edge shock waves and boundary layers can be neglected.

5. The surface temperature is constant.

Assumptions 1 and 2 are commonly accepted formalities. Assumption 3 has already been used successfully in Chap. 4 when the laminar boundary layer was treated. Assumption 4 is logical in view of the fact that turbulent boundary layers will be present downstream from the leading edge where interaction effects are usually negligible.

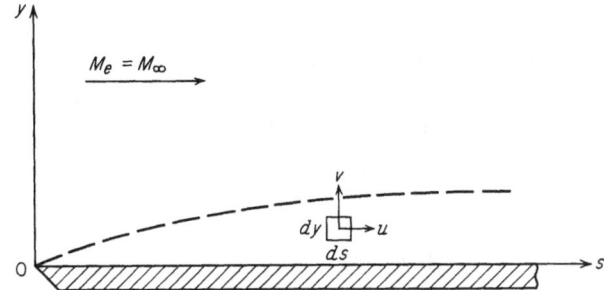

FIG. 7-3. The geometry and coordinate system used in analyzing the turbulent-boundary-layer flow over a flat plate.

Assumption 5 is necessary to the solutions of the boundary-layer equations presented later. Further assumptions will be made and, where possible, justified as the analysis proceeds.

The appropriate equations for the problems at hand can be developed from the equations already developed in the treatment of the laminar boundary layer in Chap. 2. These equations, in a slightly rearranged form, are given here as a point of departure.

The equation of state is

$$p_i = \rho_i R_i T \tag{7-1a}$$

where, according to assumption 3 and from Dalton's law,

$$p = \sum_i p_i \tag{7-1b}$$

The equation for conservation of mass is

$$\frac{\partial \rho u}{\partial s} + \frac{\partial \rho v}{\partial y} = 0 \tag{7-2}$$

where we have taken $k = 0$ in the conservation-of-mass equation for two-dimensional flow.

In presenting the equation for conservation of species we make use of Eq. (7-2) combined with Eq. (2-56) to put all dependent variables under differentiation. This is done to simplify the development of the turbulent-boundary-layer equations in that which follows. We obtain

$$\frac{\partial \rho u C_i}{\partial s} + \frac{\partial \rho v C_i}{\partial y} = -\frac{\partial \rho C_i \bar{V}_{iy}}{\partial y} + \dot{w}_i \tag{7-3a}$$

where, as before,

$$C_i = \frac{\rho_i}{\rho} \tag{7-3b}$$

and

$$\bar{V}_{iy} = -D_{12}\frac{\partial C_i}{\partial y}$$

Combining Eq. (7-2) with Eq. (2-66) (with $\partial p/\partial s = 0$) in order to put all dependent variables under differentiation, we obtain the equation for conservation of momentum in the s direction.

$$\frac{\partial \rho u u}{\partial s} + \frac{\partial \rho v u}{\partial y} = \frac{\partial}{\partial y}\left(\mu \frac{\partial u}{\partial y}\right) \tag{7-4a}$$

and, from Eq. (2-69) with $\partial p/\partial s = 0$,

$$dp = \frac{\partial p}{\partial s}ds + \frac{\partial p}{\partial y}dy = 0 \tag{7-4b}$$

Combining Eq. (7-2) with Eq. (2-72) to put all dependent variables under differentiation yields the equation for conservation of energy

$$\frac{\partial \rho u I}{\partial s} + \frac{\partial \rho v I}{\partial y} = \frac{\partial}{\partial y}\left[\frac{\mu}{P}\frac{\partial I}{\partial y} + \mu\left(1 - \frac{1}{P}\right)\frac{1}{2}\frac{\partial u^2}{\partial y}\right]$$
$$+ \frac{\partial}{\partial y}\left[\left(\frac{1}{L} - 1\right)\sum_i h_i\rho_i \bar{V}_{iy}\right] \tag{7-5}$$

where

$$I = h + \frac{u^2}{2} \tag{7-6}$$

$$h = \sum_i C_i h_i \tag{7-7}$$

$$h_i = \int_0^T C_{p_i}\,dT + h_i^0 \tag{7-8}$$

and in writing the equation for conservation of energy in the form of Eq. (7-5), it is implicitly assumed that

$$\rho_i \bar{V}_{iy} = -\rho D_{12}\frac{\partial C_i}{\partial y} \tag{7-9}$$

That is, all gas mixtures of interest are effectively binary mixtures made up of light particles (atoms) and heavy particles (molecules).

For dissociating air and reacting gas mixtures involving the components of air and foreign species including diatomic and triatomic molecules, this assumption is reasonable but must always be examined carefully depending upon the application under consideration.

Equations (7-1) through (7-8) represent our point of departure in deriving the equations for the reacting, compressible turbulent boundary layer. The procedure for derivation of those equations makes use of the concept of turbulent fluctuations about mean

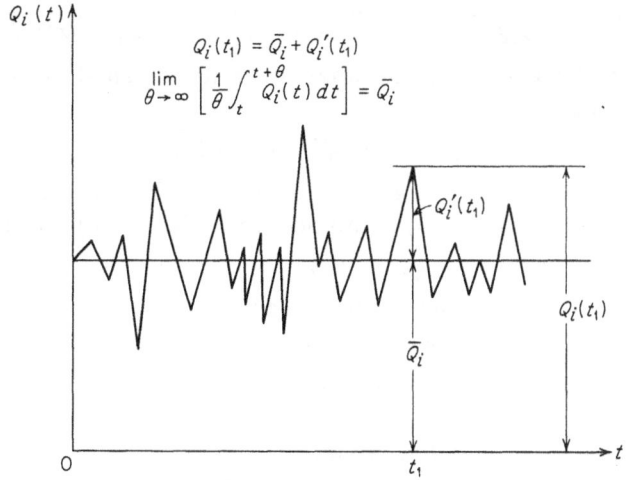

FIG. 7-4. Mean and fluctuating quantities in a turbulent boundary layer.

quantities in the boundary layer and is analogous to the well-known procedure for deriving the equations for a nonreacting turbulent boundary layer.[1] We shall not describe this procedure in great detail here but shall state all the essential assumptions and operations necessary to obtain the equations used in this chapter.

We assume that each dependent variable in the above equations can be represented by the sum of a mean quantity and a fluctuating quantity at any point within the turbulent boundary layer (see Fig. 7-4). That is,

$$Q_i(s,y,t) = \overline{Q_i(s,y)} + Q_i'(s,y,t) \tag{7-10}$$

where

$$\overline{Q_i(s,y)} = \lim_{\theta \to \infty} \left[\frac{1}{\theta} \int_t^{t+\theta} Q_i(s,y,t)\, dt \right] \tag{7-11}$$

[1] See, for example, G. M. Schubauer and C. M. Tchen, Turbulent Flow, in C. C. Lin (ed.), "Turbulent Flows and Heat Transfer," sec. B, pp. 80–90, Princeton University Press, Princeton, N.J., 1959.

thus, from Eqs. (7-10) and (7-11),[1]

$$\overline{Q_1'(s,y,t)} \equiv 0 \qquad (7\text{-}12)$$

From this point on, a bar over a quantity indicates that it is a mean quantity determined according to Eq. (7-11) and a prime indicates that it is a fluctuating quantity. Furthermore, useful rules for determining the mean of products of dependent variables can be derived using Eqs. (7-11) and (7-12) and are

$$\overline{Q_i Q_j} = \overline{Q_i}\,\overline{Q_j} + \overline{Q_i' Q_j'} \qquad (7\text{-}13)$$

and

$$\overline{Q_i Q_j Q_k} = \overline{Q_i}\,\overline{Q_j}\,\overline{Q_k} + \overline{Q_i}\,\overline{Q_j' Q_k'} + \overline{Q_j}\,\overline{Q_i' Q_k'} + \overline{Q_k}\,\overline{Q_i' Q_j'} + \overline{Q_i' Q_j' Q_k'} \qquad (7\text{-}14)$$

Generally, order-of-magnitude analyses and physical considerations will dictate that

$$|\overline{Q_i}\,\overline{Q_j}| > |\overline{Q_i' Q_j'}| \gg |\overline{Q_i' Q_j' Q_k'}| \qquad (7\text{-}15)$$

In order to proceed, we formally substitute the following into Eqs. (7-2), (7-3), (7-4a), and (7-5) combined with Eqs. (7-6) and (7-7):

$$
\begin{aligned}
u &= \bar{u} + u' \\
\rho u &= \overline{(\rho u)} + (\rho u)' \\
\rho v &= \overline{(\rho v)} + (\rho v)' \\
h &= \bar{h} + h' \\
C_i &= \overline{C_i} + C_i' \\
h_i &= \overline{h_i} + h_i' \\
\rho_i \bar{V}_{iy} &= \overline{(\rho_i \bar{V}_{iy})} + (\rho_i \bar{V}_{iy})'
\end{aligned}
\qquad (7\text{-}16)
$$

We then apply the averaging process indicated by Eq. (7-11) to each term in the resulting equations, keeping in mind the general rules for the mean of quantities and products of quantities given by Eqs. (7-12) to (7-14). Note that we assume the transport coefficients μ, k, L, P, etc., have negligible fluctuating parts compared with their mean values. The differential equations for compressible turbulent-boundary-layer flow of a reacting mixture result from the above operation.

$$\frac{\partial \overline{\rho u}}{\partial s} + \frac{\partial \overline{\rho v}}{\partial y} = 0 \qquad (7\text{-}17)$$

$$\overline{\rho u}\,\frac{\partial \overline{C_i}}{\partial s} + \overline{\rho v}\,\frac{\partial \overline{C_i}}{\partial y} = \frac{\partial}{\partial y}\left[\bar{\rho} D_{12}\,\frac{\partial \overline{C_i}}{\partial y} - \overline{(\rho v)' C_i'}\right] + w_i \qquad (7\text{-}18)$$

$$\overline{\rho u}\,\frac{\partial \bar{u}}{\partial s} + \overline{\rho v}\,\frac{\partial \bar{u}}{\partial y} = \frac{\partial}{\partial y}\left[\mu\,\frac{\partial \bar{u}}{\partial y} - \overline{(\rho v)' u'}\right] \qquad (7\text{-}19)$$

---

[1] Because of Eq. (7-12) and the steady state of the mean flow, time derivatives were omitted in the initial statement of the equations of motion.

and

$$\overline{\rho u}\,\frac{\partial \bar{h}}{\partial s} + \overline{\rho v}\,\frac{\partial \bar{h}}{\partial y} = \mu\left(\frac{\partial \bar{u}}{\partial y}\right)^2 - \overline{(\rho v)'u'}\left(\frac{\partial \bar{u}}{\partial y}\right)$$

$$+ \frac{\partial}{\partial y}\left\{ k\,\frac{\partial \bar{T}}{\partial y} - \sum_i \overline{C_i}\,\overline{(\rho v)'h_i'} - \sum_i \overline{h_i}[\overline{(\rho_i \bar{V}_{iy})} + \overline{(\rho v)'C_i'}] \right\} \quad (7\text{-}20)$$

where, in obtaining Eq. (7-20), use was made of Eqs. (7-6) to (7-9) along with the equation for frozen specific heat given below.

$$C_{p_f} = \sum_i \overline{C_i} C_{p_i} \quad (7\text{-}21)$$

The lesser terms in the inequalities listed below are discarded when they appear in the equations with the larger terms in accordance with the concept of a thin boundary layer and inequalities (7-15).

$$|\overline{\rho u}\,\bar{u}| \gg |\overline{(\rho u)'u'}|$$

$$\left|\frac{\partial \bar{u}}{\partial y}\right| \gg \left|\frac{\partial \bar{u}}{\partial s}\right|$$

$$|\overline{\rho u}\,\bar{h}| \gg |\overline{(\rho u)'h'}| \quad (7\text{-}22)$$

$$|\overline{\rho u}\,\overline{C_i}| \gg |\overline{(\rho u)'C_i'}|$$

$$|\overline{(\rho_i \bar{V}_{iy})}\,\bar{h_i}| \gg |\overline{(\rho_i \bar{V}_{iy})'h_i'}|$$

$$|\overline{C_i}\,\overline{(\rho v)'h_i'}| \gg |\overline{C_i'h_i'(\rho v)'}|$$

The Reynolds transport terms appear explicitly in Eqs. (7-18) to (7-20). These terms represent the transport of mass, momentum, and heat due to the turbulent fluctuations and are identified as follows:

$$-\overline{(\rho v)'C_i'} \sim \text{turbulent mass transfer}$$

$$-\overline{(\rho v)'u'} \sim \text{turbulent momentum transfer}$$

$$-\overline{(\rho v)'h_i'} \sim \text{turbulent heat transfer}$$

It will be convenient in that which follows to define turbulent diffusion, viscosity, and thermal-conductivity coefficients D_T, ϵ, and κ, respectively, according to the following equations:

$$\rho D_T(s,y) = -\frac{\overline{(\rho v)'C_i'}}{\partial \overline{C_i}/\partial y} \quad (7\text{-}23)$$

$$\epsilon(s,y) = -\frac{\overline{(\rho v)'u'}}{\partial \bar{u}/\partial y} \quad (7\text{-}24)$$

$$\kappa(s,y) = -\frac{\overline{(\rho v)'h_i'}}{\partial \bar{T}/\partial y} \quad (7\text{-}25)$$

Furthermore, turbulent "Lewis" number and "Prandtl" numbers can be defined, using the above, as

$$L_T = \frac{C_{p_f}\bar{\rho}D_T}{\kappa} \tag{7-26}$$

and

$$P_T = \frac{C_{p_f}\epsilon}{\kappa} \tag{7-27}$$

where, in defining D_T, ϵ, and κ as above, results in no loss in generality, since all coefficients are assumed to vary with s and y. However, assuming that L_T and/or P_T is constant restricts this definition somewhat. Combination of Eqs. (7-24), (7-25), and (7-27) with $P_T = $ const, for example, results in a relation among $-\overline{(\rho v)'u'}$, $\partial\bar{u}/\partial y$, $\overline{(\rho v)'h_i'}$, and $\partial\bar{T}/\partial y$ which is Reynolds analogy when $P_T \equiv 1$.

If Eqs. (7-23) through (7-27) are used with Eqs. (7-18) through (7-20) the following equations result (dropping the bars over mean quantities, since all dependent variables and their cross products are mean quantities from this point on):

$$\rho u \frac{\partial C_i}{\partial s} + \rho v \frac{\partial C_i}{\partial y} = \frac{\partial}{\partial y}\left(\rho D_{12}\frac{\partial C_i}{\partial y} + \rho D_T \frac{\partial C_i}{\partial y}\right) + \dot{w}_i \tag{7-28}$$

$$\rho u \frac{\partial u}{\partial s} + \rho v \frac{\partial u}{\partial y} = \frac{\partial}{\partial y}\left(\mu \frac{\partial u}{\partial y} + \epsilon \frac{\partial u}{\partial y}\right) \tag{7-29}$$

and

$$\rho u \frac{\partial h}{\partial s} + \rho v \frac{\partial h}{\partial y} = (\mu + \epsilon)\left(\frac{\partial u}{\partial y}\right)^2 + \frac{\partial}{\partial y}\left[\left(\frac{\mu}{P} + \frac{\epsilon}{P_T}\right)\frac{\partial h}{\partial y}\right]$$
$$+ \frac{\partial}{\partial y}\left\{\left[\frac{k}{C_{p_f}}(L-1) + \frac{\kappa}{C_{p_f}}(L_T-1)\right]\sum_i h_i \frac{\partial C_i}{\partial y}\right\} \tag{7-30}$$

or, making use of Eq. (7-6) with Eqs. (7-30) and (7-29),

$$\rho u \frac{\partial I}{\partial s} + \rho v \frac{\partial I}{\partial y} = \frac{\partial}{\partial y}\left[\left(\frac{\mu}{P} + \frac{\epsilon}{P_T}\right)\frac{\partial I}{\partial y}\right]$$
$$+ \frac{\partial}{\partial y}\left\{\left[\frac{k}{C_{p_f}}(L-1) + \frac{\kappa}{C_{p_f}}(L_T-1)\right]\sum_i h_i \frac{\partial C_i}{\partial y}\right\}$$
$$+ \frac{\partial}{\partial y}\left\{\left[\mu\left(1 - \frac{1}{P}\right) + \epsilon\left(1 - \frac{1}{P_T}\right)\right]\frac{1}{2}\frac{\partial u^2}{\partial y}\right\} \tag{7-31}$$

It is immediately apparent from Eqs. (7-29) and (7-31) that, if $L = L_T = P = P_T \equiv 1$, a solution to Eq. (7-31) is

$$I = au + b \tag{7-32}$$

where a and b are constants of integration. Equation (7-32) is identical with Crocco's integral developed for the laminar boundary layer for $L = P = 1$ in Sec. 2-5.

This completes our derivation of the boundary-layer equations for the reacting compressible turbulent boundary layer. Their solution for any given boundary conditions for ρ, u, v, I, and C_i is a formidable task. Note that by Eqs. (7-23) through (7-27) the turbulent transport coefficients are functions of *both* the mean and fluctuating values of their related kinematic or state variables, and it is this coupling which has defied analysis through the years except for the special cases involving restrictive assumptions such as assuming that P_T and L_T are constant. We shall proceed, now, to their application for some specific cases of interest to us.

7-3. Dissociated-turbulent-boundary-layer Equations. One of the problems of great concern to the hypersonic aerodynamicist is that of calculating the skin friction and heat transfer for a turbulent boundary layer under conditions where the air within the boundary layer is dissociating. As was seen in Chap. 4 for the laminar boundary layer, the effects of dissociation can be significant but not overwhelming, especially in heat-transfer calculations. It might be anticipated from the results obtained in Chap. 4 for the laminar boundary layer that dissociation might affect the heat transfer from a turbulent boundary layer to the cool surface over which it flows, although not to a degree resulting in an order-of-magnitude change over the heat transfer one might expect without dissociation. The purpose of this section and succeeding sections is to obtain a quantitative measure of the possible effects of dissociation upon skin friction and heat transfer of a compressible turbulent boundary layer.

To proceed we shall make an assumption used earlier in Chap. 4 when treating the dissociating laminar boundary layer, namely, that the dissociating gas is a binary mixture of atoms (O and N mostly) and molecules (O_2 and N_2 mostly) and that the differences between the transport properties of the various atoms and between the properties for the various molecules are insignificant and far less than the differences between the transport properties of the atoms and molecules. This will reduce the number of species under consideration to two (atoms and molecules) and will greatly facilitate obtaining a solution for the problem at hand. Consistent with these assumptions we write

$$\alpha = C_A = \frac{\rho_A}{\rho} = \text{mass fraction of the atoms} \qquad (7\text{-}33)$$

and

$$1 - \alpha = C_M = \frac{\rho_M}{\rho} = \text{mass fraction of the molecules} \qquad (7\text{-}34)$$

Furthermore, Eqs. (7-7) and (7-8) combined become

$$h = \sum_i C_i h_i = \alpha C_{p_A} T + \alpha h_A^0 + (1 - \alpha) C_{p_M} T$$

or
$$h = C_{p_M} T + (C_{p_A} - C_{p_M}) \alpha T + \alpha h_A^0 \tag{7-35}$$

or, using Eq. (7-21),

$$C_{p_f} = \sum_i C_i C_{p_i} = \alpha C_{p_A} + (1 - \alpha) C_{p_M} \tag{7-36}$$

whence, combining Eqs. (7-35) and (7-36),

$$h = C_{p_f} T + \alpha h_A^0 \tag{7-37}$$

where h_A^0 is the heat of dissociation of the molecules per unit mass and the subscripts A and M denote atoms and molecules, respectively, from here on. In the above we have assumed that the specific heats of the atoms and molecules are constant, thus neglecting the variation of the specific heats of both species with temperature as being of secondary consequence in the development that follows. Equations (7-33) and (7-34) can be used with Eqs. (7-28) through (7-30) to write

$$\rho u \frac{\partial \alpha}{\partial s} + \rho v \frac{\partial \alpha}{\partial y} = \frac{\partial}{\partial y} \left[(\rho D_{12} + \rho D_T) \frac{\partial \alpha}{\partial y} \right] + \rho \frac{\partial \alpha}{\partial t} \tag{7-38}$$

$$\rho u \frac{\partial u}{\partial s} + \rho v \frac{\partial u}{\partial y} = \frac{\partial}{\partial y} \left[(\mu + \epsilon) \frac{\partial u}{\partial y} \right] \tag{7-39}$$

and

$$\rho u \frac{\partial h}{\partial s} + \rho v \frac{\partial h}{\partial y} = (\mu + \epsilon) \left(\frac{\partial u}{\partial y} \right)^2 + \frac{\partial}{\partial y} \left[\left(\frac{\mu}{P} + \frac{\epsilon}{P_T} \right) \frac{\partial h}{\partial y} \right]$$
$$+ \frac{\partial}{\partial y} \left\{ \left[\frac{k}{C_{p_f}} (L - 1) + \frac{\kappa}{C_{p_f}} (L_T - 1) \right] (h_A - h_M) \frac{\partial \alpha}{\partial y} \right\} \tag{7-40}$$

or, as an alternative to Eq. (7-40), using Eq. (7-31),

$$\rho u \frac{\partial I}{\partial s} + \rho v \frac{\partial I}{\partial y} = \frac{\partial}{\partial y} \left[\left(\frac{\mu}{P} + \frac{\epsilon}{P_T} \right) \frac{\partial I}{\partial y} \right]$$
$$+ \frac{\partial}{\partial y} \left\{ \left[\frac{k}{C_{p_f}} (L - 1) + \frac{\kappa}{C_{p_f}} (L_T - 1) \right] (h_A - h_M) \frac{\partial \alpha}{\partial y} \right\}$$
$$+ \frac{\partial}{\partial y} \left\{ \left[\mu \left(1 - \frac{1}{P} \right) + \epsilon \left(1 - \frac{1}{P_T} \right) \right] \frac{1}{2} \frac{\partial u^2}{\partial y} \right\} \tag{7-41}$$

where C_{p_f} is given by Eq. (7-36). Equations (7-2), (7-17), (7-38), (7-39), and (7-40) or (7-41) represent a system of five equations which are to be solved for eight unknowns, ϵ, κ (or P_T), D_T (or L_T) ρ, u, v, α, h, or I. Exact solutions to these equations have never been obtained because of the unknown functional dependence of the turbulent transport parameters ϵ, κ, or P_T and D_T or L_T upon the mean flow variables and turbulent fluctuations as described by Eqs. (7-23) through (7-27). However, if one assumes these parameters to be constant or simple functions of the mean flow variables through the turbulent layer, some useful relations can be obtained.

7-4. Relation between C_H and C_f. We proceed to obtain an expression relating heat-transfer coefficient C_H to skin-friction coefficient C_f using Eqs. (7-39) and (7-41).

We can write Eq. (7-41) as

$$\rho u \frac{\partial I}{\partial s} + \rho v \frac{\partial I}{\partial y} = \frac{\partial G}{\partial y} \tag{7-42}$$

where G is every factor under differentiation with respect to y on the right-hand side of Eq. (7-41) and

$$G_w = -\dot{q}_w \tag{7-43}$$

since, from Eq. (7-41),

$$-\dot{q}_w = \left[k \frac{\partial T}{\partial y} + \rho D_{12}(h_A - h_M) \frac{\partial \alpha}{\partial y} \right]_w \tag{7-44}$$

and $\epsilon = \kappa = 0$ at the wall, since the velocity and fluctuations of the velocity are zero at the surface. To proceed, divide Eq. (7-42) by Eq. (7-39) to obtain

$$\frac{\rho u(\partial I / \partial s) + \rho v(\partial I / \partial y)}{\rho u(\partial u / \partial s) + \rho v(\partial u / \partial y)} = \frac{\partial G}{\partial y} \bigg/ \frac{\partial \tau}{\partial y} \tag{7-45}$$

where

$$\tau = (\mu + \epsilon) \frac{\partial u}{\partial y} \tag{7-46}$$

Now, it was shown earlier that, if $L_T = L = P_T = P \equiv 1$,

$$I = au + b \tag{7-32}$$

We shall assume that L_T, L, P_T, and P are constant and sufficiently close to the value one that Eq. (7-32) applies in the present case. (Calculations using measured temperature and velocity profiles through a turbulent boundary layer show that this assumption is a reasonable one.) This assumption amounts to assuming that the temperature and velocity boundary layers have the same thickness. Then, from Eqs. (7-32) and (7-45),

$$\frac{\partial}{\partial y} (a\tau - G) = 0 \quad \text{or} \quad G = a\tau + c \tag{7-47}$$

where a and c are constants of integration to be determined by the boundary conditions on G and τ. These are

$$G = G_w = -\dot{q}_w \qquad \text{when } \tau = \tau_w \tag{7-48}$$
$$G = 0 \qquad \text{when } \tau = 0$$

now, by definition,

$$-\dot{q}_w = C_H \rho_e u_e (I_r - h_w) \tag{7-49}$$

and

$$\tau_w = \frac{C_f}{2} \rho_e u_e^2 \tag{7-50}$$

where I_r is the recovery enthalpy defined as

$$I_r = h_e + r \frac{u_e^2}{2} \tag{7-51}$$

where r is the recovery factor which will be determined later. Applying boundary-condition equations (7-48) to Eq. (7-47) and making use of Eqs. (7-49) and (7-50) result in

$$\frac{G}{\tau} = \frac{G_w}{\tau_w} = \frac{-\dot{q}_w}{\tau_w} = \frac{C_H}{C_f/2} \frac{I_r - h_w}{u_e} \tag{7-52}$$

whence, assuming $\alpha = \alpha(u)$ and $h = h(u)$ in G, we have

$$\frac{2C_H}{C_f} = \frac{u_e}{I_r - h_w} \left[\frac{\mu/P + \epsilon/P_T}{\mu + \epsilon} \frac{dh}{du} + u \right.$$
$$\left. + \frac{(k/C_{p_f})(L - 1) + (\kappa/C_{p_f})(L_T - 1)}{\mu + \epsilon} (h_A - h_M) \frac{d\alpha}{du} \right] \tag{7-53}$$

Now, in order to obtain an equation for $2C_H/C_f$, we must integrate Eq. (7-53) across the boundary layer. In order to proceed, we shall adopt the two-layer model of the turbulent boundary layer which has proved to be an excellent model of the turbulent boundary layer at subsonic speeds[1] and a reasonable one at hypersonic speeds.[2] That is, there exist a laminar sublayer where

$$\epsilon \ll \mu \qquad \text{and} \qquad \kappa \ll k \qquad \text{for } 0 \leq u \leq u_L$$

and an outer turbulent portion of the turbulent boundary layer where

$$\epsilon \gg \mu \qquad \text{and} \qquad \kappa \gg k \qquad \text{for } u_L \leq u \leq u_e$$

The narrow region near $u = u_L$, sometimes called the buffer layer, where $\mu \simeq \epsilon$ and $\kappa \simeq k$, will be neglected in that which follows for simplicity and because taking it into account will not significantly

---

[1] See, for example, Hermann Schlichting, "Boundary Layer Theory," 4th ed., fig. 20.4, p. 507, McGraw-Hill Book Company, Inc., New York, 1960.

[2] See, for example, F. K. Hill, *Phys. Fluids*, vol. 2, no. 6, fig. 3, p. 670, 1959.

alter the results to be obtained. Now, according to our model of the boundary layer, in the laminar sublayer

$$\frac{2C_H}{C_f} = \frac{u_e}{I_r - h_w}\left[\frac{1}{P}\frac{dh}{du} + \frac{1}{P}(L-1)(h_A - h_M)\frac{d\alpha}{du} + u\right] \quad (7\text{-}54)$$

and in the turbulent portion of the boundary layer

$$\frac{2C_H}{C_f} = \frac{u_e}{I_r - h_w}\left[\frac{1}{P_T}\frac{dh}{du} + \frac{1}{P_T}(L_T - 1)(h_A - h_M)\frac{d\alpha}{du} + u\right] \quad (7\text{-}55)$$

We proceed as follows: Integrate Eq. (7-54) from the surface ($u = 0$) to the interface between the two layers ($u = u_L$). Integrate Eq. (7-55) from the interface ($u = u_L$) to the outer edge of the boundary layer ($u = u_e$). Eliminate h_L between the two equations, and solve for $2C_H/C_f$ in the resulting equation. There results

$$\frac{2C_H}{C_f} = \left[P_T + (P - P_T)\frac{u_L}{u_e}\right]^{-1}\left(1 + F_1\frac{h_A^0}{I_r - h_w}\right) \quad (7\text{-}56a)$$

with F_1 being defined as

$$F_1 = (L - 1)(\alpha_L - \alpha_w) + (L_T - 1)(\alpha_e - \alpha_L) \quad (7\text{-}56b)$$

An additional result of this calculation is

$$r = P_T + (P - P_T)\left(\frac{u_L}{u_e}\right)^2 \quad (7\text{-}56c)$$

and in Eq. (7-56a) we have used the approximation

$$h_A - h_M = (C_{p_A} - C_{p_M})T + h_A^0 \simeq h_A^0 = \text{const} \quad (7\text{-}56d)$$

Equation (7-56d) is used, since at the temperatures of concern to us here the heat of formation of the atoms far exceeds the differences in heat capacity of the molecules and atoms. Substitution of representative numerical values into exact expressions for the enthalpy of a mixture of atoms and molecules will substantiate this.

It is interesting to note that, when we assume that $u_L = u_e$ (i.e., no turbulent outer region), we obtain

$$r = P \quad (7\text{-}57)$$

and

$$\frac{2C_H}{C_f} = \frac{1}{P}\left[1 + \frac{(L-1)(\alpha_e - \alpha_w)h_A^0}{I_r - h_w}\right] \quad (7\text{-}58)$$

These are the results obtained by Clarke[1] for laminar couette flow

[1] John F. Clarke, *J. Fluid Mech.*, vol. 4, part 5, pp. 441–466, 1958.

of a dissociating mixture. Apparently, the use of Eq. (7-32) corresponds to treating the boundary layer as a quasi-couette flow. Corresponding results might be expected for turbulent couette flow where P is replaced by P_T and L by L_T in Eqs. (7-57) and (7-58) above. Of course, a fully turbulent couette flow would be impossible to obtain, since the top and bottom walls in couette flow will always damp out turbulent fluctuations, thus creating laminar sublayers at both walls such as we have postulated here for flow over a plate.

Reasonable simplifications can be made in Eqs. (7-56) without sacrificing accuracy. For example, measurements of temperature and velocity profiles in heated air jets[1] have shown that

$$P_T = \frac{C_{p_f}\epsilon}{\kappa} \simeq 0.70 \pm 10\% \tag{7-59}$$

and

$$L_T = \frac{\rho D_T C_{p_f}}{\kappa} \simeq 1.0 \pm 20\% \tag{7-60}$$

and are essentially independent of the values of the laminar-transport coefficients which were allowed to vary over wide ranges during the experiments. The geometry did not vary during this experiment, so the question remains as to the pertinence of Eqs. (7-59) and (7-60) to turbulent-boundary-layer flow. We observe, however, that semiempirical theories for turbulent flow over a flat plate incorporating the assumption that P_T is constant enjoy success when compared with experiment. The above results lead us to assume that

$$P_T = L_T = 1 \tag{7-61}$$

will be a reasonable approximation in that which follows.

Furthermore, for most values of P, L, and u_L/u_e of concern to us here, it can be shown that

$$1 + (P - 1)\frac{u_L}{u_e} \simeq P^{(u_L/u_e)} \simeq P^{2/3} \tag{7-62}$$

$$(L - 1)\frac{u_L}{u_e} \simeq L^{(u_L/u_e)} - 1 \simeq L^{2/3} - 1 \tag{7-63}$$

and

$$r = 1 - (1 - P)\left(\frac{u_L}{u_e}\right)^2 \simeq P^{1/3} \tag{7-64}$$

If $P = 0.72$, Eq. (7-62) is correct within ± 10 per cent for u_L/u_e varying from 0.40 to 0.90 and Eq. (7-64) is correct within ± 10 per

[1] Walton Forstall, Jr., and Ascher H. Shapiro, *J. Appl. Mech.*, vol. 17, no. 4, pp. 399–408, 1950.

cent for u_L/u_e varying from 0.22 to 0.83.   If $L = 1.4$, Eq. (7-63) is correct within ± 10 per cent for u_L/u_e varying from 0.56 to 0.69.   If u_L/u_e is outside these ranges, exponents other than $\frac{2}{3}$ and $\frac{1}{3}$ may be chosen to the same degree of approximation.   The values of the exponents chosen here are representative and appear most frequently in the literature.   The importance of the velocity at the edge of the laminar sublayer in determining these exponents is sharply brought out in this discussion.   Note, for example, that Eq. (7-62) reveals that, when $u_L/u_e = 1.0$ (all laminar layer), $r = P$ and, when $u_L/u_e = 0$ (all turbulent layer), $r = 1$ (since $P_T = 1$).   Note the resemblance of Eq. (7-62) to the laminar-boundary-layer recovery factor, Eq. (5-161) derived in Sec. 5-11.

One other simplification can be made.   It can be shown that

$$\alpha_L - \alpha_w \simeq (\alpha_e - \alpha_w)\frac{u_L}{u_e} \tag{7-65}$$

Equation (7-65) is exactly correct for a frozen boundary layer with $P = L = 1$.   We shall assume a priori that the use of Eq. (7-65) is justifiable in the present case and return to discussing the justification later in this chapter.   Use of Eqs. (7-61) through (7-65) in Eqs. (7-56) yields

$$\frac{2C_H}{C_f} = P^{-2/3}\left[1 + (L^{2/3} - 1)\frac{h_c}{I_r - h_w}\right] \tag{7-66a}$$

and

$$r = P^{1/3} \tag{7-66b}$$

where

$$h_c = (\alpha_e - \alpha_w)h_A^0 \tag{7-66c}$$

Equation (7-66a) has been independently derived by Lees;[1] Rose, Probstein, and Adams;[2] and Dorrance.[3]   Lees and Dorrance arrived at Eq. (7-66a) through the argument presented here.   Rose, Probstein, and Adams arrived at a generalized version of Eq. (7-66a) through analogies drawn with the laminar-boundary-layer expression presented in Chap. 4, e.g., Eq. (4-89) or (4-91).   Dorrance brought out the importance of the sublayer velocity u_L in determining the exponent on L in Eq. (7-66a).

Equation (7-66a) relates heat-transfer coefficient to skin-friction coefficient for a dissociating compressible turbulent boundary layer over a flat plate.   Our next objective will be to develop an equation

---

[1] Lester Lees, Convective Heat Transfer with Mass Addition and Chemical Reactions, paper presented in "Combustion and Propulsion, Third AGARD Colloquium," pp. 451–498, Pergamon Press, New York, 1959.

[2] Peter H. Rose, Ronald F. Probstein, and Mac C. Adams, *J. Aerospace Sci.*, vol. 25, no. 12, pp. 751–760, 1958.

[3] William H. Dorrance, *ARS J.*, vol. 31, no. 1, pp. 61–70, 1961.

for the skin-friction coefficient which, in turn, will allow heat-transfer coefficient and hence heat transfer to the plate to be calculated.

7-5. Turbulent-skin-friction Coefficient. The approach to be followed in developing an expression for skin-friction coefficient is that advanced by Dorrance.[1] Dorrance derived an expression for local-skin-friction coefficient by substituting density and velocity profiles into the well-known von Kármán momentum integral given below.

$$C_f = 2\frac{d\theta}{ds} \tag{7-67a}$$

where

$$\theta = \frac{\rho_w}{\rho_e} \int_0^1 \frac{\rho}{\rho_w} z(1-z)\left(\frac{dz}{dy}\right)^{-1} dz \tag{7-67b}$$

and

$$z = \frac{u}{u_e} \tag{7-67c}$$

It is apparent that expressions are required for dz/dy and $\rho(z)/\rho_w$. These are obtained from solving the equations of state and conservation of species, momentum, and energy, Eqs. (7-38), (7-39), and (7-40), respectively.

From Eqs. (7-17) and (7-39) it can be shown for a surface at constant surface temperature with no mass transfer present that

$$\left(\frac{\partial \tau}{\partial y}\right)_w = \left(\frac{\partial^2 \tau}{\partial y^2}\right)_w = 0$$

since u, v, $\partial u/\partial s$, and $\partial v/\partial y = 0$ at $y = 0$; thus

$$\tau = \tau_w + O\left(\frac{\partial^3 \tau}{\partial y^3}\frac{y^3}{3!}\right)$$

and $\tau = \tau_w$ is a reasonable first approximation throughout a large region of the boundary layer.[2] Therefore, following Prandtl's original mixing-length concepts we write

$$\tau = \tau_w = \epsilon\frac{du}{dy} = \rho K^2 y^2\left(\frac{du}{dy}\right)^2 \tag{7-68}$$

for the turbulent outer layer. K in Eq. (7-68) is the so-called mixing-length constant. If we define

$$E = K\left(\frac{\rho_e}{\rho_w}\frac{C_f}{2}\right)^{-1/2} \tag{7-69}$$

---

[1] *Ibid.*

[2] D. A. Spence, *J. Fluid Mech.*, vol. 8, part 3, pp. 368–387, 1960, shows through correlation of experimental data that $\tau_w \geq \tau \geq 0.9\tau_w$ for $0.81u_e \geq u \geq 0$. That is, the shear stress is within 10 per cent of the wall shear stress for values of velocity up to 81 per cent of free-stream velocity.

Eq. (7-68) can be integrated to obtain

$$\frac{1}{K} \log_e \left[\frac{\rho_w}{\mu_w} \left(\frac{\tau_w}{\rho_w} \right)^{1/2} y \right] + \phi(0) = \frac{1}{K} I_4 \qquad (7\text{-}70a)$$

where, if ξ is a variable running along the z axis, then

$$I_4 = E \int_0^z \left(\frac{\rho}{\rho_w} \right)^{1/2} d\xi \qquad (7\text{-}70b)$$

and the constant of integration $\phi(0)$ can be determined by the fact that Eq. (7-70a) becomes the velocity-distribution correlation[1] for a constant-density, low-speed, turbulent boundary layer given below as Eq. (7-70c) when $M_e = 0$ and $T = T_w = T_e$; viz.,

$$u \left(\frac{\tau_w}{\rho_w} \right)^{-1/2} = \phi(0) + \frac{1}{K} \ln \left[\frac{\rho_w}{\mu_w} \left(\frac{\tau_w}{\rho_w} \right)^{1/2} y \right] \qquad (7\text{-}70c)$$

Also, since

$$\frac{dI_4}{dz} = E \left(\frac{\rho}{\rho_w} \right)^{1/2} \qquad (7\text{-}71)$$

then, from Eqs. (7-70a) and (7-71),

$$\frac{dy}{dz} = \frac{E^2 \exp\left[-K\phi(0) \right]}{K(\rho_w u_e / \mu_w)} \exp\left(I_4 \right) \left(\frac{\rho}{\rho_w} \right)^{1/2} \qquad (7\text{-}72)$$

This is the desired expression for use in Eq. (7-67b).

In order to obtain an equation for ρ/ρ_w, we must obtain equations for concentration and temperature as a function of velocity through the boundary layer. Equations (7-38) and (7-39) are used to obtain an expression for concentration of atoms through the boundary layer as follows:

Let us assume that $\alpha = \alpha(u,t)$ and determine the conditions under which this is so using Eqs. (7-38) and (7-39). Substitute $\alpha = \alpha(u,t)$ into Eq. (7-38). We obtain

$$\left(\rho u \frac{\partial u}{\partial s} + \rho v \frac{\partial u}{\partial y} \right) \frac{\partial \alpha}{\partial u} = \frac{\partial}{\partial y} \left[\rho(D_{12} + D_T) \frac{\partial u}{\partial y} \right] \frac{\partial \alpha}{\partial u}$$
$$+ \rho(D_{12} + D_T) \left(\frac{\partial u}{\partial y} \right)^2 \frac{\partial^2 \alpha}{\partial u^2} + \rho \frac{\partial \alpha}{\partial t} \qquad (7\text{-}73)$$

Combining Eq. (7-39) with Eq. (7-73) we find that, if $\rho D_{12} = \mu$ and $\rho D_T = \epsilon$, our equation for $\alpha = \alpha(u,t)$ becomes

$$\frac{\partial^2 \alpha}{\partial u^2} + \frac{\rho(\mu + \epsilon)}{\tau^2} \frac{\partial \alpha}{\partial t} = 0 \qquad (7\text{-}74)$$

[1] G. B. Schubauer and C. M. Tchen, Turbulent Flow, in C. C. Lin (ed.), "Turbulent Flows and Heat Transfer," sec. B, pp. 122–124, Princeton University Press, Princeton, N.J., 1959.

since, in general, $\partial\alpha/\partial u \neq \infty$. It is thus apparent that the relation between the concentration of atoms α and the velocity through the boundary layer u depends upon the shear stress distribution τ, the viscosities μ and ϵ, and the rate at which atoms are created by the dissociation process. If the denominator of the second term in Eq. (7-74) is much larger than the numerator, diffusion predominates in determining the atom concentration through the boundary layer. If $\partial\alpha/\partial t = 0$, then, of course, diffusion exactly determines the concentration profile (called "frozen flow"). If the second term in Eq. (7-74) is very large, then the flow is essentially in equilibrium and Eq. (7-74) is not needed to determine the atom concentrations, which would then be determined wholly by the external stream static pressure and the temperature distribution through the boundary layer along with the equilibrium constants for the dissociation reaction.

We assume that the "frozen flow" or wholly diffusion-determined concentration profile through the boundary layer applies with equilibrium concentrations at the surface, and we shall return to justify this assumption later. Then, from Eq. (7-74),

$$\frac{d^2\alpha}{du^2} = 0$$

whence, since $\alpha = \alpha_w$ when $u = 0$ and $\alpha = \alpha_e$ when $u = u_e$, we obtain

$$\alpha = \alpha_w + (\alpha_e - \alpha_w)\frac{u}{u_e} \tag{7-75}$$

and this is the equation used to justify Eq. (7-65) used earlier.

The temperature distribution as a function of velocity is determined as follows: From Eqs. (7-39) and (7-40), assuming that $L = L_T = P = P_T = 1$ and $h = h(u)$ and using Eq. (7-39) and (7-40) in a manner analogous to the way Eqs. (7-38) and (7-39) were used to obtain Eq. (7-74) for $\alpha(u,t)$, we obtain

$$\frac{d^2h}{du^2} = -1 \tag{7-76}$$

if $dh/du \neq \infty$ as is true in general. Since $h = h_w$ when $u = 0$ and $h = h_e$ when $u = u_e$, then[1]

$$h + \frac{u^2}{2} = h_w + \left(h_e - h_w + \frac{u_e^2}{2}\right)\frac{u}{u_e} \tag{7-77}$$

---

[1] Spence, *op. cit.*, has shown that Eq. (7-77), exact for $P_T = P = 1$, can be used with good accuracy for a wide range of variation of P_T when $P = 1$. This suggests that Eq. (7-77) is a reasonable approximation when used as we use it here.

or
$$I = h_w + (I_e - h_w)\frac{u}{u_e}$$

where, from Eqs. (7-7), (7-8), (7-33), and (7-34),

$$h = \alpha\left[\int (C_{p_A} - C_{p_M})\,dT + h_A^0\right] + \int C_{p_M}\,dT \qquad (7\text{-}78)$$

Now assume that the atoms possess the translational degrees of freedom only and the molecules the translational and rotational degrees of freedom, neglecting the vibrational energy as being secondary at the temperatures of interest to us. (It could be included in a more complete analysis if desired.) Then

$$C_{p_A} = 2.5\,\frac{k}{m_A} \qquad (7\text{-}79a)$$

$$C_{p_M} = 3.5\,\frac{k}{m_M} \qquad (7\text{-}79b)$$

$$h_A^0 = \frac{D}{m_M} \qquad (7\text{-}79c)$$

and
$$2m_A = m_M \qquad (7\text{-}79d)$$

whence, from Eqs. (7-78) and (7-79),

$$h = (3.5 + 1.5\alpha)\,\frac{k}{2m_A}\,T + \alpha\,\frac{D}{2m_A} \qquad (7\text{-}80)$$

Similarly, from Eqs. (7-1a), (7-1b), (7-4b), (7-33), (7-34), and (7-79d) we obtain

$$p = p_e = \rho\,\frac{k}{2m_A}\,T(1 + \alpha) \qquad (7\text{-}81)$$

whence
$$\frac{\rho}{\rho_w} = \frac{(1 + \alpha_w)T_w}{(1 + \alpha)T} = \frac{T_w}{T_e}\frac{1 + \alpha_w}{1 + \alpha}\left(\frac{T}{T_e}\right)^{-1} \qquad (7\text{-}82)$$

The relation between T and u is obtained from Eqs. (7-75), (7-77), and (7-80) and is

$$\frac{T}{T_e} = \frac{7 + 3\alpha_w}{7 + 3\alpha}\frac{T_w}{T_e} + \left(\frac{7 + 3\alpha_e}{7 + 3\alpha} - \frac{7 + 3\alpha_w}{7 + 3\alpha}\frac{T_w}{T_e}\right)\frac{u}{u_e} + \frac{u}{u_e}\left(1 - \frac{u}{u_e}\right)\frac{1.4M_e^2}{7 + 3\alpha}$$

$$(7\text{-}83)$$

where, by definition for our purposes,

$$M_e^2 = u_e^2\left(1.4\,\frac{k}{2m_A}\,T_e\right)^{-1} \qquad (7\text{-}84)$$

When $\alpha = \alpha_e = \alpha_w = 0$, Eq. (7-83) becomes the familiar Crocco integral to the energy equation

$$\frac{T}{T_e} = \frac{T_w}{T_e} + \left(1 - \frac{T_w}{T_e}\right)\frac{u}{u_e} + \frac{u}{u_e}\left(1 - \frac{u}{u_e}\right)0.2M_e^2$$

as it should for $\gamma = 1.4$ (diatomic molecules).

Our skin-friction law now can be derived using Eqs. (7-67), (7-70b), (7-72), (7-75), (7-82), and (7-83). It is

$$R_eC_f = 0.389\left(\frac{T_w}{T_e}\right)^{0.76}I_5 \tag{7-85}$$

where $R_e = \rho_e u_e s / \mu_e$ and

$$I_4 = E\int_0^z \left(\frac{\rho}{\rho_w}\right)^{1/2}d\xi \tag{7-70b}$$

$$E = 0.557\left(\frac{T_w}{T_e}C_f\frac{1 + \alpha_w}{1 + \alpha_e}\right)^{-1/2} \tag{7-86}$$

$$I_5 = E^2\int_0^1 \left(\frac{\rho}{\rho_w}\right)^{3/2}(z - z^2)\exp(I_4)\,dz \tag{7-87}$$

$$\frac{\rho}{\rho_w} = \frac{T_w}{T_e}\frac{1 + \alpha_w}{1 + \alpha}\left(\frac{T}{T_e}\right)^{-1} \tag{7-82}$$

$T(u)/T_e$ is given by Eq. (7-83), and $\alpha(u)$ is given by Eq. (7-75). The constants $\phi(0)$ and K were chosen as 6.53 and 0.393, respectively, in order that Eq. (7-85) fit the experimental measurements of local skin-friction coefficient at

$$M_e = 0 \qquad T_w = T_e = T \qquad \alpha = \alpha_e = \alpha_w = 0$$

In obtaining Eq. (7-85) it is assumed that $\int_0^s C_f\,ds = C_f s$ for large s on the left-hand side of Eq. (7-85). This is a reasonable approximation for large s appropriate to the values for turbulent flow particularly, since the approximation is made before the empirical constants are determined by fitting the measured values of C_f at $M_e = 0$, $T_w = T_e = T$, and $\alpha = \alpha_w = \alpha_e = 0$.

The function $\mu \propto T^{0.76}$ was used in arriving at Eq. (7-85). This dependency of viscosity on temperature is in good accord with the findings of Bauer and Zlotnick[1] for high-temperature air mixtures and by Scala and Baulknight[2] for mixtures of O_2 and O and N_2 and N. Section 10-7, Calculations for Dissociated Air, discusses this in detail.

[1] Ernest Bauer and Martin Zlotnick, *ARS J.*, vol. 29, no. 10, pp. 721–728, 1959.

[2] Sinclaire M. Scala and Charles W. Baulknight, *ARS J.*, vol. 29, no. 1, pp. 39–45, 1959.

Equation (7-85) will reduce to Eq. (37) of Dorrance and Dore[1] when dissociation and mass transfer are zero and to Eq. (66) of Van Driest[2] when dissociation is zero and a series expansion of the integral I_5, Eq. (7-87), is resorted to. However, Eq. (7-85), because it retains terms neglected in Van Driest's treatment, is more precisely evaluated at supersonic Mach numbers and moderate values of T_w/T_e. Comparison with experiment will show that Eq. (7-85) agrees better with experiment over a range of values of M_e and T_w/T_e than does Van Driest's equation, possibly because of the numerical integration employed in the present case.

7-6. Calculation of Skin-friction Coefficient. Equation (7-85) was solved for C_f for a variety of combinations of R_e, M_e, T_w/T_e, α_w, and α_e using a 704 computer to perform the tedious numerical integrations and iterations involved. In preparing to present the results it became apparent that, since five parameters of C_f were involved, any method of reducing the number of plots necessary would be a contribution. Fortunately, such was suggested by expanding all expressions appearing in the integrals I_4 and I_5 involved in determining C_f into series in powers of α_w and α_e. When only first-order terms were retained, it was found that

$$\frac{C_f}{C_{f_0}} = f\left(\frac{1 + \alpha_e}{1 + \alpha_w}\right)^{1/7} - g$$

where f and g are weakly varying functions of R_e, T_w/T_e, and M_e having values near 2.0 and 1.0, respectively, and C_{f_0} is C_f evaluated at the proper value of M_e, T_w/T_e, and R_e but *with dissociation equal to zero*, that is, $\alpha_w = \alpha_e = \alpha = 0$ in Eqs. (7-82) and (7-83). Plotting the results showed that

$$\frac{C_f}{C_{f_0}} = \left[2.0\left(\frac{1 + \alpha_e}{1 + \alpha_w}\right)^{1/7} - 1\right] \begin{matrix} + 4\% \\ - 2\% \end{matrix} \tag{7-88}$$

for *all cases considered.* Figure 7-5 presents the results of correlating the calculations according to Eq. (7-88).

Equation (7-88) or the plot in Fig. 7-5 points out several interesting features regarding the effects of dissociation upon local skin-friction coefficient. First it appears that C_f is affected ± 22 per cent or less by dissociation in all cases. That is, C_f can be calculated as C_{f_0}, the value corresponding to the Mach number and temperatures present but with no dissociation and the error involved will be within 22 per

---

[1] William H. Dorrance and Frank J. Dore, *J. Aeronaut. Sci.*, vol. 21, no. 6, pp. 404–410, 1954.

[2] E. R. Van Driest, *J. Aeronaut. Sci.*, vol. 18, no. 3, pp. 145–160, 1951.

cent of the value of C_{f_0}. Second, Eq. (7-88) or Fig. 7-5 indicates that the effect of using a noncatalytic wall ($\alpha_w \neq 0$) would be such as to reduce skin-friction coefficient under the value it would have if $\alpha_w = 0$. The maximum amount of this reduction is 22 per cent for the case where $\alpha_e = 1$ and $\alpha_w = 1$ over the case where $\alpha_e = 1$ and $\alpha_w = 0$.

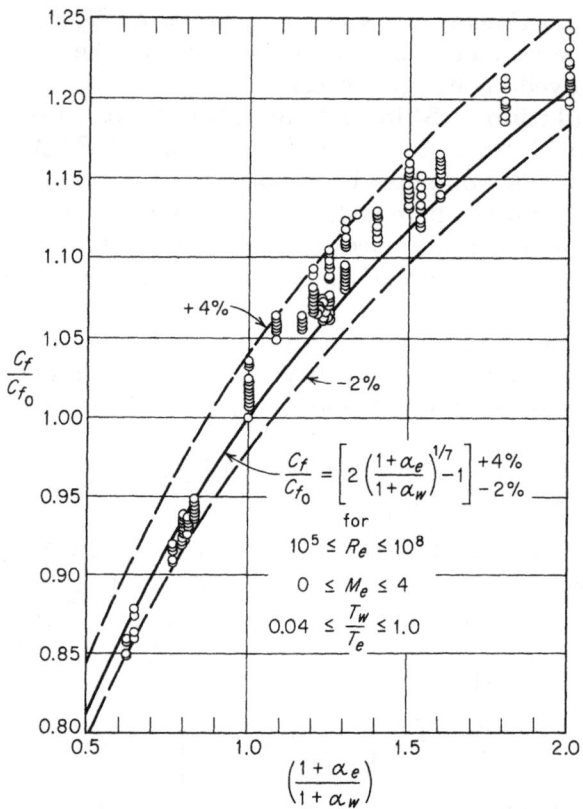

FIG. 7-5. Correlation curve for C_f/C_{f_0} given by Eq. (7-88) compared with results given by numerical integration of Eq. (7-85) for C_f and C_{f_0}.

Plots of C_{f_0} were prepared as functions of R_e, M_e, and T_w/T_e and are presented as Figs. 7-6 to 7-9. The results of the calculations are presented for local Mach numbers M_e up to 4.0 only, since it is at these low local Mach numbers that local free-stream static temperatures are high and hence equilibrium degrees of dissociation α_e are high. In the case where M_e is high, T_e is low for terrestrial flight velocities, and if the wall temperature is low so that equilibrium dissociation at the wall α_w is zero, there will be no effect of dissociation within the boundary layer upon skin friction according to the present theory. If the surface inhibits recombination, Eq. (7-88) shows that skin friction will be less than the skin friction to a catalytic wall.

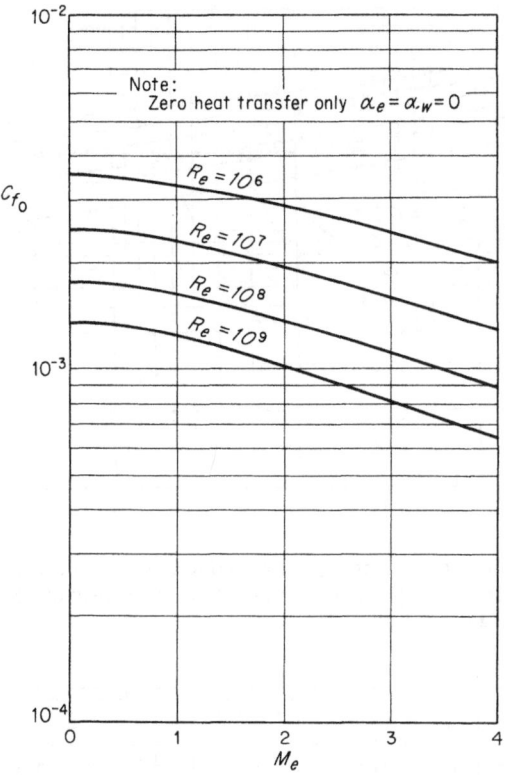

FIG. 7-6. Skin-friction coefficient C_{f_0} for zero heat transfer and zero dissociation of a local turbulent boundary layer versus local stream Mach number M_e.

All the results of the foregoing calculations can now be combined into the final equation for local heat-transfer rate. From Eqs. (7-49), (7-66), and (7-88)

$$-\dot{q}_w = \rho_e u_e P^{-2/3}(I_r - h_w)\frac{C_{f_0}}{2}\left[2\left(\frac{1 + \alpha_w}{1 + \alpha_e}\right)^{1/7} - 1\right]$$

$$\times \left[1 + (L^{2/3} - 1)\frac{h_c}{(I_r - h_w)}\right] \quad (7\text{-}89)$$

where
$$I_r = h_e + r\frac{u_e^2}{2} \quad (7\text{-}90)$$

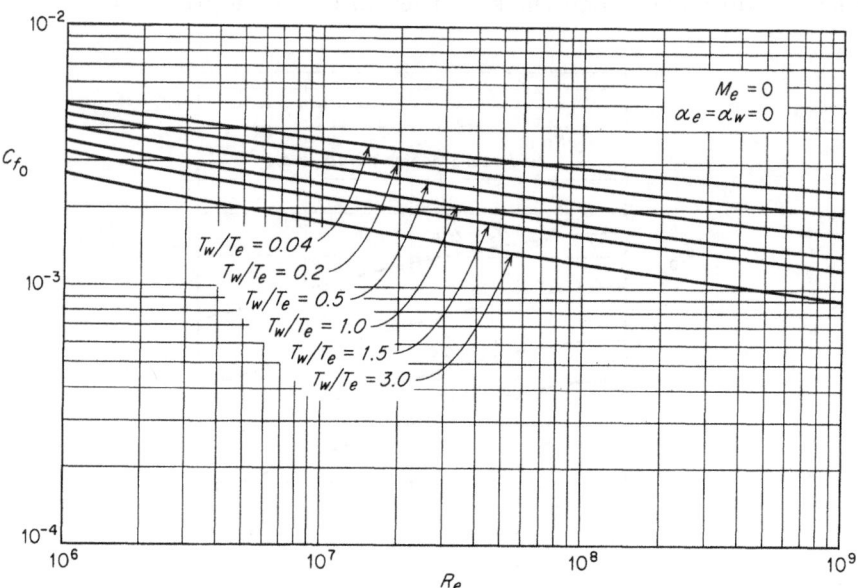

FIG. 7-7. Skin-friction coefficient C_{f_0} for zero dissociation of a turbulent boundary layer versus local stream Reynolds number R_e for a variety of temperature ratios and a local stream Mach number $M_e = 0$.

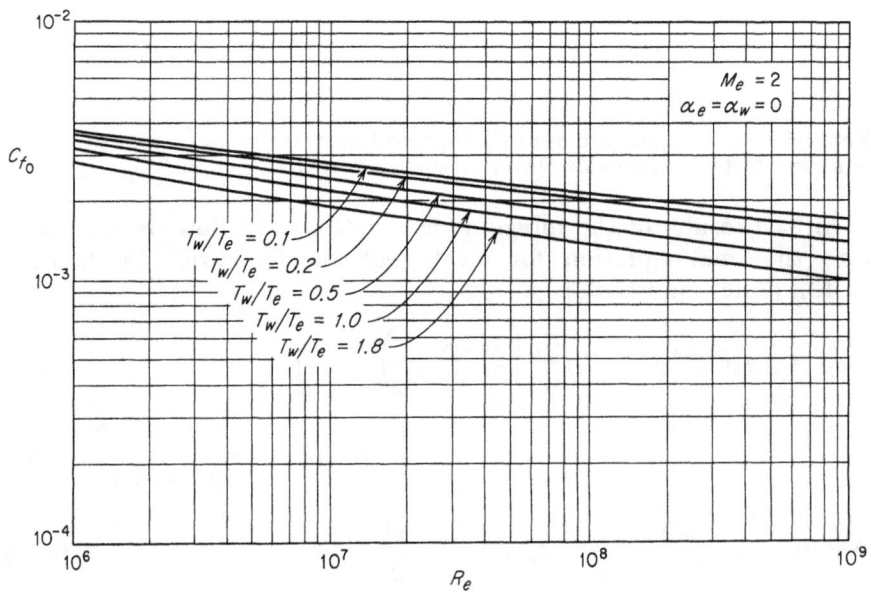

FIG. 7-8. Skin-friction coefficient C_{f_0} for zero dissociation of a turbulent boundary layer versus local stream Reynolds number R_e for a variety of temperature ratios and a local stream Mach number $M_e = 2$.

and $$r = P^{1/3} \qquad (7\text{-}66b)$$

$$h_c = (\alpha_e - \alpha_w)\frac{D}{2m_A} \qquad (7\text{-}91)$$

h_c as defined by Eq. (7-91) is strictly exact for a mixture of atoms and

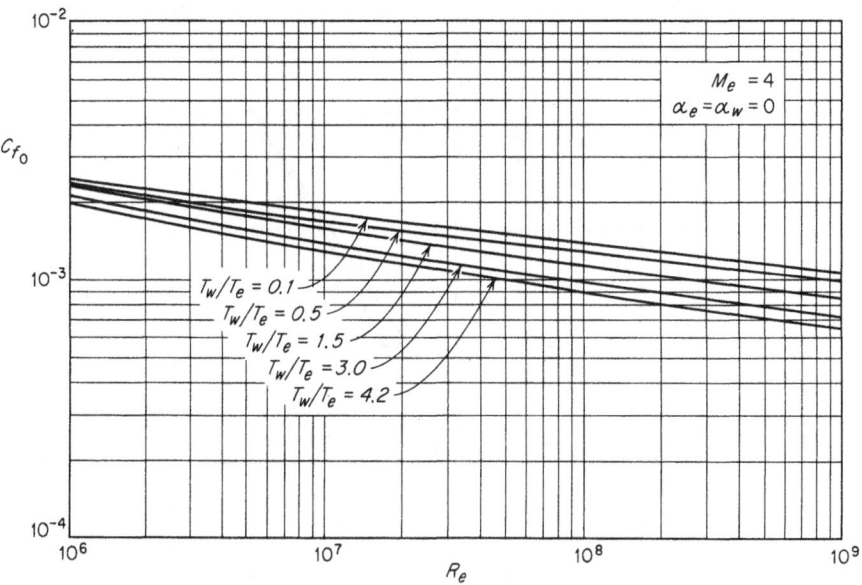

FIG. 7-9. Skin-friction coefficient C_{f_0} for zero dissociation of a turbulent boundary layer versus local stream Reynolds number R_e for a variety of temperature ratios and a local stream Mach number $M_e = 4$.

diatomic molecules of one species. For a mixture of two species such as N_2, N, O_2, and O such as for air, for example,

$$h_c = (\alpha_e - \alpha_w)\frac{\displaystyle\sum_{i=O,N}[(C_i)_e - (C_i)_w]\frac{D_i}{2m_i}}{\displaystyle\sum_{i=O,N}[(C_i)_e - (C_i)_w]} \qquad (7\text{-}92)$$

where α, as before, is the fraction by mass of the gas in the atomic state.

With respect to the assumption of using the frozen-flow concentration of atoms across the boundary layer, some conclusions can be drawn by considering Fig. 7-10. Figure 7-10 presents the diffusion-controlled distribution of atoms across a boundary layer for a typical set of external and surface conditions; that is, $p_e = 1$ atm; $\alpha_e = 0.3$; $T_e = 3600°K$; $\alpha_w = 0$; $T_w = 500°K$. Also shown is the equilibrium distribution of atoms determined by assuming that the boundary layer is locally in equilibrium throughout its extent. In both calculations the same external stream and surface conditions were assumed for a pure oxygen layer (for simplicity's sake). Equation (7-83) was used with the equilibrium constant given for oxygen dissociation to determine the equilibrium concentration of atoms through the layer.

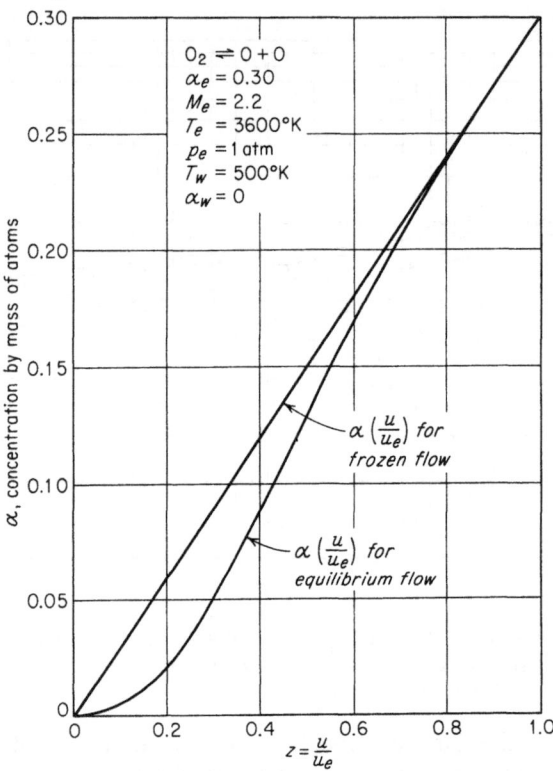

FIG. 7-10. The distribution of atom concentration through a dissociated turbulent boundary for a boundary layer in dissociation equilibrium and for a frozen boundary layer.

The intermediate case of finite reaction rates will fall near the two curves shown in Fig. 7-10. Thus, the use of Eq. (7-75) to determine the atom concentration is reasonable, since it is a great improvement over assuming that α is some constant value throughout the boundary layer. The calculation procedure for the equilibrium case involves solving Eq. (7-83), which gives $T = T(\alpha, u/u_e)$ simultaneously with the equilibrium-constant equation, which gives $\alpha = \alpha(T)$ at each value of u/u_e. The equilibrium-constant equation is

$$[K_p(T)]^{-1} = \frac{1 - \alpha^2}{\alpha^2} \frac{1}{4p_e} \qquad (7\text{-}93)$$

where K_p is given by Eq. (9-128) for oxygen dissociation. The above calculations were performed for a surface catalytic to recombination.

The concentration profiles given in Fig. 7-10 can be used in conjunction with Eq. (7-82) to obtain values of ρ/ρ_w through the boundary layer. It is this ratio which directly appears in the momentum integral, Eq. (7-67b). Figure 7-11 presents four curves

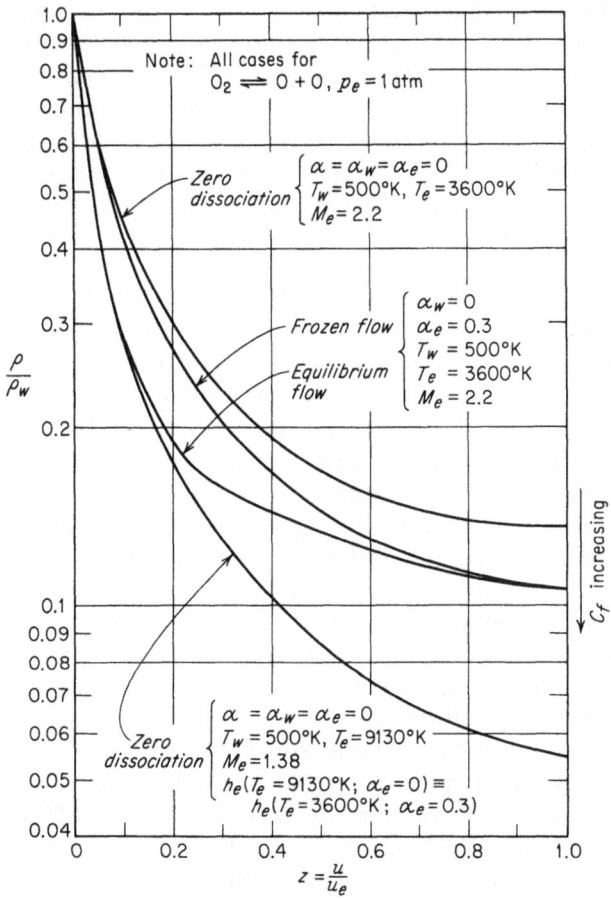

FIG. 7-11. Several density variations through a dissociated turbulent boundary layer.

of ρ/ρ_w versus u/u_e. The upper curve is calculated using Eqs. (7-82) and (7-83) assuming that $\alpha = \alpha_w = \alpha_e = 0$. It corresponds, then, to C_{f_0}. The next curve down is that appropriate to the frozen equilibrium flow case calculated using Eqs. (7-75), (7-82), and (7-83). The next lowest curve is that calculated for equilibrium flow using Eqs. (7-82), (7-83), and (7-93). The lowest curve is that calculated for the special case where the enthalpy and velocity at the wall and at the edge of the boundary layer are equal to the dissociated values with dissociation assumed to be zero. This last case corresponds to the curve assumed when C_f is evaluated as C_{f_0}, where the velocity and enthalpy at the edge of the boundary layer and at the wall are assumed as the same as when dissociation takes place. It is clear that, in view of the fact that most cases of interest will fall between the frozen-flow and dissociation equilibrium flow curves, use of the frozen-flow case for simplicity's sake is probably a reasonable simplification providing the surface is catalytic to recombination. This is further evidence that the heat transfer to the cool wall from a dissociated boundary layer is relatively insensitive to gas-phase reaction rates as long as the surface recombination rate is high so that the degree of dissociation at the surface is essentially equal to the equilibrium value there.

It is interesting to note that Kosterin and Koshmarov[1] have proposed a theory for calculating the skin friction for a dissociating turbulent boundary layer which is equivalent to using the equilibrium dissociation concentration profiles calculated using Eqs. (7-83) and (7-93) in the momentum integral equation (7-67b). According to the analysis just described, their results will not differ markedly from the present results except in so far as they may use different empirical constants $\phi(0)$ and K. In all other respects Kosterin's and Koshmarov's theory parallels the present treatment. No results of calculations were presented by Kosterin and Koshmarov.

As with any theory, the proof of validity rests with comparison with experimental results as well as with its internal consistency. The present analysis yields results which compare remarkably well with experiment as will be seen.

Figure 7-12 shows the comparison of C_{f_0} as calculated using the present theory with measurements of Schultz-Grunow,[2] Kempf,[3] Coles,[4] and Matting et al.[5] for the case of zero heat transfer and no dissociation. The excellent agreement between experiment and theory for these cases is quite remarkable.

[1] S. I. Kosterin and Yu. A. Koshmarov, *Intern. J. Heat and Mass Transfer*, vol. 1, no. 1, pp. 46–50, 1960.

[2] F. Schultz-Grunow, *NACA TM* 986, 1941.

[3] von Gunther Kempf, "Wietere Rei bungsergebnisse an ebenen glatten und Rauhen Flachen," Hydrodynamische Probleme der Schiffsantriebs, vol. 1, pp. 74–82, 1932.

[4] Donald Coles, *J. Aeronaut. Sci.*, vol. 21, no. 7, pp. 433–448, 1954.

[5] Fred W. Matting, Dean R. Chapman, Jack R. Nyholm, and Andrew G. Thomas, *Proc. 1959 Heat Transfer and Fluid Mech. Inst.*, pp. 80–93, Stanford University Press, Stanford, Calif., 1959.

When heat transfer is present, the agreement between experiment and theory is less precise but acceptable in view of the uncertainties in taking measurements with heat transfer present. Figure 7-13 presents the comparison with measurements of Winkler[1] wherein the agreement between theory and experiment gets worse as heat transfer increases (T_w/T_e decreases) but is within $+20$, -5 per cent in all cases.

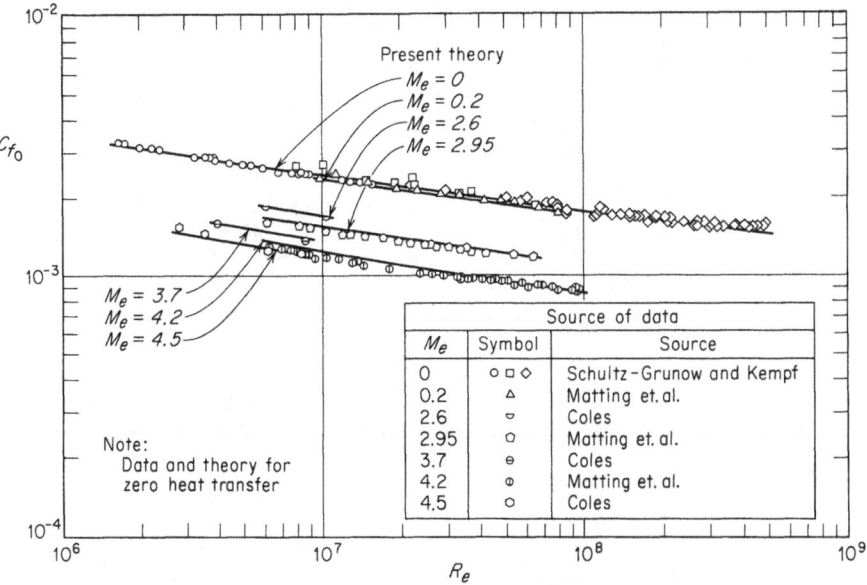

FIG. 7-12. A comparison of the theoretical nondissociated, zero-heat-transfer local skin-friction coefficient C_{f_0} given by Eq. (7-85) with measured values of nondissociated, zero-heat-transfer local skin-friction coefficient C_{f_0}.

Few data are available for heat transfer from turbulent boundary layers under high stagnation enthalpy conditions where dissociation is apt to occur. Data measured using a shock tube were reported by Rose et al.[2] and are used here for comparison with Eq. (7-89) slightly rearranged to give

$$\frac{Nu}{P} = C_H R_e = R_e \frac{C_{f_0}}{2} P^{-2/3} \left[2\left(\frac{1 + \alpha_e}{1 + \alpha_w}\right)^{1/7} - 1 \right]\left[1 + (L^{2/3} - 1)\frac{h_c}{I_r - h_w} \right]$$

$$(7\text{-}94)$$

[1] Eva M. Winkler, *Am. Rocket Soc.*, *Reprint* 856-859, American Rocket Society, New York, 1959.

[2] Peter H. Rose, Ronald F. Probstein, and Mac C. Adams, *J. Aerospace Sci.*, vol. 25, no. 12, pp. 751–760, 1958.

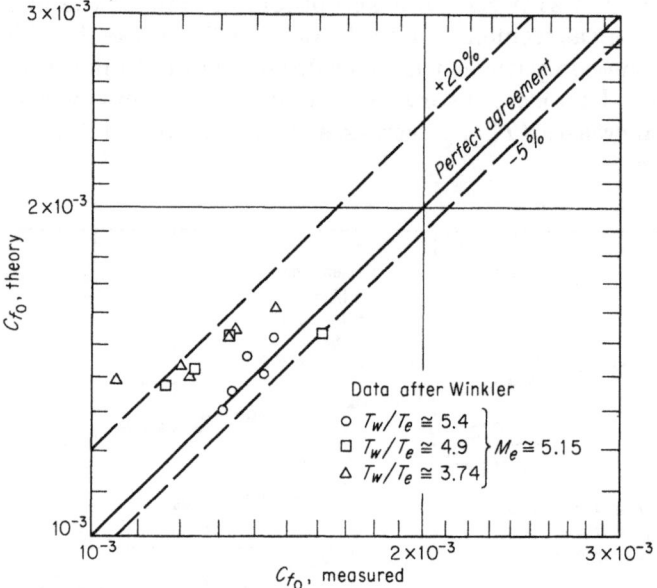

FIG. 7-13. A comparison of the theoretical, nondissociated skin-friction coefficient C_{f_0} with measured values of nondissociated local skin-friction coefficient C_{f_0}.

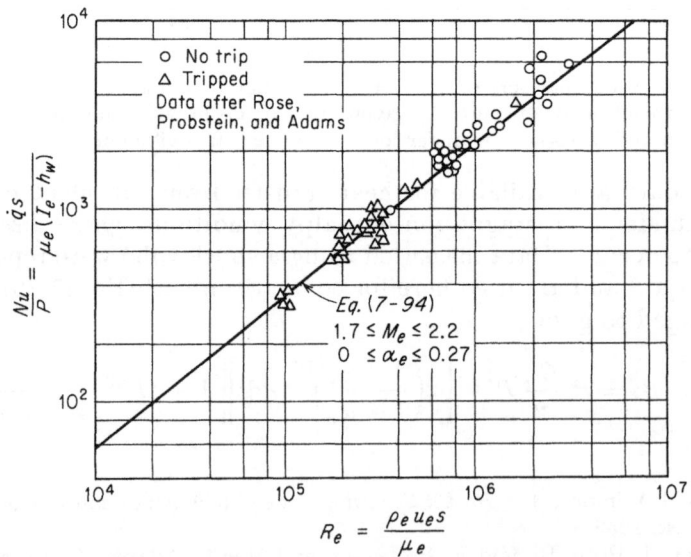

FIG. 7-14. A comparison of the theoretical heat-transfer-coefficient parameter given by Eq. (7-94) with measured values under conditions of high heat transfer and low or zero dissociation.

where C_{f_0} is interpolated using Figs. (7-7) to (7-9). Values of α_e, α_w, h_c, I_r, and h_w were calculated appropriate to the experimental or estimated values given by Rose et al. P was assumed equal to 0.72, and L equal to 1.40, reasonable values for these temperatures as the calculations for high-temperature air in equilibrium show.   See, for example, Fig. 10-11.

Figure 7-14 compares Eq. (7-94) with the data of Rose et al. Although the data scatter about the line obtained using Eq. (7-94), the agreement is satisfactory.   Note that in all cases there was considerable heat transfer and in some cases some dissociation present. ($0.14 \geq T_w/T_e \geq 0.04$; $0 \leq \alpha_e \leq 0.27$.)

7-7. Conclusions.   The following conclusions seem justified as a result of the work described in this chapter.

1. Local skin-friction coefficients for a flat-plate dissociating compressible turbulent boundary layer have been calculated using the theory of this chapter.   They agree excellently with measured values under conditions of zero heat transfer and zero dissociation and with decreasing accuracy as heat transfer is increased, although data are sparse in the latter case.

2. An equation for local heat transfer from a dissociating compressible turbulent boundary layer is derived which agrees well with measurements under conditions of strong heat transfer and low or zero dissociated boundary layer.

3. The analysis which developed the equations for both the skin-friction and heat-transfer coefficients is internally self-consistent and stems from solutions with justifiable approximations of the appropriate turbulent-boundary-layer equations.

In the next chapter we shall take up the effects of mass transfer, pressure gradient, and body shape and chemical reactions other than dissociation upon the compressible-turbulent-boundary-layer skin friction and heat transfer.

8

Mass Transfer and Chemical Reactions in the Turbulent Boundary Layer

8-1. Introduction. In Chap. 7 there was presented the derivation of the turbulent-boundary-layer equations for a reacting gas boundary layer. The equations were used to obtain expressions for the heat transfer from a dissociating turbulent boundary layer. It remains to treat a more complex situation, that of a reacting turbulent boundary layer in the presence of mass transfer. This more complex situation will be treated in this chapter.

Sections 8-2 and 8-3 will be concerned solely with the effects of mass transfer upon the skin friction and heat transfer of a non-reacting, compressible, turbulent boundary layer. Following this, we shall add to the complexity of the theory by accounting for the effects of chemical reactions along with mass-transfer effects. The chemical reactions accounted for by the theory include those involving external stream species only and reactions between external stream species and those species entering the boundary layer by means of mass transfer at the surface without regard to the mechanism of mass transfer. The results obtained will be appropriate to the flow over a two-dimensional flat plate in the absence of a pressure gradient.

The results of the turbulent-boundary-layer theory developed in Chap. 7 and the present chapter would suffer severely in their practicality if they were to be limited to application to the boundary-layer flow over flat-plate surfaces. Section 8-7 was prepared with this thought in mind. In Sec. 8-7 analyses are presented of the effects of pressure gradient and body shape upon turbulent-boundary-layer characteristics and equations are derived which can be used with the flat-plate equations of Chap. 7 and the present chapter in order to apply these equations to right circular cones and other axisymmetric bodies.

The turbulent-boundary-layer theory as presented here is

provisional at best, being limited by certain semiempirical aspects. However, it has yielded equations for skin friction and heat transfer which compare well with measurements and so offers some encouragement to those who wish to apply the theory to problems for which experimental data are not yet available. As always with this theory, such applications must be made with caution, since the theory is not completely derivable from first principles and depends upon empirically determined constants. In this respect it does not differ from any other turbulent-boundary-layer theory, however. The development of a theory for the turbulent boundary layer from first principles remains one of the major unsolved problems of gas dynamics.

8-2. Mass-transfer Effects upon $2C_H/C_f$ **and** r. This section and the next one are concerned with the effect upon skin friction and heat transfer of introducing mass into the turbulent boundary layer at the interface of the gas layer and the liquid or solid undersurface. We shall assume that the mass being introduced is a gas with the same composition as the boundary-layer gas, so that diffusion mass transfer is not important. Diffusion mass transfer will be included in Sec. 8-4, which treats the completely general case. Furthermore, we shall assume that no chemical reactions are occurring. As with the laminar-boundary-layer case taken up in Chap. 5, we anticipate that blowing will reduce heat transfer and skin friction and that sucking will have the opposite effect. Our results will bear out intuition as they did in the case of the laminar boundary layer.

Our analysis of the effects of mass transfer alone will be divided into two parts. In this section we shall determine a relationship between heat-transfer and skin-friction coefficient which depends upon the mass-transfer parameter B_5, and in the next section we shall determine the effect of mass transfer upon skin-friction coefficient. The results of the two parts can then be combined to determine the effect of mass transfer upon heat transfer to the undersurface.

Effect on $2C_H/C_f$ *and* r. Equations (7-29) and (7-31) represent our point of departure in the present case. We have, from Eq. (7-29),

$$\rho u \frac{\partial u}{\partial s} + \rho v \frac{\partial u}{\partial y} = \frac{\partial}{\partial y}\left[(\mu + \epsilon)\frac{\partial u}{\partial y}\right] = \frac{\partial \tau}{\partial y} \tag{8-1}$$

and, from Eq. (7-31) with $\partial C_i/\partial y = 0$, since we assume that the mass being introduced has the same composition as the gas layer,

$$\rho u \frac{\partial I}{\partial s} + \rho v \frac{\partial I}{\partial y} = \frac{\partial}{\partial y}\left\{\left(\frac{\mu}{P} + \frac{\epsilon}{P_T}\right)\frac{\partial I}{\partial y}\right.$$
$$\left. + \left[\mu\left(1 - \frac{1}{P}\right) + \epsilon\left(1 - \frac{1}{P_T}\right)\right]\frac{1}{2}\frac{\partial u^2}{\partial y}\right\} = \frac{\partial G}{\partial y} \tag{8-2}$$

where G is thus defined by Eq. (8-2)

Boundary conditions for the above equations are

at $y = 0$: $u = 0$ $\rho v = (\rho v)_w$ $\tau = \tau_w$ $h = h_w$ $\epsilon = 0$

at $y = \infty$: $u = u_e$ $\rho v = 0$ $\tau = 0$ $h = h_e$

We shall not attempt to solve these equations as they stand. Rather, in keeping with the provisional nature of the theory in general, we shall attempt to obtain approximate solutions the usefulness of which will depend upon their accuracy when compared with experimental results. To proceed, we assume in our thin boundary layer that for any variable Q

$$\left|\frac{\partial Q}{\partial y}\right| \gg \left|\frac{\partial Q}{\partial s}\right|$$

and

$$\rho v = \text{const} = (\rho v)_w$$

In making the latter assumption, we ignore the boundary conditions on ρv at $y = \infty$. Then Eqs. (8-1) and (8-2) become

$$\frac{d}{dy}\left[(\rho v)_w u - \tau\right] = 0$$

and

$$\frac{d}{dy}\left[(\rho v)_w I - G\right] = 0$$

whence, using the boundary conditions at $y = 0$, we obtain

$$(\rho v)_w u - \tau = -\tau_w \tag{8-3}$$

and

$$(\rho v)_w I - G = (\rho v)_w I_w - G_w \tag{8-4}$$

Now I is defined as

$$I = h + \frac{u^2}{2} \tag{8-5}$$

and at the wall $u = 0$ and $\epsilon = 0$; hence we can write

$$\left(\frac{dI}{dy}\right)_w = \left(\frac{dh}{dy} + u\frac{du}{dy}\right)_w = \left(\frac{dh}{dy}\right)_w$$

and so

$$G_w = \left(\frac{\mu}{P}\frac{dh}{dy}\right)_w = -\dot{q}_w \tag{8-6}$$

Thus, using Eq. (8-6) in Eq. (8-4) and dividing Eq. (8-4) by Eq. (8-3), where Eq. (8-5) has been used with the definition of G to express I in terms of h and u, we obtain the differential equation relating h to u.

$$\frac{dh}{du} - \frac{\mu + \epsilon}{(\mu/P) + (\epsilon/P_T)}\frac{(\rho v)_w h}{(\rho v)_w u + \tau_w}$$

$$= -\frac{\mu + \epsilon}{(\mu/P) + (\epsilon/P_T)}\left\{u + \frac{(\rho v)_w[h_w - (u^2/2)] + \dot{q}_w}{\tau_w + (\rho v)_w u}\right\} \tag{8-7}$$

Equation (8-7) can be written in two parts applicable to the laminar sublayer where $\epsilon \ll \mu$ and to the turbulent layer where $\epsilon \gg \mu$ in keeping with our two-layer model used in Chap. 7. There results

$$\frac{dh}{du} - \frac{P_i(\rho v)_w}{(\rho v)_w u + \tau_w} h = \frac{P_i(\rho v)_w[(u^2/2) - h_w] - P_i q_w}{\tau_w + (\rho v)_w u} - P_i u \qquad (8\text{-}8)$$

where the index i on P is absent in the laminar sublayer and equal to T in the turbulent layer. Equation (8-8) can be integrated as follows to obtain $h(u)$ in closed form. Let

$$h = \lambda$$

$$\bar{P}(u) = \frac{-P_i(\rho v)_w}{(\rho v)_w u + \tau_w}$$

$$Q(u) = \frac{P_i(\rho v)_w[(u^2/2) - h_w] - P_i \dot{q}_w}{\tau_w + (\rho v)_w u} - P_i u$$

Then Eq. (8-8) becomes

$$\frac{d\lambda}{du} + \bar{P}(u)\lambda = Q(u) \qquad (8\text{-}9)$$

the solution of which is

$$\lambda = \frac{1}{R}\left(\int RQ\,du + C_3\right) \qquad (8\text{-}10)$$

where

$$R = \exp\left(\int \bar{P}\,du\right) \qquad (8\text{-}11)$$

and

$$C_3 = \text{const of integration}$$

Obtain the solution to Eq. (8-8) in the laminar sublayer by making use of Eq. (8-10), where C_3 is eliminated by integrating between the definite limits $u = 0$ and $u = u_L$. Obtain the solution to Eq. (8-8) in the turbulent layer by making use of Eq. (8-10), where C_3 is eliminated by integrating between the definite limits $u = u_L$ and $u = u_e$. Eliminate h_L between the resulting two equations, and solve for $2C_H/C_f$. There results, for $P_T = 1$,

$$\frac{2C_H}{C_f} = \frac{B_5}{(1 + B_5)[1 + B_5(u_L/u_e)]^{P-1} - 1} \qquad (8\text{-}12)$$

and

$$r = \frac{1 + B_5}{B_5^2}\left\{\frac{1 + B_5(u_L/u_e)}{2 - P}\left[P - 2\left(1 + B_5 \frac{u_L}{u_e}\right)^{P-2}\right]\right.$$

$$\left. + \left[\frac{1}{1 + B_5} - B_5\left(\frac{u_L}{u_e} - 1\right)\right]\right\} \qquad (8\text{-}13)$$

where

$$B_5 = \frac{(\rho v)_w u_e}{\tau_w} = \frac{(\rho v)_w}{\rho_e u_e (C_f/2)} \qquad (8\text{-}14)$$

and
$$r = \frac{2}{u_e^2}(I_r - h_e) \tag{8-15}$$

When $B_5 \to 0$, Eqs. (8-12) and (8-13) reduce to

$$\frac{2C_{H_0}}{C_{f_0}} = \frac{1}{1 + (P-1)(u_L/u_e)} \tag{8-16}$$

and
$$r_0 = 1 + (P-1)\left(\frac{u_L}{u_e}\right)^2 \tag{8-17}$$

which are in complete agreement with Eqs. (7-56a) and (7-56c) for zero mass transfer when $P_T = 1$ and $\alpha = 0$ in those equations corresponding to the case where no mass transfer or dissociation occurs. In this section the subscript 0 denotes zero mass-transfer values.

Equation (8-12) gives an explicit relation between C_H and C_f which depends upon mass-transfer parameter B_5 and on u_L/u_e. Equation (8-13) gives the dependency of recovery factor upon the mass-transfer parameter B_5 and u_L/u_e. The importance of the sublayer velocity u_L becomes apparent. These equations were first obtained by Rubesin.[1]

Figure 8-1 presents $2C_H/C_f$ plotted versus B_5 for different values of u_L/u_e. Figure 8-2 presents r plotted versus B_5 for different values of u_L/u_e. These figures can also be used in conjunction with equations to be derived in the next section as will be shown.

The Velocity Ratio u_L/u_e. Some question remains as to the value of u_L/u_e to use when Eqs. (8-12) and (8-13) are applied. In general we can state that u_L/u_e is bounded by the inequalities

$$0 \leq \frac{u_L}{u_e} \leq 1$$

and that experimental results indicate that it is relatively insensitive to variations in Reynolds number and Mach number *when mass transfer is absent.* This latter statement is verified by recognizing that, over a wide range of Mach numbers and Reynolds numbers, the turbulent-boundary-layer recovery factor r_0 is relatively constant and equal to about 0.88. Thus, from Eq. (8-17) for $P = 0.72$,

$$0.28\left(\frac{u_L}{u_e}\right)^2 = 1 - 0.88 = 0.12$$

or
$$\frac{u_L}{u_e} = 0.65 \tag{8-18}$$

[1] Morris W. Rubesin, *NACA TN* 3341, 1954.

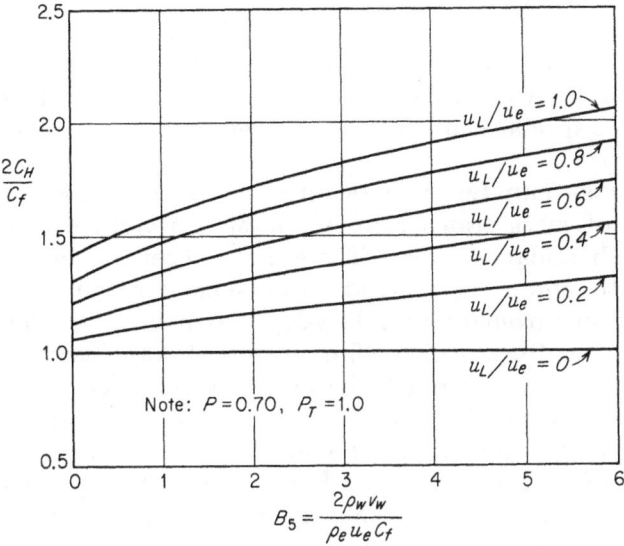

FIG. 8-1. $2C_H/C_f$ versus blowing parameter B_5 calculated using Eq. (8-12) for $P = 0.70$ and $P_T = 1.0$.

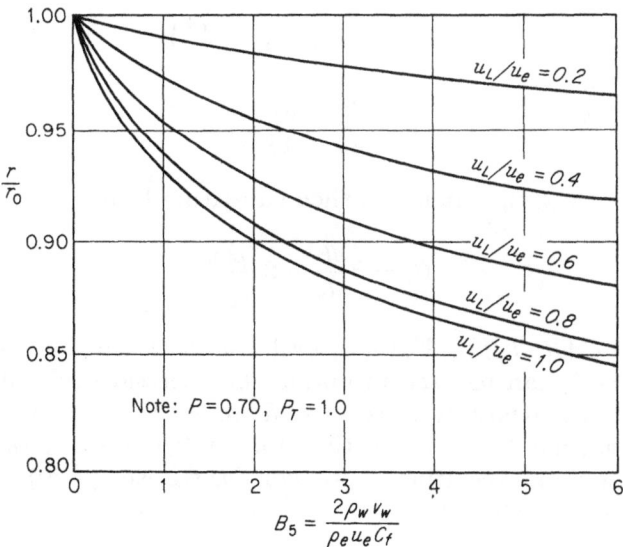

FIG. 8-2. Variation of recovery factor r with blowing parameter B_5 calculated using Eq. (8-13) for $P = 0.70$ and $P_T = 1.0$.

and this appears to be the proper magnitude from examination of measured velocity profiles for a turbulent boundary layer. However, because of the idealization of the two-layer model used here, and because it would appear that u_L/u_e should vary with heat transfer (and probably with mass transfer), we shall resist accepting Eq. (8-18) as the specification of u_L/u_e and retain this ratio as a parameter of the theory.

8-3. Mass-transfer Effects on C_f. We turn now to evaluating the effect of mass transfer upon the skin-friction coefficient for a turbulent boundary layer. There have been several interesting attempts to obtain theoretical expressions for the effect of mass transfer upon turbulent-boundary-layer skin friction. The simplest approach, sometimes called "film theory,"[1] was described in detail in Sec. 3-5 and was found to be unsatisfactory because it made no distinction between the laminar and turbulent boundary layer and neglected the thickening of the boundary layer with mass addition.

Dorrance and Dore[2] used the Prandtl mixing-length concept with the degenerate form of the momentum equation given as Eq. (8-19) to obtain a velocity distribution law which was then used with the appropriate form of the von Kármán momentum integral to obtain a local skin-friction law which depended upon M_e, T_w/T_e, R_e, and $(\rho v)_w$. This procedure consists in using

$$\tau = \rho K^2 y^2 \left(\frac{du}{dy}\right)^2$$

with

$$(\rho v)_w \frac{du}{dy} = \frac{d\tau}{dy} \qquad (8\text{-}19)$$

to obtain $u = u(y)$, which was then substituted into

$$C_f = 2\frac{d\theta}{ds} - 2\frac{(\rho v)_w}{\rho_e u_e} \qquad (8\text{-}20)$$

to obtain $C_f[R_e, T_w/T_e, M_e, (\rho v)_w/\rho_e u_e]$. θ is given by Eq. (7-67b), and Eq. (7-32) can be used to obtain the necessary relation between ρ and u. Two arbitrary constants [K above and a constant of integration resulting from integrating Eq. (8-19)] were determined by assuming that the resulting expression for C_f, Eq. (8-20), must agree with experiment when $M_e = 0$, $T_w = T_e$, and $(\rho v)_w = 0$. In essence,

---

[1] H. S. Mickley, R. C. Ross, A. L. Squyers, and W. E. Stewart, *NACA TN* 3208, 1954.

[2] W. H. Dorrance and F. J. Dore, *J. Aeronaut. Sci.*, vol. 21, no. 6, pp. 404–410, 1954.

this theory assumed that a turbulent velocity profile could be applied across the entire boundary layer and that any error involved in neglecting the exact velocity distribution in the laminar sublayer when evaluating the integral for θ would be compensated for when the arbitrary constants were determined by matching Eq. (8-20) with experiment. It has been shown that Dorrance's and Dore's theory does not agree quantitatively with experiment except at $M_e = 0$, where agreement might reasonably be expected, since it was at $M_e = 0$ that the arbitrary constants were evaluated.[1]

Rubesin[2] presented a theory which agreed with that of Dorrance and Dore in most details except for the method of evaluating the arbitrary constants. Rubesin used an expression for the velocity distribution in the laminar sublayer which identified one of the arbitrary constants with the thickness of the laminar sublayer. He then considered three alternative values for y_L, finally choosing one which agreed best with the $M_e = 0$ data of Mickley et al.[3] He chose to take $u_L = 13.1[(C_f/2)(T_w/T_e)]^{1/2}$ and to determine y_L by integrating Eq. (8-1) across the laminar sublayer, assuming constant properties in the sublayer. Rubesin's theory agrees best with experiment[2] at $M_e = 0$ as might be expected, again, since it was at $M_e = 0$ that his empirical constants were determined. The results of Dorrance's and Dore's theory and Rubesin's theory do not differ appreciably at any Mach number and agree with experiment best at $M_e = 0$ as might be anticipated from a detailed examination of their basic assumptions. One may conclude that these approaches could be improved at supersonic Mach numbers if velocity-distribution measurements in turbulent boundary layers were available under conditions of supersonic external stream Mach numbers and various conditions of heat and mass transfer. The numerical computations using these theories are tedious, and the work involved is not commensurate with the accuracy one obtains.

Effect of Mass Transfer on C_f. In this section we take the position that the rather complex theories described above meet with only fair success in matching experiment and hence one is justified in seeking a more direct approach to the problem of determining the effect of mass transfer upon turbulent-boundary-layer skin friction. In proceeding, we shall take note of the observation that the turbulent outer portion of the incompressible turbulent boundary layer, when plotted as u versus log y, has a slope which is insensitive to blowing

[1] Constantine C. Pappas and Arthur F. Okuno, *J. Aerospace Sci.*, vol. 27, no. 5, pp. 321–333, 1960.

[2] Rubesin, *op. cit.*

[3] Mickley, Ross, Squyers, and Stewart, *op. cit.*

rate. Leadon[1] plotted the data of Mickley and Davis[2] taken at $M_e = 0$ in the form of u versus $\log y$ in order to draw this conclusion. We shall make use of this observation to show that it indicates that the shear in the turbulent portion of the boundary layer is equal to that present in the absence of mass transfer if none of the mass entering the boundary layer at the surface reaches that portion of the turbulent core of the boundary layer which exhibits the above-mentioned characteristic. These conclusions can then be used to arrive at the effect of mass transfer upon turbulent-boundary-layer skin friction for low rates of mass transfer. We proceed as follows: Assume that

$$\left| \frac{\partial Q}{\partial s} \right| \ll \left| \frac{\partial Q}{\partial y} \right|$$

and that, throughout the sublayer,

$$\mu = \mu_w$$

and

$$(\rho v) \simeq (\rho v)_w$$

in Eq. (8-1) as applied to the laminar sublayer. Then Eq. (8-1) becomes

$$\frac{d}{dy} \left[(\rho v)_w u - \mu_w \frac{du}{dy} \right] = 0 \tag{8-21}$$

whence, integrating Eq. (8-21),

$$\tau_w + (\rho v)_w u = \mu_w \frac{du}{dy} \tag{8-22}$$

since $\tau = \mu_w (du/dy)_w = \tau_w$ at $u = 0$. In the turbulent outer layer assume that

$$(\rho v) \simeq (\rho v)_L$$

$$\epsilon \gg \mu$$

and

$$\tau = \epsilon \frac{du}{dy} = \rho K^2 y^2 \left(\frac{du}{dy} \right)^2$$

Then Eq. (8-1) becomes for the turbulent boundary layer

$$\frac{d}{dy} \left[(\rho v)_L u - \rho K^2 y^2 \left(\frac{du}{dy} \right)^2 \right] = 0 \tag{8-23}$$

[1] Bernard M. Leadon, *Am. Rocket Soc. Reprint* 857-59, paper presented at the June, 1959, Semiannual meeting, American Rocket Society, New York, N.Y.
[2] H. S. Mickley and R. S. Davis, *NACA TN* 4017, 1957.

whence, using the boundary conditions at the sublayer edge that

$$u = u_L$$

$$\rho = \rho_L = \text{const}$$

$$y = y_L$$

and, from Eq. (8-22), the additional condition that

$$\left(\frac{du}{dy}\right)_L = \frac{\tau_w + (\rho v)_w u_L}{\mu_w} \tag{8-24}$$

then Eq. (8-23) can be integrated to obtain, with the help of Eq. (8-24),

$$(\rho v)_L(u - u_L) + \rho_L K^2 y_L^2 \left[\frac{\tau_w + (\rho v)_w u_L}{\mu_w}\right]^2 = \rho K^2 y^2 \left(\frac{du}{dy}\right)^2$$

or

$$\frac{du}{d(\log y)} = \frac{1}{K} A_1^{1/2} \tag{8-25a}$$

where

$$A_1 = \rho^{-1}\left\{(\rho v)_L(u - u_L) + \rho_L K^2 y_L^2 \left[\frac{\tau_w + (\rho v)_w u_L}{\mu_w}\right]^2\right\} \tag{8-25b}$$

Now, if $du/d(\log y) = \text{const}$ and is independent of u and $(\rho v)_w$, then we conclude according to Eqs. (8-25a) and (8-25b) that a possible explanation is that

1.
$$(\rho v)_L = 0 \tag{8-26}$$

that is, none of the mass introduced at the wall enters the region where $du/d(\log y) = \text{const}$.

2.
$$\tau_w + (\rho v)_w u_L = (\tau_w)_0 \tag{8-27}$$

a constant independent of $(\rho v)_w$, where $\tau_w = (\tau_w)_0$ when $(\rho v)_w = 0$.

Equation (8-27) is our desired expression for the effect of mass transfer upon skin friction.   That is,

$$\frac{\tau_w}{(\tau_w)_0} = \frac{C_f}{C_{f_0}} = \frac{1}{1 + B_5(u_L/u_e)} \tag{8-28}$$

where, as before,

$$B_5 = \frac{(\rho v)_w u_e}{\tau_w} = \frac{2(\rho v)_w}{\rho_e u_e C_f} \tag{8-14}$$

An alternative to Eq. (8-28) can be obtained from Eq. (8-27) and is, for $B_1 \to 0$,

$$\frac{\tau_w}{(\tau_w)_0} = 1 - B_1 \frac{u_L}{u_e} \tag{8-29}$$

where, as defined in Chap. 3, Eq. (3-56),

$$B_1 = \frac{(\rho v)_w u_e}{(\tau_w)_0} = \frac{2(\rho v)_w}{\rho_e u_e C_{f_0}} \tag{8-30}$$

Again, the parameter u_L/u_e appears and remains unknown except that we know that $0 \leq u_L/u_e \leq 1$ and, when $(\rho v)_w = 0$, $u_L/u_e \simeq 0.65$.

The Velocity Ratio u_L/u_e. It is not clear, a priori, that u_L/u_e as used in Eqs. (8-28) and (8-29) above is identical with the laminar sublayer velocity ratio. Rather, as the preceding development shows, u_L/u_e is the velocity ratio for the velocity at that point in the boundary layer above which the slope of the curve of u versus log y is unaffected by mass transfer. Clearly, for large blowing rates it is possible for some of the fluid introduced at the wall to enter the turbulent core of the boundary layer and thus compromise the above analysis. It would be fortunate indeed if the ratio u_L/u_e turned out to be the velocity ratio at the edge of the sublayer when large blowing rates are considered.

It is possible to determine the velocity ratio u_L/u_e from experimental measurements of velocity through a turbulent boundary layer under various blowing conditions. At the time of writing this book such data were available for $M_e = 0$ only, the data of Mickley and Davis. These data can be used to determine u_L/u_e for the case of $M_e = 0$ and $T_w/T_e = 1$. In proceeding, we take note of the work of Turcotte,[1] who pointed out that among the dimensionless velocity ratios which can be used in correlating such velocity data are

$$u^* = \frac{u}{U_\tau} \tag{8-31}$$

$$v_w^* = \frac{v_w}{U_\tau} \tag{8-32}$$

where
$$\rho_w U_\tau^2 = \tau_{w_0} = \frac{C_{f_0}}{2} \rho_e u_e^2 \tag{8-33}$$

Thus Eq. (8-29) becomes

$$\frac{C_f}{C_{f_0}} = 1 - v_w^* u_L^* \tag{8-34}$$

and when C_f/C_{f_0} is plotted versus v_w^*, the slope of the resulting curve (hopefully a straight line) will be u_L^*. Using the data of Mickley and Davis for $M_e = 0$ and $\rho_w = \rho_e = \rho$, one can plot C_f/C_{f_0} versus v_w^*, and the results are correlated by Eq. (8-34) when

$$u_L^* = 12.4 \tag{8-35}$$

providing values of v_w^* are restricted to the range

$$0 \leq v_w^* \leq 0.031$$

[1] Donald L. Turcotte, *J. Aerospace Sci.*, vol. 27, pp. 675–678, 1960.

For the values of v_w^* beyond 0.031, the data of Mickley and Davis depart from the simple linear expression for C_f/C_{f_0} given by Eq. (8-34) presumably owing to penetration of the mass introduced at the surface into the turbulent core of the boundary layer and the consequent breakdown of the model used in this analysis.

The value for u_L^* given by Eq. (8-35) is quite close to the value of u^* at the laminar sublayer edge when no mass transfer is present. Most investigators accept this latter value of u^* as being between 11.5 and 13.5.   According to Eqs. (8-31), (8-33), and (8-35)

$$\frac{u_L}{u_e} = 12.4\left(\frac{\rho_e}{\rho_w}\frac{C_{f_0}}{2}\right)^{1/2} \tag{8-36}$$

at least for $M_e = 0$.   It appears that for blowing rates such that $v_w^* \leq 0.031$ the value of u_L/u_e to be used with the model developed here is close to the sublayer velocity ratio in the absence of blowing, at least at $M_e = 0$.   Note that the expected weak dependence of u_L/u_e upon Reynolds number is borne out by Eq. (8-36), which suggests that u_L/u_e is proportional to $(R_e)^{-1/10}$, since C_{f_0} is roughly proportional to $(R_e)^{-1/5}$.

Comparison with Experiment.   At the time of writing this book there were no experimental data available where C_H, C_f, and r were all measured during the same experiment at supersonic Mach numbers for turbulent-boundary-layer flow under mass-transfer conditions. However, two separate experiments were reported having the same external flow Mach number of 3.20 which provide such data.   Pappas and Okuno[1] report measurements of C_F on a cone for an external conical flow Mach number of 3.20.   Bartle and Leadon[2] report measurements of r and C_H for flow over a flat plate for Mach number 3.20 external flow.   Accepting the assumption that $C_f/C_{f_0} = C_F/C_{F_0}$ and that these ratios for a cone at a given Mach number are the same as for a plate at the same Mach number, a complete set of experimental data at one supersonic Mach number, $M_e = 3.20$, is thus available for comparison with our theory as presented here.

Since, at best, the analysis of this section must be considered provisional, let us concern ourselves with asymptotic values of C_H/C_{H_0} and r/r_0 for small values of the blowing parameters.   That is, for B_1 small,

$$\frac{C_f}{C_{f_0}} \simeq 1 - \frac{u_L}{u_e} B_1 \tag{8-29}$$

[1] Pappas and Okuno, *op. cit.*

[2] E. Roy Bartle and Bernard M. Leadon, *J. Aerospace Sci.*, vol. 27, no. 1, pp. 78–80, 1960.

and making use of Eqs. (8-12), (8-16), and (8-28) for $B_5 \to 0$, we obtain

$$\frac{C_H}{C_{H_0}} \simeq \frac{1}{1 + B_5(u_L/u_e)} \tag{8-37}$$

However, from Eqs. (8-14), (8-16), and (8-28) and the definition for B_2 given below,

$$B_5 = B_2 \left\{ 1 + [(P-1) - B_2]\frac{u_L}{u_e} \right\}^{-1} \tag{8-38}$$

where
$$B_2 = \frac{(\rho v)_w}{\rho_e u_e C_{H_0}}$$

We can obtain from Eqs. (8-37) and (8-38) for B_2 small

$$\frac{C_H}{C_{H_0}} \simeq 1 - \frac{B_2(u_L/u_e)}{1 + (P-1)u_L/u_e} \tag{8-39}$$

From Eqs. (8-13), (8-17), and (8-38) for B_2 small, we find that

$$\frac{r}{r_0} \simeq 1 - \left\{ \frac{(1-P)(u_L/u_e)^2[1 - (3-P)/3(u_L/u_e)]}{[1 - (1-P)(u_L/u_e)^2][1 - (1-P)(u_L/u_e)]} \right\} B_2 \tag{8-40}$$

In using Eqs. (8-29), (8-39), and (8-40) to calculate values to compare with the data of Pappas and Okuno and Bartle and Leadon, one might begin by assuming that $P = 0.72$, as is reasonable for air, and use the value of u_L/u_e as determined by Eq. (8-36). When this is done, it is immediately found that poor agreement with the experiment results. The alternative was followed of matching Eq. (8-39) with the data of Bartle and Leadon to determine the value of u_L/u_e to use then with Eqs. (8-29) and (8-40). This latter procedure provides a check on the mutual consistency of these three equations but appears to suggest that Eq. (8-36) gives improper values of u_L/u_e at supersonic Mach numbers. We find that, if $u_L/u_e = 0.305$ in Eqs. (8-29), (8-39), and (8-40), these equations compare with measurements favorably, as Figs. 8-3, 8-4, and 8-5 demonstrate. These equations are strictly valid for $B_1 \to 0$ and $B_2 \to 0$, and it is in the vicinity of $B_1 = 0$ and $B_2 = 0$ that the sought-for agreement is approached. One can conclude that the Eqs. (8-24), (8-39), and (8-40) are consistent with one another as are the two sets of data but that Eq. (8-36) fails to give the proper value of u_L/u_e to use in these equations.

Some improvement over using Eq. (8-36) to predict the value of u_L/u_e results if an unspecified reference density ρ' is used in defining U_τ in Eq. (8-33) where $\rho' > \rho_w$. That is, let

$$\rho' U_\tau^2 = \tau_{w_0} = \frac{C_{f_0}}{2} \rho_e u_e^2 \tag{8-41}$$

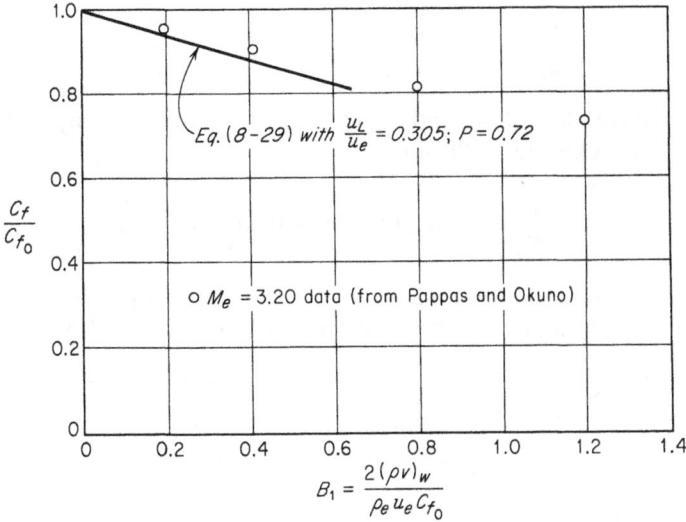

Fig. 8-3. Comparison of Eq. (8-29) with results of measurements of average skin friction on a cone for $M_e = 3.20$.

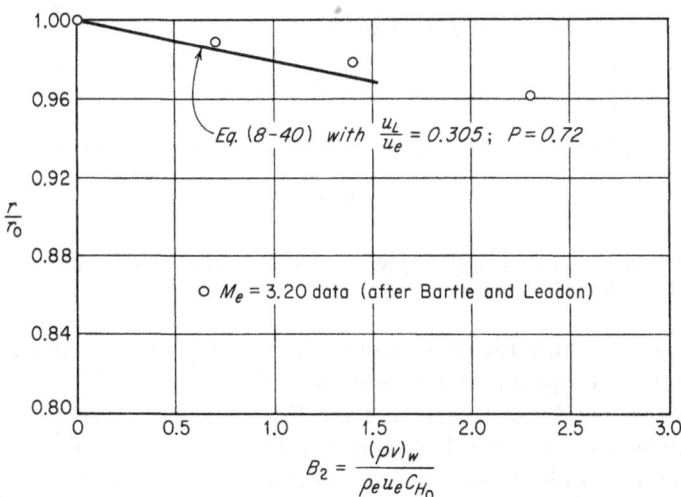

Fig. 8-4. Comparison of Eq. (8-40) with measured recovery factors on a flat plate for $M_e = 3.20$.

Then by Eqs. (8-31), (8-35), and (8-41), using 11.5 in Eq. (8-35),

$$\frac{u_L}{u_e} = 11.5\left(\frac{\rho_e}{\rho'}\frac{C_{f_0}}{2}\right)^{1/2} = 11.5\left(\frac{T'}{T_e}\frac{C_{f_0}}{2}\right)^{1/2} \tag{8-42}$$

and T' might be identified with the "reference temperature" of Eckert,[1] for example. This is

$$\frac{T'}{T_e} = 1 + 0.038M_e^2 + 0.50\left(\frac{T_w}{T_e} - 1\right) \tag{8-43}$$

If Eq. (8-42) is used instead of Eq. (8-36) to determine a value of u_L/u_e for $M_e = 3.20$, $T_w/T_e = 2.85$, and $R_x \simeq 6 \times 10^6$, we obtain

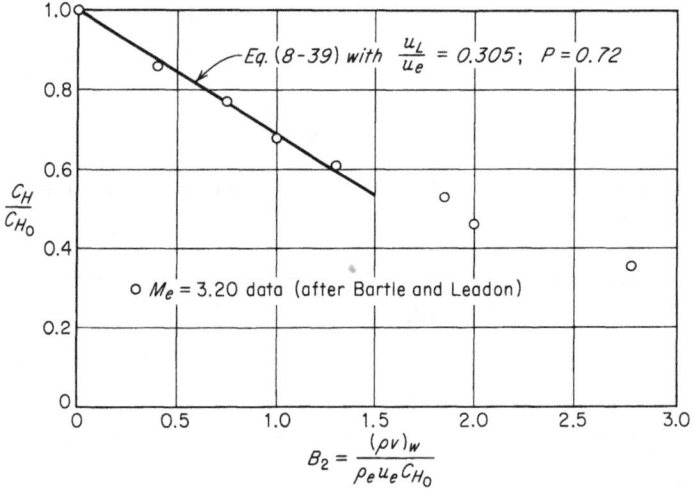

FIG. 8-5. Comparison of Eq. (8-39) with measured local heat-transfer coefficients on a flat plate for $M_e = 3.20$.

$u_L/u_e = 0.48$. This result is more satisfactory and lends support to the reference temperature concept as used herein.

In conclusion, it is suggested that Eq. (8-12) (or Fig. 8-1) be used to calculate the relation between C_H and C_f when mass transfer is present in a turbulent boundary layer and that Eq. (8-13) (or Fig. 8-2) be used to calculate the recovery factor r where the required value of u_L/u_e is calculated using Eqs. (8-42) and (8-43). It should be kept in mind that the results obtained using Eqs. (8-12), (8-13), and (8-29) are sensitive to the value of u_L/u_e chosen and that it is uncertain that Eq. (8-42) will give accurate values of u_L/u_e except for $M_e = 0$.

[1] E. R. G. Eckert, *J. Aeronaut. Sci.*, vol. 22, no. 8, pp. 585–587, 1955.

8-4. Effects of Chemical Reactions with Mass Transfer. In this section we shall treat the cases where mass transfer, diffusion of species, and chemical reactions take place in the compressible turbulent boundary layer. The effects of both exothermic and endothermic reactions can be treated within the framework of the methods presented in this section. An example calculation will be given to illustrate the use of the equations derived.

It will be recalled that the results of Chap. 7, wherein the effects of dissociation upon the compressible turbulent boundary layer were treated, indicated that, providing the wall was catalytic to recombination, the heat transfer to the wall was relatively insensitive to whether the recombination took place within the boundary layer (chemical equilibrium in the boundary layer represents an extreme in this direction) or at the surface (frozen flow in the boundary layer with equilibrium dissociation at the wall, the other extreme for gas-phase reaction rates). A similar result was found for the laminar boundary layer case treated in Chap. 4. These results suggest that for at least one gas-phase chemical reaction, that of dissociation, the heat transfer to the wall is relatively insensitive to where the reaction takes place providing that equilibrium conditions exist at the interface and in the external stream. We shall take the position here that this conclusion holds for any gas-phase chemical reaction, exothermic or endothermic, under conditions for which the boundary-layer equations are appropriate. This latter restriction rules out consideration of combustion situations incorporating a flame zone normal to or nearly normal to the direction of flow within the boundary layer. However, several important situations of gas-phase chemical reactions occurring at or near the surface over which the gas is flowing are included within the framework of the theory to be presented and are compatible with the assumptions made in developing it.

Keeping the foregoing discussion in mind, we shall develop a theory for the reacting compressible turbulent boundary layer which has incorporated within it the assumptions that all chemical reactions take place at the surface or interface between the gas layer and the solid or liquid undersurface and that no reactions occur within the gas layer itself. The products of the chemical reactions and the reactants will be distributed throughout the gas layer by diffusion alone according to these assumptions. It will be found that these assumptions result in removing the species source term $\dot{w}_i$ from the conservation-of-species equation, which will simplify the obtaining of solutions to the boundary-layer equations to a considerable extent. To proceed, we begin with the thin-boundary-layer assumptions used in the previous section. That is, for any dependent variable Q within the boundary layer

$$\left| \frac{\partial Q}{\partial y} \right| \gg \left| \frac{\partial Q}{\partial s} \right|$$

and

$$\rho v = (\rho v)_w$$

Our boundary-layer equations then become the equation for conservation of species[1]

$$\frac{d}{dy}[(\rho v)_w C_i - F_2] = 0 \qquad (8\text{-}44a)$$

the equation for conservation of momentum

$$\frac{d}{dy}[(\rho v)_w u - \tau] = 0 \qquad (8\text{-}45a)$$

and the equation for conservation of energy

$$\frac{d}{dy}[(\rho v)_w I - G] = 0 \qquad (8\text{-}46a)$$

where

$$F_2 = \rho(D_{12} + D_T)\frac{dC_i}{dy} \qquad (8\text{-}44b)$$

$$\tau = (\mu + \epsilon)\frac{du}{dy} \qquad (8\text{-}45b)$$

and

$$G = \left(\frac{\mu}{P} + \frac{\epsilon}{P_T}\right)\frac{dI}{dy} + \left[\mu\left(1 - \frac{1}{P}\right) + \epsilon\left(1 - \frac{1}{P_T}\right)\right]u\,\frac{du}{dy}$$
$$+ \left[\frac{k}{C_{p_f}}(L - 1) + \frac{\kappa}{C_{p_f}}(L_T - 1)\right]\sum_i h_i \frac{dC_i}{dy} \qquad (8\text{-}46b)$$

with boundary conditions

$C_i = (C_i)_w \quad I = I_w \quad F_2 = (F_2)_w \quad \tau = \tau_w \quad G = -\dot{q}_w$ when $u = 0$
$C_i = (C_i)_e \quad I = I_e \quad F_2 = 0 \quad \tau = 0 \quad G = 0$ when $u = u_e$

Let us make use of the concept of a partial enthalpy I_f introduced in Chap. 5. That is, let

$$I_f = I - \sum_i C_i h_i^0 \qquad (8\text{-}47)$$

---

[1] The simplified turbulent-boundary-layer equations given by Eqs. (8-44a) through (8-46b) and the simplified laminar-boundary-layer equations represented by Eq. (5-24) of Sec. 5-4 are examples of the generalization of the equations for the steady convective mass transfer given by D. B. Spalding, *Intern. J. Heat and Mass Transfer*, vol. 1, nos. 2, 3, pp. 192–207, 1960. It is important to note that the generalization applies only where the transport parameters are constant, $L = L_T = 1$, and viscous dissipation and pressure gradient can be neglected.

then define
$$g_f = \frac{I_f}{I_{f_e}} \tag{8-48}$$

and define the dimensionless chemical enthalpy z as

$$z = \frac{\sum_i h_i^0 C_i}{\left(\sum_i h_i^0 C_i\right)_e} \tag{8-49}$$

Then it can be shown that

$$-\dot{q}_w = -(\dot{q}_w)_c - (\dot{q}_w)_d \tag{8-50}$$

where the convective-heat-transfer rate is

$$-(\dot{q}_w)_c = (I_f)_e\left(\frac{\mu}{P}\frac{dg_f}{dy}\right)_w \tag{8-51}$$

and the diffusion-heat-transfer rate is

$$-(\dot{q}_w)_d = \left(\sum_i h_i^0 C_i\right)_e\left(\rho D_{12}\frac{dz}{dy}\right)_w \tag{8-52}$$

Equations (8-47), (8-48) and (8-44a) through (8-52) will now be combined to obtain relations among $-\dot{q}_{w_c}$, $\dot{q}_{w_d}$, and τ_w under conditions of mass transfer and chemical reactions at the surface.

Equations for $2C_{H_c}/C_f$ and r.   Multiply Eq. (8-44a) by h_i^0 and sum over all species i. Subtract the resulting differential equation from Eq. (8-46a) combined with Eqs. (8-47) and (8-48). There results

$$\frac{d}{dy}\left\{(\rho v)_w g_f - \left(\frac{\mu}{P} + \frac{\epsilon}{P_T}\right)\frac{dg_f}{dy} - \left[\mu\left(1 - \frac{1}{P}\right) + \epsilon\left(1 - \frac{1}{P_T}\right)\right]\frac{u}{(I_f)_e}\frac{du}{dy}\right\} = 0 \tag{8-53}$$

where, in arriving at Eq. (8-53), it is taken that

$$\sum_i h_i\frac{dC_i}{dy} - \sum_i h_i^0\frac{dC_i}{dy} = \sum_i\left(\int_0^T C_{p_i}dT\right)\frac{dC_i}{dy} = 0 \tag{8-54}$$

This latter result is obtained for most atomic and molecular species likely to be of interest to us because $C_{p_i} \simeq \text{const} = C_p$, whence

$$\sum_i\left(\int C_{p_i}dT\right)\frac{dC_i}{dy} = \left(\int_0^T C_p\, dT\right)\sum_i\frac{dC_i}{dy} = 0$$

since $\sum_i C_i = 1$.

Furthermore, from Eqs. (8-44a) and (8-45a),

$$\frac{d}{dy}\left[(\rho v)_w z - \rho(D_{12} + D_T)\frac{dz}{dy}\right] = 0 \tag{8-55}$$

Integrate Eqs. (8-45a), (8-53), and (8-55) once and apply the boundary conditions to eliminate the constants of integration. There result

$$(\rho v)_w u + \tau_w = \tau = (\mu + \epsilon)\frac{du}{dy} \tag{8-56}$$

$$(\rho v)_w g_f - \left(\frac{\mu}{P} + \frac{\epsilon}{P_T}\right)\frac{dg_f}{dy} - \left[\mu\left(1 - \frac{1}{P}\right) + \epsilon\left(1 - \frac{1}{P_T}\right)\right]\frac{u}{(I_f)_e}\frac{du}{dy}$$
$$= (\rho v)_w (g_f)_w + \frac{1}{(I_f)_e}(\dot{q}_w)_c \tag{8-57}$$

and $\qquad (\rho v)_w z - \rho(D_{12} + D_T)\dfrac{dz}{dy} = (\rho v)_w z_w + \dfrac{1}{\left(\sum\limits_i C_i h_i^0\right)_e}(\dot{q}_w)_d \tag{8-58}$

When Eq. (8-57) is divided by Eq. (8-56), there results a differential equation of the form of Eq. (8-9) obtained earlier. In Eq. (8-9) we let

$$\lambda = g_f \tag{8-59}$$

$$\bar{P}(u) = -\frac{P_i(\rho v)_w}{(\rho v)_w u + \tau_w} \tag{8-60}$$

and $\qquad Q(u) = -\left\{\dfrac{P_i(\rho v)_w(g_f)_w + [P_i/(I_f)_e](\dot{q}_w)_c}{(\rho v)_w u + \tau_w} + \dfrac{(P_i - 1)}{(I_f)_e}u\right\} \tag{8-61}$

where, in Eqs. (8-60) and (8-61) for the laminar sublayer,

$$0 \leq u \leq u_L \qquad \epsilon = \kappa = 0 \qquad \text{and} \qquad P_i = P$$

and, for the turbulent outer layer,

$$u_L \leq u \leq u_e \qquad \mu = k = 0 \qquad \text{and} \qquad P_i = 1$$

Using Eqs. (8-59) through (8-61) with Eqs. (8-9) through (8-11), we obtain two equations for $g_f(u)$ applicable in the sublayer and turbulent core. Evaluate the constant of integration C_3 by using the boundary condition $g_f = (g_f)_w$ at $u = 0$ with the sublayer equation for $g_f(u)$ and the boundary condition $g_f = (g_f)_e$ at $u = u_e$ with the turbulent-outer-layer equation for $g_f(u)$. Eliminate $(g_f)_L = g_f(u_L)$ between the two equations, and solve for $2C_{H_c}/C_f$. There result

$$\frac{2C_{H_c}}{C_f} = \frac{B_5}{(1 + B_5)[1 + B_5(u_L/u_e)]^{P-1} - 1} \tag{8-12}$$

and

$$r = \frac{1 + B_5}{B_5^2}\left\{\left[\frac{1 + B_5(u_L/u_e)}{2 - P}\right]\left[P - 2\left(1 + B_5\frac{u_L}{u_e}\right)^{P-2}\right] + \left[\frac{1}{1 + B_5} - B_5\left(\frac{u_L}{u_e} - 1\right)\right]\right\} \tag{8-13}$$

where, as before,

$$B_5 = \frac{(\rho v)_w u_e}{\tau_w} = \frac{2(\rho v)_w}{\rho_e u_e C_f} \tag{8-14}$$

$$r = \frac{2}{u_e^2}[(I_f)_r - (I_f)_e] + 1 = \frac{2}{u_e^2}(I_r - h_e) \tag{8-62}$$

and we have used the definition for C_{H_c}

$$-(\dot{q}_w)_c = C_{H_c}\rho_e u_e[(I_f)_r - (I_f)_w] \tag{8-63}$$

These results are entirely consistent with those derived in Sec. 8-2 for the effect of mass transfer upon the nonreacting compressible turbulent boundary layer and, in effect, amount to an independent derivation of Eqs. (8-12) and (8-13).

Equation for $2C_{H_d}/C_f$. Consider, now, the derivation of a relation for $2C_{H_d}/C_f$. When Eq. (8-58) is divided by Eq. (8-56), there results a differential equation of the form of Eq. (8-9) except that for this case

$$\lambda = z \tag{8-64}$$

$$\bar{P}(u) = -\frac{S_i(\rho v)_w}{(\rho v)_w u + \tau_w} \tag{8-65}$$

and

$$Q(u) = -\frac{S_i(\rho v)_w z_w + \left[S_i \Big/ \left(\sum_i C_i h_i^0\right)_e\right](\dot{q}_w)_d}{(\rho v)_w u + \tau_w} \tag{8-66}$$

where, in Eqs. (8-65) and (8-66), in the laminar sublayer

$$0 \leq u \leq u_L \qquad \epsilon = \kappa = 0 \qquad \text{and} \qquad S_i = S$$

and in the turbulent core

$$u_L \leq u \leq u_e \qquad \mu = k = 0 \qquad \text{and} \qquad S_i = 1$$

Use Eqs. (8-64) through (8-66) with Eqs. (8-9) through (8-12) to obtain two equations for $z(u)$: one applicable in the sublayer ($0 \leq u \leq u_L$), the other applicable in the turbulent core ($u_L \leq u \leq u_e$). Eliminate the constant C_3 in the former by using the boundary condition that $z = z_w$ when $u = 0$, and eliminate the constant C_3 in the latter by using the boundary condition $u = u_e$ when $z = z_e$. Eliminate $z_L = z(u_L)$ between the two equations, and solve for $2C_{H_d}/C_f$ in the resulting equation. There results

$$\frac{2C_{H_d}}{C_f} = \frac{B_5}{(1 + B_5)[1 + B_5(u_L/u_e)]^{S-1} - 1} \tag{8-67}$$

$$S = \text{Schmidt number} = \frac{\mu}{\rho D_{12}}$$

where C_{H_d} is defined by the equation

$$-(\dot{q}_w)_d = C_{H_d}\rho_e u_e \left\{ \sum_i h_i^0[(C_i)_e - (C_i)_w] \right\} \tag{8-68}$$

Boundary-layer Heat Transfer. Since, by Eq. (8-50),

$$-\dot{q}_w = -(\dot{q}_w)_c - (\dot{q}_w)_d \tag{8-50}$$

then we obtain, using Eqs. (8-12), (8-63), (8-67), and (8-68) with Eq. (8-50),

$$-\dot{q}_w = \frac{C_f}{2}\rho_e u_e \frac{2C_{H_c}}{C_f}\left\{ (I_f)_r - (I_f)_w + \frac{C_{H_d}}{C_{H_c}}\sum_i h_i^0[(C_i)_e - (C_i)_w] \right\} \tag{8-69}$$

or, since by Eq. (8-47),

$$I = I_f + \sum_i h_i^0 C_i \tag{8-47}$$

we can write Eq. (8-69) as

$$-\dot{q}_w = \frac{C_f}{2}\rho_e u_e \frac{2C_{H_c}}{C_f}(I_r - h_w)\left\{ 1 + \left(\frac{C_{H_d}}{C_{H_c}} - 1\right)\frac{\sum_i h_i^0[(C_i)_e - (C_i)_w]}{I_r - h_w} \right\} \tag{8-70}$$

where

$$\frac{2C_{H_c}}{C_f} = \frac{B_5}{(1 + B_5)[1 + B_5(u_L/u_e)]^{P-1} - 1} \tag{8-12}$$

$$\frac{C_{H_d}}{C_{H_c}} = \frac{(1 + B_5)[1 + B_5(u_L/u_e)]^{P-1} - 1}{(1 + B_5)[1 + B_5(u_L/u_e)]^{S-1} - 1} \tag{8-71}$$

and from Eq. (8-13)

$$r = \frac{1 + B_5}{B_5^2}\left\{ \frac{1 + B_5(u_L/u_e)}{2 - P}\left[P - 2\left(1 + B_5\frac{u_L}{u_e}\right)^{P-2}\right] + \left[\frac{1}{1 + B_5} - B_5\left(\frac{u_L}{u_e} - 1\right)\right] \right\} \tag{8-13}$$

Let us check the consistency of Eq. (8-70) with Eq. (7-89) for the reacting boundary layer without mass transfer. In the limit as $B_5 \to 0$, it can be shown that (if $u_L/u_e = \frac{2}{3}$)

$$\frac{2C_{H_c}}{C_f} \to \frac{1}{1 + (P - 1)(u_L/u_e)} \simeq P^{-2/3} \tag{8-72}$$

$$\frac{C_{H_d}}{C_{H_c}} \to \frac{1 + (P - 1)(u_L/u_e)}{1 + (S - 1)(u_L/u_e)} \simeq \frac{P^{2/3}}{S^{2/3}} = L^{2/3} \tag{8-73}$$

$$r \to 1 + (1 - P)\left(\frac{u_L}{u_e}\right)^2 \simeq P^{1/3} \tag{8-74}$$

and thus

$$-q_w \to \frac{C_f}{2}\, \rho_e u_e P^{-2/3}(I_r - h_w)\left[1 + (L^{2/3} - 1)\frac{h_c}{I_r - h_w}\right] \quad (8\text{-}75)$$

which is exactly Eq. (7-89) combined with Eqs. (7-88) and (7-92).

Figures 8-1 and 8-2 present plots of r and $2C_{H_c}/C_f$ versus B_5 for representative values of u_L/u_e and $P = 0.70$. Figure 8-6 presents the ratio C_{H_d}/C_{H_c} plotted versus the parameter B_6, where

$$B_6 = B_3(C_{H_c}/C_{H_d})$$

The variable B_6 was chosen because of its appearance in a useful expression for $\dot{q}_s$ yet to be derived. In preparing Fig. 8-6 it was assumed

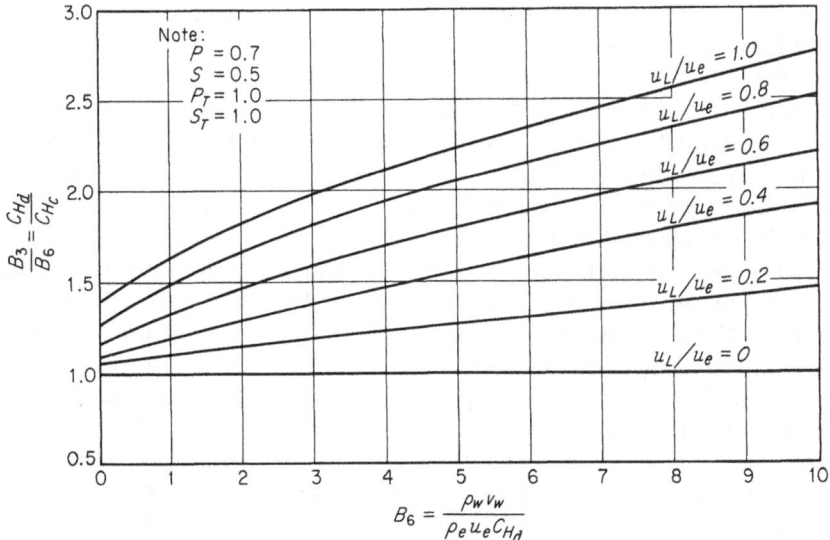

FIG. 8-6. C_{H_d}/C_{H_c} versus blowing parameter B_6 calculated using Eq. (8-71). B_6 is related to B_5 through Eq. (8-113). Calculations for $P = 0.70$, $S = 0.5$, $P_T = 1.0$, and $S_T = 1.0$.

that $P = 0.70$ and $S = 0.50$, values which are not unreasonable for gas mixtures composed of molecules and atoms with molecular and atomic weights near those of the components of air or simple molecules like CO_2, CO, and BO_2. When light gases like helium or hydrogen are used, the choice of P and S will depend upon the fractional proportions of the light gas to the remaining species in the boundary layer and will become variables through the boundary layer. Rubesin and Pappas[1] have treated the effect of injecting a light gas such as He or H_2 into an

[1] Morris W. Rubesin and Constantine C. Pappas, *NACA TN* 4149, 1958.

isothermal air boundary layer, which clearly demonstrates the dependency of P and S upon the fractional composition of He or H_2 in the boundary layer. This case will not be dealt with here except to point out that it can be handled in a manner analogous to the development given in this section, with the additional requirement that Prandtl numbers and Schmidt numbers used are dependent upon the mass fractions of the light gas species present through the boundary layer.

Before proceeding to the development of an expression for $\dot{q}_s$, the net heat-transfer rate to the liquid or solid undersurface, it is pointed out here that the unknown parameter in Eq. (8-69) is u_L/u_e as it has been previously. u_L/u_e is a weighting factor which enters in as a determination of that portion of the boundary layer over which the laminar-flow transport properties predominate. If $u_L/u_e = 1$, the boundary layer is entirely laminar and the above equations for $u_L/u_e = 1$ would reduce to those appropriate to laminar couette flow with mass transfer and chemical reactions present at the boundary of the flow.

Lapin[1] treated the effect of chemical reactions upon skin friction and heat transfer of the compressible turbulent boundary layer in a manner somewhat similar to that method outlined in Sec. 8-3 and the present section. The effect of mass transfer on skin-friction coefficient is treated in a manner analogous to the theory of Dorrance and Dore described in Sec. 8-3. The chemical reactions are assumed to occur at an infinitely thin reaction surface within the boundary layer, so that the reactants are supplied to this reaction surface in stoichiometric proportions and one of the reactants is entirely consumed there. This assumption approximates the assumption of chemical equilibrium at the reaction surface. Since the effects of chemical reactions on skin friction and heat transfer are independent of the location of the reaction plane to a first approximation, the method of Lapin and those of this book for $L = P = 1$ should give equivalent results. Lapin does not cite any numerical results.

8-5. Net Heat Transfer into a Surface. Consider, now, the net heat transfer from the boundary layer into the vaporizing or sublimating liquid or solid undersurface. It can be shown at the interface that, in the absence of radiation,

$$\dot{q}_s = -\dot{q}_w + (\rho v)_w[h_E(s)]_w - (\rho v)_w h_w \qquad (8\text{-}76)$$

The first term on the right-hand side of Eq. (8-76) represents the diffusive, convective, and conductive heat transfer from the reacting boundary layer to the interface. The second term is the counter-current of heat carried toward the interface because of melting,

[1] Yu. V. Lapin, *Zhur. Tekh. Fiz.*, vol. 30, no. 10, pp. 1227–1237, 1960; translation appears in *Soviet Phys. JETP*, vol. 5, p. 1162, 1961.

chemical reactions, or vaporization. The last term represents the heat carried away from the interface into the boundary layer by the mass transfer occurring at the interface. If mass transfer through a solid, nonmelting, porous material is being considered, the second term represents the heat being carried toward the interface by the fluid being used as a coolant.

Equation for Heat Transfer. If vaporization or sublimation is occurring at the interface, then for our vaporizing material with the chemical symbol E,

$$(h_E)_w = [h_E(g)]_w = [h_E(s)]_w + L_v \tag{8-77}$$

where L_v is the heat of vaporization. Combining Eqs. (8-69), (8-76), and (8-77) results in

$$\dot{q}_s = \rho_e u_e C_{H_c}\left((I_f)_r - (I_f)_w + \frac{C_{Hd}}{C_{H_c}}\left\{\sum_i h_i^0[(C_i)_e - (C_i)_w]\right.\right.$$
$$\left.\left. - B_6 h_w + B_6(h_E)_w\right\} - B_3 L_v\right) \tag{8-78}$$

where B_6 is related to B_3 by the equation

$$B_6 = B_3 \frac{C_{H_c}}{C_{Hd}} = \frac{(\rho v)_w}{\rho_e u_e C_{Hd}} \tag{8-79}$$

Now, since $C_p \simeq$ const, then, according to Eq. (8-54) used earlier,

$$\sum_i h_i^0[(C_i)_e - (C_i)_w] \simeq \sum_i (h_i)_w[(C_i)_e - (C_i)_w] \tag{8-80}$$

and Eq. (8-78) can be written

$$\dot{q}_s = \rho_e u_e C_{H_c} \Delta I\left[1 + \left(\frac{C_{Hd}}{C_{H_c}} - 1\right)\frac{h_c}{\Delta I} - B_3 \frac{L_v}{\Delta I}\right] \tag{8-81}$$

where

$$\Delta I = (I_f)_r - (I_f)_w + h_c \tag{8-82}$$

and the chemical enthalpy potential h_c is defined by

$$h_c = \sum_{i \neq E} (h_i)_w[(C_i)_e - (1 + B_6)(C_i)_w] + (h_E)_w[B_6 - (1 + B_6)(C_E)_w] \tag{8-83}$$

Equation (8-81) is analogous in every detail to Eq. (5-53) for the effect of surface mass transfer and chemical reactions upon net heat transfer from a laminar boundary layer to the solid or liquid undersurface. Using the results of Sec. 3-3, it can be shown that

$$h_c = \sum_{J=O,N} \Delta Q_{EJ}[(\bar{C}_J)_e - (1 + B_6)(\bar{C}_J - r_{J,EJ}C_{EJ})_w]$$
$$+ \sum_{J=O} \Delta Q_{EJ_2}[(\bar{C}_J)_e - (1 + B_6)(\bar{C}_J - r_{J,EJ_2}C_{EJ_2})_w]$$
$$- \sum_{J=O,N} \frac{1}{2}\Delta Q_{J_2}[(C_{J_2})_e - (1 + B_6)(C_{J_2})_w] \tag{8-84}$$

where, for most cases of interest, the concentrations of E and the products of chemical reaction will be zero at the edge of the boundary layer and where the ΔQ's in Eq. (8-84) are obtained using Eq. (3-8).

Effect of Mass Transfer with Unlike Gases on C_f. Before the use of Eq. (8-81) is illustrated with an example calculation, a question arises regarding the variation of C_f with mass transfer when unlike gases are injected into the boundary layer. It will be recalled that the results of the previous section dealt exclusively with the effect of mass transfer upon C_f when a like gas is injected into a like gas. For these conditions it was found that, approximately, for $B_1 \rightarrow 0$,

$$\frac{C_f}{C_{f_0}} = 1 - B_1 \frac{u_L}{u_e} \tag{8-29}$$

where the principal unknown is the value of the velocity ratio u_L/u_e to use. Since the analysis upon which Eq. (8-29) rests is provisional at best, it is unlikely that an expression similar in form to Eq. (8-29), which accounts for the fact than an unlike gas is being introduced into the boundary layer, will be any more accurate than Eq. (8-29) is under the conditions for which it is reasonable to expect Eq. (8-29) to apply. It is probably best to rely upon correlations established upon experimental measurements and what few theoretical calculations have been made for the turbulent boundary layer when an unlike gas is injected into the boundary layer. Such a correlation has been presented by Hartnett, Masson, Gross, and Gazley[1] and can be represented by the following equation. If C_f/C_{f_0} is the same for two cases, one for like gas blowing into like gas and the other for unlike gas blowing into unlike gas, then correlation of data and calculations shows that

$$\frac{B_1(M_2)}{B_1(M_1)} = \left(\frac{M_2}{M_1}\right)^{0.46} \pm 28\% \tag{8-85}$$

for
$$2 \leq M_1; \quad M_2 \leq 120$$

where M_1 is the molecular weight of the boundary-layer gas and M_2 is the molecular weight of the injected gas. Using Eq. (8-85) in Eq. (8-29) we obtain

$$\frac{C_f}{C_{f_0}} = 1 - B_1(M_2)\left(\frac{M_1}{M_2}\right)^{0.46} \frac{u_L}{u_e} \tag{8-86}$$

When $M_2 = M_1$, Eq. (8-86) becomes Eq. (8-29). It is apparent that for very light gases being injected into a gas like air ($M_2 < M_1$) C_f for a given value of the blowing parameter B_1 will be somewhat less

[1] J. P. Hartnett, D. J. Masson, J. F. Gross, and Carl Gazley, Jr., *J. Aerospace Sci.*, vol. 27, no. 8, pp. 623–625, 1960.

than that which would result if air was being blown into air.   This can be accounted for by the realization that light gases usually have a lower viscosity than air and blowing a light gas into the boundary layer results in a high fraction of light gas near the surface, thus lowering the skin friction and hence C_f.   Quite the opposite conclusion would apply when a heavy gas is blown into air $(M_2 > M_1)$.

8-6. A Sample Calculation.   We shall now illustrate the use of Eq. (8-81) by an example.   Consider the case of a (hot) atomic oxygen stream flowing over a sublimating graphite surface.   Assume that only three species are present: C, O, and CO.   Such a case is hypothetical but is nevertheless helpful to illustrate use of the derived equations.   Our purpose in choosing a hypothetical case is simply to hold down the number of species involved, since increasing the number of species will not intrinsically alter the method used to solve such problems.   We desire to calculate the effect of the exothermic chemical reaction

$$C(g) + O(g) \to CO(g) + \Delta Q_{CO} \tag{8-87}$$

upon the net heat transfer to the graphite surface.   We wish to obtain values for the ratio $\dot{q}_s / (\dot{q}_s)_0$ formed using Eq. (8-81), where $(\dot{q}_s)_0$ is the net heat transfer to the graphite in the absence of the reaction described by Eq. (8-87).   That is,

$$\frac{\dot{q}_s}{(\dot{q}_s)_0} = \left\{ \frac{C_{H_e}}{(C_{H_e})_0} \right\} \left\{ \frac{\Delta I}{\Delta I_0} \right\} \left\{ 1 + \left(\frac{C_{H_d}}{C_{H_e}} - 1 \right) \frac{h_c}{\Delta I} - B_3 \frac{L_v}{\Delta I} \right\} \tag{8-88}$$

where the subscripts 0 denote quantities evaluated with no chemical reactions or mass transfer.

We proceed, now, to evaluate the various terms in this equation using the input data given in Table 8-1.

TABLE 8-1. Physical-Chemical Constants and Parameters Used in Example Calculation

(Assume: $(I_f)_e - (I_f)_w = 6500$ Btu/lb. K_3 and K_6 are in atmospheres; T is in degrees Rankine)

$p_e = 1$ atm	$\Delta Q_{CO} = 28{,}800$ Btu/lb O
$T_w = 6000°$R	$L_v = 25{,}500$ Btu/lb C
$L = 1.4$	$M_C = 12$
$P = 0.70$	$M_O = 16$
$S = 0.50$	$M_{CO} = 28$
$\log\left(\dfrac{K_3}{3.0172 \times 10^7}\right) = -\dfrac{2.37607 \times 10^5}{T}$	$K_3 = \dfrac{p_C p_O}{P_{CO}}$
$\log\left(\dfrac{K_6}{1.3596 \times 10^8}\right) = -\dfrac{1.55372 \times 10^5}{T}$	$K_6 = p_C$

From Eq. (8-84), where $E = C$, $J = O$, and J_2 is absent, we have

$$h_c = \Delta Q_{CO}[(\overline{C_O})_e - (1 + B_6)(\overline{C_O} - r_{0,CO}C_{CO})_w] \tag{8-89a}$$

and it is apparent that we must obtain B_3, B_6, $(C_O)_w$, and $(C_{CO})_w$ in order to proceed. C_{CO} and C_O are related to $\overline{C}_O$ by the equations

$$(\overline{C_O})_e = (C_O)_e + r_{0,CO}(C_{CO})_e \tag{8-89b}$$

and
$$(\overline{C_O})_w = (C_O)_w + r_{0,CO}(C_{CO})_w \tag{8-89c}$$

The Species Mass Fractions.   In that which follows we assume that the reaction goes to completion at the surface and that the equilibrium concentrations determine the mass-transfer parameters B_3 and B_6. We can make use of the equilibrium concentrations at the surface to determine our species mass fractions as shown in the following.

For the chemical reaction given by Eq. (8-87) the equilibrium constant is, as was defined in Sec. 5-7,

$$\frac{p_C p_O}{p_{CO}} = K_3(T) \tag{8-90}$$

Furthermore, for the sublimation of graphite

$$p_C = K_6(T) \tag{8-91}$$

and Dalton's law states that

$$\sum_i p_i = p_O + p_{CO} + p_C = p_e \tag{8-92}$$

since the pressure is constant across the boundary layer.   Also, since we assume that all the gas species behave as perfect gases, then

$$p_i = \rho_i \frac{k}{m_i} T \tag{8-93}$$

Furthermore, conservation of mass requires that

$$\sum_i C_i = 1 = C_O + C_{CO} + C_C \tag{8-94}$$

where
$$C_i = \frac{\rho_i}{\rho} \tag{8-95}$$

Combining Eqs. (8-90) through (8-95) it can be shown that the species mass fractions are

$$C_i = p_i M_i \chi \tag{8-96}$$

where, if N_0 is Avagadro's number and $M_i = m_i N_0$, then

$$\chi = \left(\sum_i p_i M_i \right)^{-1} \tag{8-97}$$

and the necessary partial pressures are determined by solving Eqs. (8-90) through (8-92). That is,

$$p_C = K_6$$

$$p_{CO} = \frac{K_6 p_e - K_6^2}{K_3 + K_6} \tag{8-98}$$

$$p_O = \frac{K_3 p_e - K_3 K_6}{K_3 + K_6}$$

The above equations determine C_O, C_C, and C_{CO} at the surface once the pressure p_e and surface temperature is specified if it is assumed that the reaction goes to completion at the surface.

The results of the calculations for $p_e = 1$ atm and $T_w = 6000°R$ are

$$(C_{CO})_w \simeq 1 \qquad (C_{CO})_e = 0$$
$$(C_O)_w \simeq 0 \qquad (C_O)_e = 1$$
$$(C_C)_w \simeq 0 \qquad (C_C)_e = 0$$

The Mass-transfer Parameter B_6. In order to determine the blowing parameter B_6, we must work with the equation for conservation of species at the interface. That is, in order to conserve mass at the interface, we have, from Eq. (3-19) which applies in the present case,

$$\left(\rho D_{12} \frac{d\bar{C}_C}{dy} \right)_w + (\rho v)_w = (\rho v)_w (\bar{C}_C)_w \tag{8-99}$$

Equation (8-99) can be combined with a relation between z and $\bar{C}_C$ to obtain B_6. From Eqs. (8-44a) and (8-44b) with the index i changed to k,

$$\frac{d}{dy} \left[(\rho v)_w C_k - \rho(D_{12} + D_T) \frac{dC_k}{dy} \right] = 0$$

or integrating with respect to y and evaluating the constant of integration at $y = 0$, we obtain

$$(\rho v)_w C_k - \rho(D_{12} + D_T) \frac{dC_k}{dy} = (\rho v)_w (C_k)_w - \left(\rho D_{12} \frac{dC_k}{dy} \right)_w \tag{8-100}$$

but

$$\bar{C}_i = \sum r_{i,k} C_k \tag{8-101}$$

hence, combining Eqs. (8-100) and (8-101), we find that

$$(\rho v)_w \bar{C}_i - \rho(D_{12} + D_T) \frac{d\bar{C}_i}{dy} = (\rho v)_w (\bar{C}_i)_w - \left(\rho D_{12} \frac{d\bar{C}_i}{dy} \right)_w \tag{8-102}$$

From Eqs. (8-52) and (8-58),

$$(\rho v)_w z - \rho(D_{12} + D_T) \frac{dz}{dy} = (\rho v)_w z_w - \left(\rho D_{12} \frac{dz}{dy} \right)_w \tag{8-103}$$

hence it follows that an integral to Eq. (8-102) is

$$(\overline{C}_i) = z \frac{(\overline{C}_i)_e - (\overline{C}_i)_w}{z_e - z_w} + \frac{(\overline{C}_i)_w z_e - (\overline{C}_i)_e z_w}{z_e - z_w} \tag{8-104}$$

which can be verified by substitution into Eq. (8-102). Thus, differentiating Eq. (8-104) and evaluating the result at $y = 0$ for $i = C$, we obtain

$$\left(\rho D_{12} \frac{d\overline{C}_C}{dy} \right)_w = \frac{(\overline{C}_C)_e - (\overline{C}_C)_w}{z_e - z_w} \left(\rho D_{12} \frac{dz}{dy} \right)_w \tag{8-105}$$

and making use of Eqs. (8-52) and (8-68) in Eq. (8-105) results in

$$\left(\rho D_{12} \frac{d\overline{C}_C}{dy} \right)_w = [(\overline{C}_C)_e - (\overline{C}_C)_w] C_{H_d} \rho_e u_e \tag{8-106}$$

Combining Eq. (8-99) with Eq. (8-106), we obtain the sought-after expression for B_6 in terms of the species concentrations at the wall. This is

$$B_6 = \frac{(\overline{C}_C)_w}{1 - (\overline{C}_C)_w} = \frac{(C_C + r_{C,CO} C_{CO})_w}{1 - (C_C + r_{C\ CO} C_{CO})_w} \tag{8-107}$$

and using the concentrations previously determined and the values of M_C and M_{CO} given in Table 8-1, we obtain for our example problem

$$B_6 = \frac{\frac{12}{28}}{1 - \frac{12}{28}} = \tfrac{3}{4} \tag{8-108}$$

It is interesting to note that Eq. (8-104) can be obtained directly from Eqs. (2-107) and (2-114) and applies in general to the laminar and turbulent-boundary-layer equations with the only restriction that $\dot{w}_i = 0$ (frozen flow).

The Chemical Enthalpy Potential h_c. Using the results obtained for $(C_{CO})_w$, $(C_C)_w$, $(C_O)_w$, and B_6 with the values of ΔQ_{CO}, M_O, and M_{CO} given in Table 8-1 and Eq. (8-89a) we obtain, since for our example $(C_O)_w \simeq 0$,

$$h_c = 28{,}800 \text{ Btu/lb} \tag{8-109}$$

and, since Table 8-1 gives $(I_f)_e - (I_f)_w$ as input data, then

$$\Delta I = 6500 + 28{,}800 = 35{,}300 \text{ Btu/lb} \tag{8-110}$$

and

$$\Delta I_0 = 6500 \text{ Btu/lb} \tag{8-111}$$

Next we shall show how the mass-transfer parameter B_3 is calculated. B_3 is needed in order to apply Eq. (8-88) in our example problem.

The Mass-transfer Parameter B_3. The parameter B_3 is found using the definitions for B_3, B_5, and B_6. That is, from Eq. (8-79),

$$B_3 = \frac{C_{Hd}}{C_{H_c}} B_6 \qquad (8\text{-}112)$$

and from Eqs. (8-14) and (8-79),

$$B_5 = \frac{C_H}{C_f/2} B_3 = \frac{C_{Hd}}{C_f/2} B_6 \qquad (8\text{-}113)$$

Equations (8-12), (8-71), (8-112), and (8-113) can all be manipulated algebraically to obtain B_3 and B_5 once B_6 is known. There result

$$B_6 = \left[\left(1 + B_3 \frac{C_{H_c}}{C_f/2}\right) \left(1 + B_3 \frac{C_{H_c}}{C_f/2} \frac{u_L}{u_e}\right)^{S-1} - 1 \right] \qquad (8\text{-}114)$$

and

$$B_3 = \left[\left(1 + B_3 \frac{C_{H_c}}{C_f/2}\right) \left(1 + B_3 \frac{C_{H_c}}{C_f/2} \frac{u_L}{u_e}\right)^{P-1} - 1 \right] \qquad (8\text{-}115)$$

Given B_6, Eq. (8-114) is solved for $2B_3 C_{H_c}/C_f = B_5$. This value of $2B_3 C_{H_c}/C_f$ is used in Eq. (8-115) to obtain B_3. The ratio

$$B_3/B_6 = C_{H_d}/C_{H_c}$$

is plotted as Fig. 8-6 for $P = 0.70$ and $S = 0.50$ for various values of u_L/u_e and can be used in lieu of Eqs. (8-114) and (8-115) when these values of P and S are appropriate to the problem at hand. In the case of our problem, $P = 0.70$ and $S = 0.50$ are quite appropriate and it is found that if we assume that $u_L/u_e = 0.60$, and use $B_6 = 0.75$, we obtain

$$B_3 = 0.98 \qquad (8\text{-}116)$$

The Ratio $C_{H_c}/(C_{H_c})_0$. We need to calculate the ratio $C_{H_c}/(C_{H_c})_0$. This is

$$\frac{C_{H_c}}{(C_{H_c})_0} = \frac{2C_{H_c}/C_f}{2(C_{H_c}/C_f)_0} = \frac{2C_H/C_f}{(2C_{H_c}/C_f)_0}\left(1 - B_1 \frac{u_L}{u_e}\right) \qquad (8\text{-}117)$$

making use of Eq. (8-29), since $M_C \simeq M_O$. $2C_H/C_f$ is given by Eq. (8-12) once B_5 is known. $B_1 = B_5 (C_f/C_{f_0})$. Hence by Eq. (8-29)

$$\frac{C_f}{C_{f_0}} = 1 - B_5 \frac{C_f}{C_{f_0}} \frac{u_L}{u_e} \qquad (8\text{-}118)$$

and C_f/C_{f_0} is determined once B_5 is determined. B_5 is related to B_3 by Eq. (8-113) and is determined by Eq. (8-114) once B_6 has been found. In the present example it is found that

$$B_5 = \frac{2C_H}{C_f} B_3 = 1.38$$

for $B_6 = 0.75$, $u_L/u_e = 0.6$, and $S = 0.5$ in Eq. (8-113). For

$$B_5 = 1.38 \quad \text{and} \quad u_L/u_e = 0.6$$

Eq. (8-118) gives

$$\frac{C_f}{C_{f_0}} = 0.547 \qquad (8\text{-}119)$$

Equation (8-12) or Fig. 8-1 gives $2C_H/C_f$ for $B_5 = 1.38$, $P = 0.7$, and $u_L/u_e = 0.6$ and $(2C_H/C_f)_0$ for $B_5 = 0$, $P = 0.7$, and $u_L/u_e = 0.6$. We find, for this case, that

$$\frac{2C_{H_e}/C_f}{(2C_{H_e}/C_f)_0} = \frac{1.41}{1.21} = 1.163 \qquad (8\text{-}120)$$

Hence, from Eqs. (8-117), (8-119), and (8-120) for the present example

$$\frac{C_{H_e}}{(C_{H_e})_0} = 1.163 \times 0.547 = 0.635 \qquad (8\text{-}121)$$

and this value is to be substituted into Eq. (8-88).

Results of All Calculations. Using Eqs. (8-108), (8-110), (8-111), (8-112), (8-116), and (8-121) and the value of L_v given in Table 8-1 with Eq. (8-88) results in

$$\frac{\dot{q}_s}{(\dot{q}_s)_0} = \{0.635\}\{5.43\}\{0.545\} = 1.87 \qquad (8\text{-}122)$$

The braces in Eq. (8-122) correspond sequentially to the braces of Eq. (8-88). We see that the mass-transfer term acts to reduce the heat transfer, the term involving h_c acts to increase the heat transfer because of the exothermic reaction given by Eq. (8-87), and the last term acts to reduce heat transfer because the reaction $C(s) \rightarrow C(g)$ is endothermic. It is apparent that the middle term predominates and that in the present example the exothermic reaction

$$C(g) + O(g) \rightarrow CO(g)$$

which occurs in the gas layer liberates enough heat to offset the reduction in heat transfer due to the mass-transfer effect and the heat absorbed by the reaction $C(s) \rightarrow C(g)$.

The method developed in this chapter depends implicitly upon the assumption that all chemical reactions occurring go to completion at the wall and are not rate dependent at the wall. If the wall in some way inhibits or slows down the chemical reactions, then reaction-rate-dependent terms must be taken into account and the present method does not apply.

Denison[1] presented a method for treating the effects of chemical

[1] M. R. Denison, *J. Aerospace Sci.*, vol. 28, no. 26, pp. 471–479, 1961.

reactions upon heat transfer and skin friction of the compressible turbulent boundary layer when $P = P_T = L = L_T = 1$. Denison's method can be obtained from the treatment developed in Secs. 8-4 and 8-5 if P and L are taken equal to 1 and the relations among $2C_{H_c}/C_f$, $2C_{H_d}/C_f$, and r and the blowing parameters given by Eqs. (8-12), (8-67), and (8-13), respectively, are taken at their values for zero blowing with $P = P_T = L = L_T = 1$. (That is, if $2C_{H_c}/C_f = 1$, $2C_{H_d}/C_f = 1$ and $r = 1$.) Denison also used an approximation to the method of Dorrance and Dore[1] to account for the effects of mass transfer upon skin friction. When the results of Denison's calculations are compared with his reported experimental findings, the differences between experiment and theory found are in the direction of the magnitude as those to be expected if one considers the differences which result in making calculations with $P = 1.0$ and $L = 1.0$ rather than the more accurate values $P = 0.70$ and $L = 1.4$ in Eqs. (8-12), (8-67), and (8-13) with blowing parameters zero rather than the correct finite value in those equations. The experimental results reported by Denison suggest that the analyses developed in Secs. 8-4 and 8-5 give results in good agreement with measurement, although additional experimental data are desirable before broad conclusions can be drawn. The experiments referred to by Denison consisted in blowing varied mixtures of O_2 and N_2 through an inductively heated graphite cylinder while measuring graphite surface temperatures and the amount of mass lost at differing times during the experiments. The experiments are similar to the hypothetical model selected for an example calculation in the present section.

8-7. Effects of Body Shape and Pressure Gradient. The previous sections of this chapter have all dealt with the turbulent boundary layer under various conditions flowing over a two-dimensional flat plate. It is natural to inquire as to what modifications to the derived equations must be made in order to adapt them to use with three-dimensional surfaces. It is the purpose of this section to deal with that question.

It seems reasonable to rule out the necessity of considering a turbulent boundary layer at the stagnation point of a blunt body. Experimental evidence indicates that turbulent boundary layers over blunt bodies are preceded by a run of laminar flow before transition to turbulence occurs providing the boundary layer flows over a reasonably smooth aerodynamic surface. The data of Stetson[2]

---

[1] W. H. Dorrance and F. J. Dore, *J. Aeronaut. Sci.*, vol. 21, no. 6, pp. 404–410, 1954.

[2] Kenneth F. Stetson, *J. Aerospace Sci.*, vol. 27, no. 2, pp. 81–91, 1960.

indicate that the transition point occurs at or slightly before the sonic point on a blunt body even under the most unfavorable conditions for laminar boundary layer over a smooth surface. This being so, it seems necessary to consider only the effects of body shape and pressure gradient upon turbulent-boundary-layer characteristics in regions somewhat downstream from the stagnation point of a blunt body. We shall develop a theory for treating such cases in that which follows.

A useful approximate method of calculating turbulent-boundary-layer heat transfer at the point of highest heat transfer, the sonic point, has been developed by Sibulkin.[1] Sibulkin's method is based upon using skin-friction and heat-transfer formulas for an incompressible turbulent boundary layer applied to axisymmetric bodies with the fluid properties evaluated at the value of the Eckert's reference enthalpy at the sonic point. The procedure used is analogous to the procedure for calculating heat transfer and skin friction of a flat-plate compressible laminar boundary layer which was described in Sec. 5-10.

Our theory for accounting for the effects of pressure gradient and body shape at and beyond the sonic point begins with the momentum integral for steady boundary-layer flow over an axisymmetric body as given by Young,[2] for example. This is

$$\frac{\tau_w}{\rho_e u_e^2} = \frac{C_f}{2} = \frac{d\theta}{ds} + \theta\left(\frac{2 + H_f}{u_e}\frac{du_e}{ds} + \frac{1}{\rho_e}\frac{d\rho_e}{ds} + \frac{1}{r_0}\frac{dr_0}{ds}\right) \qquad (8\text{-}123)$$

where

$$H_f = \frac{\delta^*}{\theta} \qquad (8\text{-}124)$$

$$\delta^* \simeq \int_0^\delta \left(1 - \frac{\rho u}{\rho_e u_e}\right) dy \qquad (8\text{-}125)$$

and

$$\theta \simeq \int_0^\delta \frac{\rho u}{\rho_e u_e}\left(1 - \frac{u}{u_e}\right) dy \qquad (8\text{-}126)$$

Equation (8-123) can be rearranged for our purposes to read

$$\frac{C_f}{2} = \frac{1}{\rho_e u_e^2 r_0}\frac{d}{ds}(\rho_e u_e^2 r_0 \theta) + \theta\left(\frac{H_f}{u_e}\frac{du_e}{ds}\right) \qquad (8\text{-}127)$$

[1] Merwin Sibulkin, *Jet Propulsion*, vol. 28, no. 8, pp. 548–554, 1958.

[2] A. D. Young, "Modern Developments in Fluid Dynamics: High Speed Flow," in L. Howarth (ed.), sec. X, Boundary Layers, p. 395, Oxford University Press, New York, 1953.

Now most hypersonic bodies with which we are concerned are apt to be characterized by one or both of the following considerations:

1. The wall temperature is less than or the same order of magnitude as the local free-stream temperature T_e.

2. Favorable pressure gradients predominate if there is an appreciable pressure gradient at all.

Both of the above conditions make their effect in such a way as to lead to relatively high values of the momentum thickness θ and low values of the displacement thickness δ^* and thus can lead to low values of the form factor H_f. If H_f has a small magnitude, its influence upon the solution of the differential equation (8-127) is negligible and the last term of Eq. (8-127) might reasonably be neglected for $H_f \simeq 0$. We shall assume that $H_f \simeq 0$ for cases of interest to us here. [Use of flat-plate values of δ^* and θ to calculate H_f for $T_w \ll T_e$ and $M_e \leq 3$ leads to the result that $H_f \simeq 0$ compared with values of $1 < H_f < 5$ when adverse pressure gradients are present and $T_w > T_e$. For the latter conditions the term involving H_f could not reasonably be neglected in solving Eq. (8-127), of course.]

Accepting the above discussion, Eq. (8-127) becomes

$$\rho_e u_e^2 r_0 \frac{C_f}{2} = \frac{d}{ds} \left(\rho_e u_e^2 r_0 \theta \right) \tag{8-128}$$

Making use of Eqs. (7-67b) and (7-72), we find that

$$\theta = \frac{\mu_w}{K \rho_e u_e} \exp\left[-K\phi(0)\right] I_5 \tag{8-129}$$

where, as before, from Eqs. (7-67b), (7-70b), and (7-87),

$$I_5 = E^2 \int_0^1 \left(\frac{\rho}{\rho_w} \right)^{3/2} (z - z^2) \exp\left(I_4\right) dz \tag{8-130}$$

$$I_4 = E \int_0^z \left(\frac{\rho}{\rho_w} \right)^{1/2} d\xi \tag{8-131}$$

and for this development only

$$z = \frac{u}{u_e} \tag{8-132}$$

Use of Eq. (8-129) with Eq. (8-128) yields, if μ_w is constant,

$$C_f \rho_e u_e^2 r_0 \, ds = \frac{2}{K} \mu_w \exp\left[-K\phi(0)\right] d(u_e r_0 I_5) \tag{8-133}$$

Whence, integrating the right- and left-hand sides from $s = 0$ to $s = s$ and from $u_e r_0 I_5 = 0$ to $u_e r_0 I_5 = u_e(s) r_0(s) I_5(s)$ and using $\mu \propto T^{0.76}$, $K = 0.393$, and $\phi(0) = 6.53$ as before, we arrive at

$$C_f \bar{R}_e = 0.389 \left(\frac{T_w}{T_e} \right)^{0.76} I_5 \qquad (8\text{-}134)$$

where

$$\bar{R}_e = \frac{\displaystyle\int_0^s \rho_e u_e^2 r_0 \, ds}{\mu_e u_e r_0} \qquad (8\text{-}135)$$

and, as before, we assume the variation of C_f with s of secondary importance in evaluating the integral on the left-hand side of Eq. (8-133) as was done in deriving Eq. (7-85).

Note that, when ρ_e, u_e, and r_0 are all constant (flat plate or circular cylinder), then

$$\bar{R}_e = R_e = \frac{\rho_e u_e s}{\mu_e}$$

and Eq. (8-134) becomes Eq. (7-85) derived for the flat-plate case.

For a right circular cone u_e and ρ_e are constant and $r = s \sin \omega$, where ω is the cone semivertex angle, and for this case Eq. (8-135) yields

$$\bar{R}_e = \frac{R_e}{2}$$

a result obtained earlier by Van Driest.[1]

Equations (8-134) and (8-135) are remarkably useful in that they allow the previously derived and calculated values of skin-friction coefficient as plotted in Figs. 7-5 through 7-9 to be used with any axisymmetric body of revolution providing that the assumptions used in arriving at Eqs. (8-134) and (8-135) are respected. It is apparent that these equations can be used with Eq. (7-88) to account for the effects of dissociation in the boundary layer over an axisymmetric body in a manner completely analogous to the manner previously described for the flat-plate case in Chap. 7. That is, given a point on an axisymmetric body, Eq. (8-135) is used to calculate the value of $\bar{R}_e$ for that point. The appropriate figure among Figs. 7-6 through 7-9 is then entered to obtain C_{f_0} for the point. Given the equilibrium values of dissociation at the outer edge of the boundary layer and at the surface, α_e and α_w, respectively, Eq. (7-88) or Fig. 7-5 can then be used to obtain C_f/C_{f_0}. C_f then is the product of the C_{f_0} determined by the former step and the ratio C_f/C_{f_0} determined by the latter step.

[1] E. R. Van Driest, *J. Aeronaut. Sci.*, vol. 19, no. 1, pp. 55–57, 1952.

Before turning to the calculation of local heat-transfer rates, it is appropriate to restate the assumptions involved in arriving at Eqs. (8-134) and (8-135). They are as follows:

1. $H_f \to 0$. This is appropriate when a favorable pressure gradient and/or a cool wall ($T_w < T_e$) is present.

2. The expression for $\theta(s)$ derived for a flat plate in the absence of a pressure gradient, Eq. (8-129), applies in the presence of a pressure gradient. In essence, this assumption implies that the velocity-distribution correlation, Eq. (7-20c), should correlate velocity-distribution data in the presence of a pressure gradient with the empirical constants unchanged. Experimental data in this regime appear to support this assumption.[1]

For flow conditions under which $H_f \gg 0$ and/or strong pressure gradients are present, there is no reason to believe that Eqs. (8-134) and (8-135) would be applicable, nor are they intended to be so.

A relation between C_H and C_f was derived in Chap. 7 under conditions of zero pressure gradient and zero mass transfer. This relation is Eq. (7-66a) and is

$$\frac{C_H}{C_f/2} = P^{-2/3}\left[1 + (L^{2/3} - 1)\frac{h_c}{I_r - h_w}\right] \tag{8-136}$$

where, from Eqs. (7-49), (7-51), and (7-66b)

$$-\dot{q}_w = C_H \rho_e u_e (I_r - h_w) \tag{8-137}$$

$$I_r = h_e + r\frac{u_e^2}{2} \tag{8-138}$$

$$r = P^{1/3} \tag{8-139}$$

and from Eq. (7-66c), for the dissociation reaction,

$$h_c = (\alpha_e - \alpha_w)h_A^0 \tag{8-140}$$

We shall assume that this expression is adequate when favorable pressure gradients are present and rely upon comparisons with experiment to justify its use. Equation (8-136) is proper only in the limit as $dp_e/ds \to 0$.

8-8. Conclusions. In this chapter we have derived several useful expressions for calculating compressible turbulent-boundary-layer skin friction and heat transfer under a variety of conditions. As a starting point, the equations for the reacting, compressible turbulent boundary layer which were derived in Sec. 7-2 were used. It was found that, when i chemical species are present, there were $i + 7$

[1] See, for example, Francis H. Clauser, *J. Aeronaut. Sci.*, vol. 21, no. 1, pp. 91–108, 1954.

unknowns and only $i + 4$ equations available with which to determine the unknowns. The remaining 3 equations necessary were found by suggesting the plausible and empirically supported assumptions that $L_T = 1$, $P_T = 1$, and $\epsilon = \rho K^2 y^2 (du/dy)$.

The question of the effect of mass transfer upon the nondissociating compressible turbulent boundary layer was taken up in Secs. 8-2 and 8-3. It was found that useful expressions for $2C_H/C_f$, r, and C_f could be derived in a fairly simple manner which include the effect of a mass-transfer parameter but which also contain a parameter u_L/u_e. It was pointed out that velocity-distribution data for a turbulent boundary layer taken with mass transfer present at supersonic Mach numbers may cast some light upon the parameter u_L/u_e.

Sections 8-4, 8-5, and 8-6 dealt with the effects of chemical reactions, either exothermic or endothermic, upon heat transfer in a compressible turbulent boundary layer. An example solution for turbulent heat transfer to a sublimating surface with exothermic reactions taking place near the surface was presented to illustrate the use of the equations which were derived in this section. It was found for the conditions of the example problem that the heat released in the assumed exothermic reaction predominated over the mass-transfer effects and heat of vaporization.

All the above work was devoted to the flow of a compressible turbulent boundary layer over a two-dimensional flat plate. In Sec. 8-7 methods were derived which enable the flat-plate results to be used with axisymmetric bodies under certain circumstances.

In closing this chapter it seems appropriate to the author to state that, while he is not completely satisfied with the assumptions which must be made in deriving useful expressions for calculation of skin friction and heat transfer for a reacting, compressible turbulent boundary layer, nevertheless it is gratifying to note that the expressions which result agree with experiment under widely different conditions. We believe that this chapter has presented something useful for the gas dynamicist who desires to calculate engineering quantities with some confidence and also some underlying fundamentals from which the research worker can depart in the never-ending struggle to improve turbulent-boundary-layer theory.

9

Thermodynamic Properties of Dilute
Gas Mixtures

9-1. Introduction. Those concerned with hypersonic gas dynamics deal with gas mixtures at high temperatures. Under some conditions the mixture is that of high-temperature air. Under more complex conditions, such as are present in a number of situations described in earlier chapters of this book, the gas mixture consists of products and reactants including air species and species injected into the boundary layer under diverse circumstances. Chemical reactions may occur within this gas mixture, and the specification of the thermodynamic properties for the equilibrium mixture becomes difficult. A method is required to determine the state variables: internal energy, enthalpy, entropy, and equilibrium composition of high-temperature gas mixtures.

As the gas temperature increases far beyond room temperature, the gas particles of a dilute gas mixture begin to assert their individuality. Depending upon the elements making up the gas molecules, various internal modes of energy storage begin to become activated. The atoms of the oxygen and nitrogen molecules, for example, begin to vibrate along the axis through the two atoms making up these diatomic molecules. Increase in temperature also results in disturbance of orbital electrons until temperatures are reached where electrons are induced to leave the molecules completely; that is, ionization occurs. It is not difficult to imagine that other, more complex, internal energy modes exist for polyatomic molecules.

It was early recognized that the methods of classical physics did not describe correctly the behavior of the thermodynamic properties of gas mixtures under the circumstances just described. Kinetic theory, which treats all molecules as simple mass points distinguished only by their differing masses, adequately described the thermodynamic properties of gas mixtures near room temperature. The treatment of gas mixtures at temperatures high enough that a significant portion of the molecules had their internal energy modes activated could not be adequately treated by kinetic theory alone. It remained for quantum theory and statistical mechanics to describe the behavior of gas mixtures at temperatures of interest to us here.

This chapter will describe how the thermodynamic properties and equilibrium constants for equilibrium gas mixtures can be obtained using the methods of statistical mechanics and quantum theory. This chapter is not intended to represent an exhaustive treatment of this branch of science. Rather, the objective is to illustrate how the methods of statistical mechanics are employed, where quantum theory enters into the calculations, and to present expressions for the thermodynamic properties and equilibrium constants which can be applied in problems such as those described in earlier chapters of this book.

9-2. Boltzmann Statistics. Air at conditions of standard temperature and pressure (STP) has a molecular number density of about 2.7×10^{19} molecules per cubic centimeter. Despite this enormous number of particles in such a small volume, the individual particles are relatively far apart at any given instant because of the tremendous concentration of matter at the position of any given molecule. Because the particles, on the average, travel distances between collisions far exceeding the effective range of interparticle potentials, these interparticle potentials have little bearing on the motion of the particles and can be neglected. Such would not be true, of course, for dense fluids or ionized gases for which the range of interparticle potentials is a significant fraction of the mean free path. Although it is conceptually possible to follow the motion of any given molecule in such a force-free field and to describe its behavior according to the laws of classical mechanics, the description of the aggregate on a macroscopic scale by this means would be a stupifying task even using modern computing machines. The situation seems made to order for a statistical approach.

The correct statistical approach to the problem of specifying the macroscopic behavior of such a large number of particles in a closed system is to consider each particle indistinguishable. In that which follows we shall knowingly use a statistical approach somewhat easier to follow which incorporates the concept that the particles of each species making up a closed system are distinguishable; that is, by some means we can tag each particle and keep track of its position and behavior if we choose to do so. While we know for quantum-theory reasons that this concept leads to serious errors at very low temperatures and at higher temperatures for particles with low mass (protons, electrons, and sometimes hydrogen), nevertheless the statistical method here described, Boltzmann statistics, can be shown to correspond to the formally correct statistical approach at temperatures of interest for particles of interest to us here.

Consider that we are dealing with a closed system of N_0 particles such that

$$N_0 = \sum_i n_i \qquad (9\text{-}1)$$

where we have arbitrarily chosen the total number of particles as N_0, Avogadro's number, and where the particles n_i have energy ϵ_i. Then the total energy for 1 mole of gas is

$$E = \sum_i n_i \epsilon_i \qquad (9\text{-}2)$$

where, because the total number of particles is N_0, the energy E is the energy for 1 g mole of the gas mixture.

Now the probability W that the system exists in any given configuration or state is proportional to the number of different ways the N_0 distinguishable particles can be arranged in the specified state, that is, arranged so that n_1 particles have energy ϵ_1, n_2 particles have energy ϵ_2, n_3 particles have energy ϵ_3, etc. This is the probability that N_0 objects can be separated into a set of boxes such that there are n_1 in the first box, n_2 in the second, etc. Thus the total probability is

$$W = (\text{const}) \frac{N_0!}{n_1! \, n_2! \, n_3! \cdots ! \, n_i!} \qquad (9\text{-}3)$$

where the constant in Eq. (9-3) is determined by the finite total number of possible arrangements including those not satisfying the requirement that n_1 be in the first box, n_2 be in the second box, etc.

For example, let us assume that $N_0 = 4$ and that $n_1 = 1$, $n_2 = 2$, and $n_3 = 1$. Then the number of different ways this distribution can be achieved is given by

$$\frac{N_0!}{n_1! \, n_2! \, n_3!} = \frac{1 \cdot 2 \cdot 3 \cdot 4}{1 \cdot 1 \cdot 2 \cdot 1} = 12$$

and the alternatives are as shown in Fig. 9-1.

Now, Eq. (9-3) can be written

$$\log W = \log N_0! - \sum_i \log (n_i!) + \text{const}$$

but from Stirling's formula,[1] for large numbers n,

$$\log n! \simeq n \log n - n$$

hence, since $\sum_1 n_i = N_0$,

$$\log W = N_0 \log N_0 - \sum_i n_i \log n_i + \text{const} \tag{9-4}$$

The actual state of the system will be close to that state having

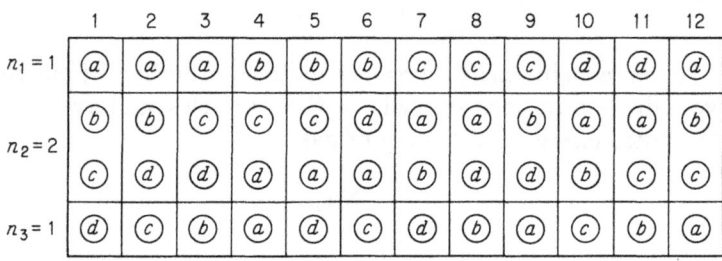

Fig. 9-1. The number of ways four distinguishable particles can be arranged in three separate piles such that one particle is in the first pile, two in the second, and one in the third.

maximum probability. When the system is in the state of maximum probability, the variation $\delta(\log W)$ must be zero; that is,

$$\delta(\log W) = - \sum_i \delta(n_i \log n_i) = - \sum_i \delta n_i (1 + \log n_i) = 0 \tag{9-5}$$

But not all δn_i are permissible, since, from Eqs. (9-1) and (9-2),

$$\delta N_0 = \sum_i \delta n_i = 0 \tag{9-6}$$

and

$$\delta E = \sum_i \epsilon_i \, \delta n_i = 0 \tag{9-7}$$

Equations (9-6) and (9-7) provide constraints on the variations δn_i permissible. Thus, making use of Lagrange's method of multipliers,[2] we obtain for our δn_i

$$\sum_i \delta n_i (1 + \log n_i + \lambda_1 + \lambda_2 \epsilon_i) = 0 \tag{9-8}$$

where λ_1 and λ_2 are the Lagrangian multipliers. Since the δn_i are independent of each other, we find that the solution to Eq. (9-8) is

$$n_i = \alpha \exp(-\beta \epsilon_i) \tag{9-9a}$$

where

$$\alpha = \exp[-(\lambda_1 + 1)] \tag{9-9b}$$

$$\beta = \lambda_2 \tag{9-9c}$$

[1] See, for example, I. S. Sokolnikoff and E. S. Sokolnikoff, "Higher Mathematics for Engineers and Physicists," pp. 508–511, McGraw-Hill Book Company, Inc., New York, 1941.

[2] *Ibid.*, pp. 163–167.

Equation (9-9a) is the classical or Maxwell-Boltzmann distribution law. The multiplier β will be identified with the kinetic theory temperature in the next section.

9-3. The Multiplier β. While Eq. (9-9a) is perfectly general, let us apply it to a system of particles possessing kinetic energy only. Such a system might represent a monatomic gas at room temperature, for example. Then if we assume that each gas particle has mass m, we have

$$\epsilon = \frac{m}{2}\,(v_x^2 + v_y^2 + v_z^2) \tag{9-10}$$

and Eq. (9-9a) becomes

$$n_i = \alpha \exp\left[-\frac{\beta m}{2}\,(v_x^2 + v_y^2 + v_z^2)\right] \tag{9-11}$$

where we have made the classical theory assumption that the particle energies ϵ_i form a continuum; that is, ϵ_i can take any value from 0 to ∞. Under these circumstances Eq. (9-11) is recognized as the classical distribution law which gives the number of particles n_i with velocity components v_x, v_y, and v_z in the interval between v_x and $v_x + dv_x$, v_y and $v_y + dv_y$, and v_z and $v_z + dv_z$, respectively. Thus Eq. (9-1) for this case is replaced by the integral

$$N_0 = \int_{-\infty}^{\infty} \int_{-\infty}^{\infty} \int_{-\infty}^{\infty} n_i\, dv_x\, dv_y\, dv_z \tag{9-12}$$

and Eq. (9-2) is replaced by the integral

$$E = \int_{-\infty}^{\infty} \int_{-\infty}^{\infty} \int_{-\infty}^{\infty} n_i \epsilon(v_x,v_y,v_z)\, dv_x\, dv_y\, dv_z \tag{9-13}$$

Then, using Eq. (9-11) in Eq. (9-12), there results

$$N_0 = \alpha(I_1)^3 \tag{9-14}$$

where
$$I_1 = \int_{-\infty}^{\infty} \exp\left(-\frac{\beta m}{2}\,v^2\right) dv = \left(\frac{2\pi}{\beta m}\right)^{1/2} \tag{9-15}$$

or, combining Eqs. (9-14) and (9-15),

$$\alpha = N_0 \left(\frac{\beta m}{2\pi}\right)^{3/2} \tag{9-16}$$

Thus, from Eqs. (9-11) and (9-16),

$$n_i \epsilon\, dv_x\, dv_y\, dv_z = N_0 \left(\frac{\beta m}{2\pi}\right)^{3/2} \epsilon \exp\left(-\beta\epsilon\right) dv_x\, dv_y\, dv_z \tag{9-17}$$

where $\epsilon(v_x,v_y,v_z)$ is given by Eq. (9-10). Using Eq. (9-17) in Eq. (9-13) gives

$$E = \frac{3}{2} mN_0I_2 \tag{9-18}$$

where

$$I_2 = \left(\frac{\beta m}{2\pi}\right)^{1/2} \int_{-\infty}^{\infty} v^2 \exp\left(-\frac{\beta m}{2} v^2\right) dv = \frac{1}{\beta m} \tag{9-19}$$

Hence, from Eqs. (9-18) and (9-19),

$$E = \frac{3}{2} \frac{N_0}{\beta} \tag{9-20}$$

Kinetic theory gives[1]

$$E = \frac{3}{2} N_0kT \tag{9-21}$$

Comparing Eq. (9-20) with (9-21) we find that the Lagrangian multiplier β is identified with the temperature according to the equation

$$\beta = \frac{1}{kT} \tag{9-22}$$

and thus Eq. (9-9a) becomes

$$n_i = \alpha \exp\left(-\frac{\epsilon_i}{kT}\right) \tag{9-23}$$

Let us now assume that there are a number of particles n_0 which possess a minimum of energy ϵ_0. Then, from Eq. (9-23),

$$\alpha = n_0 \exp \frac{\epsilon_0}{kT} \tag{9-24}$$

from which, combining Eqs. (9-23) and (9-24), we obtain

$$n_i = n_0 \exp \frac{-(\epsilon_i - \epsilon_0)}{kT} \tag{9-25}$$

Equation (9-25) gives the number of particles n_i which possess energy ϵ_i in terms of the number of particles n_0 having the ground level of energy ϵ_0 at any temperature T.

9-4. The Partition Functions. To this point no use has been made of quantum theory. Now we shall turn aside from classical theory and introduce the quantum-theory postulate that the energy levels ϵ_i do not form a continuum and, in fact, take discrete values specifically determined by the formal application of quantum mechanics. That is, there are discrete quantum states of energy ϵ_i

[1] See, for example, R. D. Present, "Kinetic Theory of Gases," pp. 12–20, McGraw-Hill Book Company, Inc., New York, 1958.

populated according to the Maxwell-Boltzmann distribution law, Eq. (9-25). There may be one or more quantum states at each energy level; that is, the energy levels are degenerate. The number of quantum states at an energy level ϵ_i is g_i. Thus the sum over all quantum states having energy ϵ_i, referred to the ground-state energy ϵ_0, is

$$Q = \sum_i g_i \exp \frac{-(\epsilon_i - \epsilon_0)}{kT} \qquad (9\text{-}26)$$

and Q is called the partition function. The summation in Eq. (9-26) is sometimes written as

$$Q = \sum_i \exp \frac{-(\epsilon_i - \epsilon_0)}{kT} \qquad (9\text{-}27)$$

where it is understood that the summation is made over all quantum states of the molecules.

The quantum states of interest to us here include the states of translational energy, rotational energy, vibrational energy, and electronic energy. These energy levels are mentioned in the decreasing order that they are significantly occupied at ordinary temperatures. It will be shown later that the translational-energy states are determined by a set of quantum numbers l, the rotational-energy states by quantum numbers j, the vibrational-energy states by quantum numbers v, and the electronic-energy states by quantum numbers which we shall represent by s. If, as is usually true, the various modes do not significantly interact with one another, then it can be written that

$$\epsilon_i = \epsilon(l) + \epsilon(j) + \epsilon(v) + \epsilon(s)$$

and, in terms of energy above the ground-state energy,

$$\epsilon_i - \epsilon_0 = [\epsilon(l) - \epsilon_0(l)] + [\epsilon(j) - \epsilon_0(j)]$$
$$+ [\epsilon(v) - \epsilon_0(v)] + [\epsilon(s) - \epsilon_0(s)] \quad (9\text{-}28)$$

The writing of Eq. (9-28) neglects the effect of such interactions as vibration upon rotation, for example. As a result of this assumption, we can treat the energy stored in the various modes of a diatomic molecule separately. That is, the energy of a translating, rotating, vibrating, and electronically disturbed diatomic molecule is the sum of the kinetic energy of a mass point, the kinetic and potential energy of a harmonic oscillator, the rotational energy of a rigid rotator, and the energy stored in disturbed electronic orbits of a fixed molecule, all particles, of course, having masses related to the mass of the diatomic molecule. Each of these energy modes can then be treated separately when quantum theory is used to obtain the partition functions as later developments will show. When Eq. (9-28) is substituted into Eq. (9-26), there results

$$Q = \sum_l \sum_j \sum_v \sum_s g_{l,j,v,s}$$
$$\times \exp \frac{-[\epsilon(l) - \epsilon_0(l)] - [\epsilon(j) - \epsilon_0(j)] - [\epsilon(v) - \epsilon_0(v)] - [\epsilon(s) - \epsilon_0(s)]}{kT}$$

however, when the identity $\exp (x + y) = \exp x \exp y$ is used with the above equation, we obtain

$$Q = Q_l Q_j Q_v Q_s \tag{9-29}$$

where
$$Q_n = \sum_n g_n \exp \frac{-[\epsilon(n) - \epsilon_0(n)]}{kT} \qquad n = l, j, v, \text{ or } s \tag{9-30}$$

and the subscripts l, j, v, and s refer to the translational, rotational, vibrational, and electronic-energy states, respectively. In the next section we shall derive equations for the thermodynamic functions in terms of the partition functions. Following that we shall show how the partition functions are derived using the methods of quantum mechanics.

9-5. The Thermodynamic Functions. The thermodynamic relations for 1 mole of molecules in a gaseous state will now be derived in terms of the partition functions. From Eqs. (9-1) and (9-25), for 1 mole of molecules having degenerate quantum-energy levels,

$$N_0 = n_0 \sum_i g_i \exp \frac{-(\epsilon_i - \epsilon_0)}{kT} \tag{9-31}$$

or, in view of Eq. (9-26),

$$n_0 = \frac{N_0}{Q} \tag{9-32}$$

Now, 1 mole of molecules possesses an energy $E - E_0$ referred to the energy of 1 mole of molecules in the ground state given by

$$E - E_0 = n_0 \sum_i g_i \epsilon_i \exp \frac{-(\epsilon_i - \epsilon_0)}{kT} - N_0 \epsilon_0 \tag{9-33}$$

where the summation is taken over all quantum-energy levels. Making use of Eqs. (9-26), (9-31), and (9-32), Eq. (9-33) becomes

$$E - E_0 = \frac{N_0}{Q} \left[\sum_i g_i (\epsilon_i - \epsilon_0) \exp \frac{-(\epsilon_i - \epsilon_0)}{kT} \right] \tag{9-34}$$

or
$$E - E_0 = N_0 k T^2 \left[\frac{d}{dT} (\log Q) \right]_V$$

where the subscript V indicates differentiation at constant volume. Thus, since $N_0 k = R$,

$$\frac{E - E_0}{RT} = T\left[\frac{d}{dT}(\log Q)\right]_V \tag{9-35}$$

Equation (9-35) gives a relation between the energy of 1 mole of molecules and the partition function Q. Further relations are obtained as follows:

$$\bar{C}_v = \left(\frac{\partial E}{\partial T}\right)_V = \left\{\frac{\partial}{\partial T}\left[RT^2 \frac{\partial}{\partial T}(\log Q)\right]\right\}_V \tag{9-36}$$

or

$$\frac{\bar{C}_v}{R} = 2T\left[\frac{\partial}{\partial T}(\log Q)\right]_V + T^2\left[\frac{\partial^2}{\partial T^2}(\log Q)\right]_V \tag{9-37}$$

also, from thermodynamics,

$$\bar{C}_p = \bar{C}_v + R \tag{9-38}$$

Equations (9-37) and (9-38) give relations between the molar heat capacities and the partition function.

Now, from thermodynamics,

$$H = E + pV \tag{9-39}$$

and for a perfect gas such as we assume here, from kinetic theory,

$$pV = RT \tag{9-40}$$

Hence, using Eqs. (9-35), (9-39), and (9-40),

$$H - H_0 = E - E_0 + RT$$

or

$$\frac{H - H_0}{RT} = 1 + T\left[\frac{\partial}{\partial T}(\log Q)\right]_V \tag{9-41}$$

Equation (9-41) relates enthalpy to the partition function Q.

The entropy S is given by the relation from thermodynamics

$$S = \int_0^T \bar{C}_v \, d(\log T)$$

or, using Eq. (9-36),

$$S = \int_0^T \left\{\frac{\partial}{\partial T}\left[RT^2 \frac{\partial}{\partial T}(\log Q)\right]\right\}_V d(\log T)$$

from which we obtain

$$S = \left\{\int_0^T \frac{1}{T} d\left[RT^2 \frac{\partial}{\partial T}(\log Q)\right]\right\}_V$$

or, integrating by parts,

$$S = \left\{ RT \frac{\partial}{\partial T} (\log Q) + \int_0^T R \frac{\partial}{\partial T} (\log Q)\, dT \right\}_V$$

Thus

$$\frac{S - S_0}{R} = \left\{ T \left[\frac{\partial}{\partial T} (\log Q) \right] + \log Q \right\}_V$$

and so

$$\frac{S - S_0}{R} = \left[\frac{\partial}{\partial T} (T \log Q) \right]_V \tag{9-42}$$

where

$$\frac{S_0}{R} = \log Q_0 \tag{9-43}$$

Equation (9-42) gives the entropy in terms of the partition function. The subscript 0 on all thermodynamic variables will be taken to indicate the value of the quantity at absolute zero temperature.

Another useful thermodynamic function is the free energy F defined by the equation

$$F = E + RT - TS \tag{9-44}$$

where $F_0 = E_0$. Using Eqs. (9-35) and (9-42) with Eq. (9-44) yields

$$\frac{F - H_0}{RT} = 1 - \log Q \tag{9-45}$$

Equations (9-35), (9-37), (9-38), (9-41), (9-42), and (9-45) relate $E - E_0$, $\bar{C}_v$, $\bar{C}_p$, $H - H_0$, $S - S_0$, and $F - H_0$ to the partition function Q. It remains to determine the partition functions for the various energy modes using the methods of quantum mechanics.

9-6. Partition Functions from Quantum Mechanics. In Sec. 9-5 we found how the thermodynamic functions are related to the partition functions, which are as yet unspecified. In this section we shall illustrate how the partition functions for a simple gas are obtained using the methods of quantum mechanics.

Let us postulate a function Ψ, called a wave function, such that the product of the complex conjugate Ψ^* with Ψ gives the probability that a particle exists at the point in phase space at which Ψ and Ψ^* are evaluated.[1] Then for a single particle the integral of $\Psi^*\Psi$ over the entire space (x,y,z) must be unity; that is,

$$\int_{-\infty}^{\infty} \int_{-\infty}^{\infty} \int_{-\infty}^{\infty} \Psi^*\Psi\, dx\, dy\, dz = 1 \tag{9-46}$$

We speak of Eq. (9-46) as the requirement that Ψ be integrable squared. Furthermore, in quantum mechanics the dynamical variables assigned to the particle are identified with operators so that the

---

[1] The complex conjugate of $x + iy$ is $x - iy$. The complex conjugate of $= \psi_x + i\psi_y$ is $\psi_x - i\psi_y$, where ψ_x and ψ_y are functions of real numbers only.

following correspondence between dynamical variables and operators is followed:[1]

Dynamical variable		Operator
x	$\rightarrow$	x
y	$\rightarrow$	y
z	$\rightarrow$	z
mv_x	$\rightarrow$	$\dfrac{h}{2\pi i}\dfrac{\partial}{\partial x}$
mv_y	$\rightarrow$	$\dfrac{h}{2\pi i}\dfrac{\partial}{\partial y}$
mv_z	$\rightarrow$	$\dfrac{h}{2\pi i}\dfrac{\partial}{\partial z}$
$\phi(x)$	$\rightarrow$	$\phi(x)$
$\phi(y)$	$\rightarrow$	$\phi(y)$
$\phi(z)$	$\rightarrow$	$\phi(z)$
w(total energy)	$\rightarrow$	$-\dfrac{h}{2\pi i}\dfrac{\partial}{\partial t}$

where h is the Planck constant.

The average value of any dynamical variable which corresponds to an operator q is then given by

$$\bar{q} = \int_{-\infty}^{\infty} \int_{-\infty}^{\infty} \int_{-\infty}^{\infty} \Psi^* q \Psi \, dx \, dy \, dz \qquad (9\text{-}47)$$

where by average value we mean the average of a large number of measurements. Furthermore, if the equation for the total energy of a particle is the sum of the kinetic and potential energy and is

$$w = \frac{1}{2} m(v_x^2 + v_y^2 + v_z^2) + \phi(x,y,z)$$

where ϕ represents the potential energy of the particle, then substitution of the operators into the same equation according to the rules outlined above gives

$$-\frac{h}{2\pi i}\frac{\partial}{\partial t} = w = -\frac{h^2}{8m\pi^2}\left(\frac{\partial^2}{\partial x^2} + \frac{\partial^2}{\partial y^2} + \frac{\partial^2}{\partial z^2}\right) + \phi(x,y,z)$$

---

[1] See, for example, Chalmers W. Sherwin, "Introduction to Quantum Mechanics," pp. 13–19, Holt, Rinehart and Winston, Inc., New York, 1959. The correspondences shown are only one of many possible.

Operating on Ψ we obtain the Schrödinger wave equation for our single particle; viz.,

$$\frac{\partial^2 \Psi}{\partial x^2} + \frac{\partial^2 \Psi}{\partial y^2} + \frac{\partial^2 \Psi}{\partial z^2} + \frac{8\pi^2 m}{h^2}(w - \phi)\Psi = 0 \qquad (9\text{-}48)$$

where Ψ is required to be single-valued, continuous, and finite throughout the configuration space (x,y,z).

Now we are in a position to interpret the wave equation (9-48). If we assume that the operator q in Eq. (9-47) is given by

$$q = \frac{h^2}{8m\pi^2}\left(\frac{\partial^2}{\partial x^2} + \frac{\partial^2}{\partial y^2} + \frac{\partial^2}{\partial z^2}\right) + (w - \phi)$$

then we find that application of Eq. (9-47) with subsequent substitution back from operator language to dynamical variables gives

$$\bar{w} = \tfrac{1}{2}m(\overline{v_x^2} + \overline{v_y^2} + \overline{v_z^2}) + \bar{\phi}$$

That is, quantum mechanics says that the average value of the total energy is equal to the sum of the average value of the kinetic energy and the average value of the potential energy but makes no explicit statement about the relation among total energy, kinetic energy, and the potential energy at any time at any point in configuration space. Rather, the best we can do is specify·that the equation for the Ψ function is satisfied and, having been satisfied, yields a function Ψ which can be used to obtain average values of the variables. That is, the position of a particle and its velocity cannot be simultaneously specified in accordance with Heisenberg's uncertainty principle. Thus is illustrated the statistical nature of quantum mechanics.

Let us use Eq. (9-48) to obtain our quantum-energy values. Consider first a simple mass point moving in a field-free space. Such an idealization fits the description of a monatomic gas particle moving freely in a box. Since the interparticle potentials have short range relative to the mean free paths of such particles, we set $\phi \equiv 0$ in Eq. (9-48) and obtain

$$\frac{\partial^2 \Psi}{\partial x^2} + \frac{\partial^2 \Psi}{\partial y^2} + \frac{\partial^2 \Psi}{\partial z^2} + \frac{8\pi^2 m}{h^2} w\Psi = 0 \qquad (9\text{-}49)$$

If the particle is assumed to be contained in a cubic box of dimensions (a,a,a) with volume a^3, then Ψ must be zero whenever x, y, or z are zero or a, assuming that the coordinate system has an origin at a corner of the box and the x, y, and z axes run along the edges of the box. The only solution to Eq. (9-49) which has Ψ continuous, finite, and single-valued within the box and zero at the walls of the box occurs for values of w given by

$$w = \frac{h^2}{8mV^{2/3}} (l_x^2 + l_y^2 + l_z^2) = \epsilon(l) - \epsilon_0(l) \tag{9-50}$$

since $\epsilon_0(l) = \epsilon(l = 0) = 0$. The equation for Ψ for this case is[1]

$$\Psi = \left(\frac{2}{a}\right)^{3/2} \sin\left(\pi \frac{l_x x}{a}\right) \sin\left(\pi \frac{l_y y}{a}\right) \sin\left(\pi \frac{l_z z}{a}\right) \exp \frac{-2\pi i w t}{h} \tag{9-51}$$

where $V = a^3$, $l^2 = l_x^2 + l_y^2 + l_z^2$, and l_x, l_y, and l_z are integers and are the quantum numbers for translational motion. Thus we find that the energy levels allowed in translational motion are discrete and defined by Eq. (9-50). As the numbers l_x, l_y, and l_z become large, the energy levels $\epsilon(l)$ are closer together, so that Eq. (9-50) approaches (but never becomes) a continuum as is assumed by classical theory.

Equation (9-50) can be used with Eq. (9-30) to obtain .

$$Q_l = \left[\sum_{l=1}^{\infty} \exp -\left(\frac{h^2 l^2}{8V^{2/3} mkT}\right) \right]^3 \tag{9-52}$$

where the translational levels corresponding to each degree of freedom are nondegenerate ($g_i = 1$) and $\epsilon_0(l) = 0$. Since the argument of the exponent changes very slowly with l for large l, we replace the summation with an integration and obtain

$$Q_l = 8V\left(\frac{2mkT}{h^2}\right)^{3/2} (I_3)^3 \tag{9-53}$$

where $\quad I_3 = \frac{h}{2V^{1/3}(2mkT)^{1/2}} \int_0^{\infty} \exp\left(-\frac{h^2 l^2}{8V^{2/3} mkT}\right)^2 dl = \left(\frac{\pi}{4}\right)^{1/2} \tag{9-54}$

finally, combining Eqs. (9-53) and (9-54),

$$Q_l = V\left(\frac{2\pi mkT}{h^2}\right)^{3/2} = \frac{RT^{5/2}}{p} \left(\frac{2\pi mk}{h^2}\right)^{3/2} \tag{9-55}$$

Equation (9-55) is the partition function for translational energy.

For most cases of interest we can represent the rotational mode of a linear molecule as a simple, rigid rotator. That is,

$$\epsilon(j) = \tfrac{1}{2} I\omega^2 \tag{9-56}$$

where I is the moment of inertia of the system about the axis of rotation and ω is the angular velocity. The axis of rotation for a

---

[1] See, for example, F. K. Richtmyer, E. H. Kennard, and T. Lauritsen, "Introduction to Modern Physics," 5th ed., pp. 199–200, McGraw-Hill Book Company, Inc., New York, 1955.

rigid dumbbell model is the center of mass the location of which is at a distance

$$r_1 = \frac{m_2}{m_1 + m_2} r \qquad \text{from } m_1 \qquad (9\text{-}57)$$

and a distance
$$r_2 = \frac{m_1}{m_1 + m_2} r \qquad \text{from } m_2 \qquad (9\text{-}58)$$

where r is the distance between the two masses m_1 and m_2. Then, because

$$I = m_1 r_1^2 + m_2 r_2^2 \qquad (9\text{-}59)$$

substitution of (9-57) and (9-58) into (9-59) gives

$$I = \bar{m} r^2 \qquad (9\text{-}60)$$

where
$$\bar{m} = \frac{m_1 m_2}{m_1 + m_2} \qquad (9\text{-}61)$$

is the reduced mass of the system which then acts as a single mass $\bar{m}$ rotating about the center of gravity of the two masses m_1 and m_2 with a radius r_1. For such a rotator situated so that the center of gravity of the two masses is at the origin, the Schrödinger equation (9-48) becomes

$$\frac{\partial^2 \Psi}{\partial x^2} + \frac{\partial^2 \Psi}{\partial y^2} + \frac{\partial^2 \Psi}{\partial z^2} + \frac{8\pi^2 \bar{m}}{h^2} w \Psi = 0 \qquad (9\text{-}62)$$

where $x^2 + y^2 + z^2 = r^2$ is a constant. Ψ is finite for all x, y, and z, so that $x^2 + y^2 + z^2 \leq r^2$ and zero elsewhere. $\partial \Psi / \partial r = 0$, and the problem is usually solved in the spherical coordinates, so that $\Psi(x,y,z) \to \Psi(\theta,\phi)$ independent of r. A solution to Eq. (9-62) which is single-valued, finite, and continuous for $x^2 + y^2 + z^2 \leq r^2$ occurs only for values of w given by[1]

$$w = \frac{j(j+1)h^2}{8\pi^2 I} = \epsilon(j) - \epsilon_0(j) \qquad (9\text{-}63)$$

since $\epsilon_0(j) = \epsilon(j = 0) = 0$. It can be shown that the rotational-energy states are degenerate with degeneracy

$$g_j = (2j + 1) \qquad (9\text{-}64)$$

Furthermore, it can be shown that using Eqs. (9-63) and (9-64) in Eq. (9-30) and replacing the summation by integration give

$$Q_j = \frac{8\pi^2 I k T}{\alpha h^2} \qquad (9\text{-}65)$$

[1] See, for example, Samuel Glasstone, "Theoretical Chemistry," pp. 35–47, D. Van Nostrand Company, Inc., Princeton, N.J., 1944.

where α is a symmetry number[1] which takes integer values depending upon the structure of polyatomic molecules. For nonlinear polyatomic molecules,

$$Q_j = \left(\frac{8\pi^2 kT}{h^2}\right)^{3/2} \frac{(\pi I_x I_y I_z)^{1/2}}{\alpha}$$ (9-66)

where I_x, I_y, and I_z are the moments of inertia about the three axes of the polyatomic molecule and α is again a symmetry number.

For a simple harmonic, two-particle oscillator the Schrödinger equation is equivalent to that for a single particle of reduced mass $\bar{m}$ vibrating on a spring. The equation is

$$\frac{d^2\Psi}{dx^2} + \frac{8\pi^2 \bar{m}}{h^2}\left(w - \frac{1}{2}\bar{k}^2 x^2\right)\Psi = 0$$ (9-67)

where $\bar{k}$ is the spring constant of the oscillator. The solutions to this equation which are single-valued, finite, and continuous and which vanish at infinity exist only for values of w given by[2]

$$w = \epsilon(v) = \left(\frac{h}{2\pi}\frac{\bar{k}}{\bar{m}}\right)^{1/2}\left(v + \frac{1}{2}\right)$$

hence, since $\epsilon_0(v) = \epsilon_0(v = 0)$, then

$$\epsilon(v) - \epsilon_0(v) = \frac{h}{2\pi}\left(\frac{\bar{k}}{\bar{m}}\right)^{1/2} v$$ (9-68)

Substituting Eq. (9-68) into Eq. (9-30) and evaluating the resulting series give

$$Q_v = \left(1 - \exp\frac{h\nu'}{kT}\right)^{-1}$$ (9-69)

where the frequency of vibration ν', later shown to be the frequency of emitted light, is given by

$$\nu' = \frac{1}{2\pi}\left(\frac{\bar{k}}{\bar{m}}\right)^{1/2}$$ (9-70)

Thus are obtained the simplest expressions for the partition functions for the translational, rotational, and vibrational modes of diatomic molecules and the translational and rotational modes of polyatomic molecules under the assumption that the energy modes do not interact with one another. More complex cases are treated in the cited references, to which the interested reader is referred.

---

[1] α is the number of ways a polyatomic molecule can be superimposed upon itself by rotation of the molecule. For example, for the linear molecule CO_2 or OCO, α is 2; for N_2O or NNO, α is 1; for O_2 or OO, α is 2; and for NO, α is 1.
[2] Glasstone, *op. cit.*, pp. 28–35.

Before going on to give explicit expressions for the partition functions of diatomic gases, we shall explain briefly how spectroscopic measurements can be used to find the rotational moment of inertia I and the vibrational frequency ν'. These values are required before our equations for the rotational and vibrational partition function, Eqs. (9-65) and (9-69), can be applied.

9-7. Using Spectroscopic Data. The rotational moments of inertia I for a rigid rotator which appears in the rotational mode partition function is determined using spectroscopic data. According to quantum theory, when a rotator changes from one allowed energy level to another, light is emitted or absorbed with a frequency ν' given by the following remarkable equation attributed to Bohr:

$$h\nu' = \epsilon_u - \epsilon_l \tag{9-71}$$

where h is Planck's constant, ν' is the frequency of emitted or absorbed light, and the subscripts u and l refer to upper and lower energy levels, respectively. The emitted light can be discriminated spectroscopically according to its wavelength. Now, for a rigid rotator we have

$$\epsilon(j) = \frac{h^2 j(j+1)}{8\pi^2 I} \tag{9-63}$$

Hence, combining Eqs. (9-63) and (9-71), we obtain

$$h\nu' = \frac{h^2}{8\pi^2 I} [j_u(j_u + 1) - j_l(j_l + 1)] \tag{9-72}$$

where the subscripts u and l on j refer to upper and lower energy levels, respectively.

Now quantum-mechanical selection rules allow energy-level changes such that

$$\Delta j = \pm 1$$

only. Hence, letting $j_u = j_l + 1$ in Eq. (9-72), we obtain

$$h\nu' = \frac{h^2}{4\pi^2 I} (1 + j) \tag{9-73}$$

and the rotational spectrum should be equally spaced in units of $h/4\pi^2 I$. Furthermore, Eq. (9-73) yields a value for I, the moment of inertia. The value of the factor dividing T in the equation for the rotational partition function, Eq. (9-65), is, in terms of the separation of spectral lines $\Delta \nu'$,

$$\frac{\alpha h^2}{8\pi^2 I k} = \frac{h\alpha}{2k} \Delta \nu' \tag{9-74}$$

where h = Planck's constant

k = Boltzmann's constant

α = symmetry number

and $\Delta v'$ represents the spacing between successive spectral lines in terms of frequency of the emitted light. It is also possible, using Eqs. (9-56), (9-60), (9-61), and (9-74), to express the rotational frequency ω and the interparticle distance r in terms of the separation of spectral lines $\Delta v'$.

Equation (9-71) is quite general and applies for a harmonic oscillator as well. Substituting Eq. (9-68) into (9-71), we obtain

$$hv' = \frac{h}{2\pi}\left(\frac{\bar{k}}{\bar{m}}\right)^{1/2}\left(v_u + \frac{1}{2} - v_l - \frac{1}{2}\right) \qquad (9\text{-}75a)$$

where $\bar{k}$ is the spring constant of the oscillator and all other terms are as used before. The quantum-theory selection rules require that

$$\Delta v = \pm 1$$

so, letting $v_u = 1 + v_l$ in Eq. (9-75a), we obtain

$$hv' = \frac{h}{2\pi}\left(\frac{\bar{k}}{\bar{m}}\right)^{1/2} \qquad (9\text{-}75b)$$

Now, for a simple harmonic oscillator with spring constant $\bar{k}$ and $\bar{m}$, the classical theory frequency of oscillation v_0 is

$$v_0 = \frac{1}{2\pi}\left(\frac{\bar{k}}{\bar{m}}\right)^{1/2} \qquad (9\text{-}76)$$

and hence the frequency of light emitted is equal to the classical theory frequency of vibration. The term multiplying T^{-1} in the equation for the vibrational partition function, Eq. (9-69), is

$$\frac{hv'}{k} = \frac{1}{2\pi k}\left(\frac{\bar{k}}{\bar{m}}\right)^{1/2} \qquad (9\text{-}77)$$

and is directly determined using spectroscopic data.

The frequency of the emitted light is related to the wavelength λ of the emitted light by the equation

$$v' = \left(\frac{c}{\lambda}\right) \qquad |\sec^{-1}| \qquad (9\text{-}78)$$

where c is the velocity of light. Spectroscopists prefer to use the wave number v, where v is related to v' and λ according to the equation

$$v = \frac{v'}{c} = \frac{1}{\lambda} \qquad |\text{cm}^{-1}| \qquad (9\text{-}79)$$

Wave numbers of atomic and molecular spectra are tabulated in standard references.[1] The rotational spectrum for diatomic molecules shows up in the far infrared (wavelengths near 2×10^{-2} cm, frequencies near 1.5×10^{12} sec^{-1}, wave numbers near 50 cm^{-1}). The vibrational spectra show up in the near infrared (wavelengths near 1.2×10^{-3} cm, frequencies near 2.5×10^{13} sec^{-1}, wave numbers near 835 cm^{-1}). Both emission and absorption spectra are used in spectrographic analyses, the latter technique being most effective in the infrared.

9-8. The Equilibrium Constant. By no means have we exhausted the possibilities for using the relations yielded by the methods of statistical mechanics and quantum theory. One useful equation yet to be presented is that for the equilibrium constant for a chemical reaction which can be obtained using Eqs. (9-44) and (9-45) presented earlier. Consider the chemical reaction

$$a_1 A_1 + a_2 A_2 \rightarrow b_1 B_1 + b_2 B_2 \qquad (9\text{-}80)$$

where the lower-case letters a_i and b_i are the stoichiometric coefficients and the capital letters A_i and B_i denote chemical elements or compounds. The proper quotient of pressures for this reaction is here defined as

$$K_p = \frac{(p_{B_1})^{b_1}(p_{B_2})^{b_2}}{(p_{A_1})^{a_1}(p_{A_2})^{a_2}} \qquad (9\text{-}81)$$

When each partial pressure in Eq. (9-81) is that which would exist in the thermodynamic equilibrium for reaction equation (9-80), then Eq. (9-81) gives the equilibrium constant.

We shall now show how the equilibrium constant for the reaction given by Eq. (9-80) can be related to the change of free energy accompanying the reaction and, hence, to the partition functions of the participating species.

From Eqs. (9-39), (9-40), and (9-44) for a perfect gas,

$$F = H - TS \qquad (9\text{-}82)$$

hence $\qquad dF = dH - T\,dS - S\,dT$

However, for a reversible process from thermodynamics

$$dH = T\,dS + V\,dp$$

hence $\qquad dF = V\,dp = RT d(\log p)$

---

[1] See, for example, C. E. Moore, Atomic Energy Levels, *Natl. Bur. Standards Circ.* 467, 1948; and Gerhard Herzberg, "Molecular Spectra and Molecular Structure," vol. I, Spectra of Diatomic Molecules, D. Van Nostrand Company, Inc., Princeton, N.J., 1950.

if our process proceeds at constant temperature $(dT = 0)$

and $$pV = RT$$

Thus, if the process begins in a thermodynamic state at pressure p^e and ends in a thermodynamic state at pressure p^0, then integration of the above equation gives

$$F^0 - F^e = RT \log \frac{p^0}{p^e} \tag{9-83}$$

Now, for the chemical reaction given by Eq. (9-80), we can write

$$\Delta F = b_1 F_{B_1} + b_2 F_{B_2} - a_1 F_{A_1} - a_2 F_{A_2} \tag{9-84}$$

and this expression can be written for any thermodynamic state of the participating species. Let us write the equation for two different thermodynamic states of the participating species. We shall use the superscript 0 when the participating species appear in the thermodynamic state appropriate to pressures $p^0_{A_i}$ and $p^0_{B_i}$ and use the superscript e when the participating species appear in the thermodynamic state appropriate to pressures $p^e_{A_i}$ and $p^e_{B_i}$. That is, from Eq. (9-84), for the state designated by the superscript 0,

$$\Delta F^0 = b_1 F^0_{B_1} + b_2 F^0_{B_2} - a_1 F^0_{A_1} - a_2 F^0_{A_2} \tag{9-85}$$

and for the state designated by the superscript e

$$\Delta F^e = b_1 F^e_{B_1} + b_2 F^e_{B_2} - a_1 F^e_{A_1} - a_2 F^e_{A_2} \tag{9-86}$$

whence, combining Eqs. (9-85) and (9-86),

$$\Delta F^0 - \Delta F^e = b_1(F^0_{B_1} - F^e_{B_1}) + b_2(F^0_{B_2} - F^e_{B_2})$$
$$- a_1(F^0_{A_1} - F^e_{A_1}) - a_2(F^0_{A_2} - F^e_{A_2}) \tag{9-87}$$

or making use of Eq. (9-83), we obtain

$$\Delta F^0 - \Delta F^e = RT(\log K^0_p - \log K^e_p) \tag{9-88}$$

where K_p is defined by Eq. (9-81). Now, let us assume that thermodynamic state denoted by the superscript 0 is the condition where $p^0_{A_1} = p^0_{A_2} = p^0_{B_1} = p^0_{B_2} = 1$ atm. Furthermore, identify the thermodynamic state denoted by the superscript e as the state of thermodynamic equilibrium at the given temperature. Then from Eq. (9-81),

$$K^0_p = 1 \tag{9-89}$$

and according to our criterion for equilibrium,[1]

$$\Delta F^e = 0 \tag{9-90}$$

---

[1] A criterion for the existence of a state of equilibrium in a system is that the change of free energy is zero for any reversible change of that system.

and so Eq. (9-88) becomes

$$\Delta F^0 = -RT \log K_p \qquad (9\text{-}91)$$

where we drop the superscript e from K_p, since when K_p is used from here on, we refer to the value of the ratio given by Eq. (9-81) at thermodynamic equilibrium.

Equation (9-91), along with Eqs. (9-29) and (9-30) for the partition functions, Eq. (9-45) for the free energy of each species, and the spectroscopically determined atomic and molecular constants, can be used to determine the equilibrium constant for any chemical reaction described according to Eq. (9-80). In practice the equilibrium constants of few of the large number of chemical reactions possible have been tabulated. Rather, the more convenient and flexible constants for the formation of various chemical compounds from the elements are tabulated which can then be used to determine the equilibrium constants for many chemical reactions according to the following scheme. For the reaction

$$\sum_{i=1}^{m} a_i A_i \rightarrow \sum_{i=1}^{n} b_i B_i \qquad (9\text{-}80)$$

we can write $\quad \log K_p = \sum_{i=1}^{n} b_i \log K_f(B_i) - \sum_{i=1}^{m} a_i \log K_f(A_i) \qquad (9\text{-}92)$

where, for example, $\quad K_f(B_1) = \dfrac{p_{B_1}}{\displaystyle\prod_{i=1}^{m} p_{A_i}} \qquad (9\text{-}93)$

appropriate to the reaction describing the formation of B_i from its atomic elements. That is, the reaction

$$\sum_{i=1}^{m} A_i \rightarrow B_1 \qquad (9\text{-}94)$$

where all A_i are chemical elements. For example, if we have

$$C + O \rightarrow CO \qquad (9\text{-}95)$$

then $\quad\quad\quad K_f(CO) = \dfrac{p_{CO}}{p_C p_O} \qquad (9\text{-}96)$

Let us demonstrate the validity of Eq. (9-92) with an example. Consider the chemical reaction

$$C_2 + 2CO_2 \rightarrow 4CO \qquad (9\text{-}97)$$

ccording to Eq. (9-81) the equilibrium constant is

$$K_p = \frac{(p_{CO})^4}{p_{C_2}(p_{CO_2})^2}$$

or $\quad\quad\quad \log K_p = 4 \log p_{CO} - \log p_{C_2} - 2 \log p_{CO} \qquad (9\text{-}98)$

Application of Eq. (9-92) gives

$$\log K_p = 4 \log K_f(\text{CO}) - 2 \log K_f(\text{CO}_2) - \log K_f(\text{C}_2) \qquad (9\text{-}99)$$

However, from Eq. (9-93),

$$K_f(\text{C}) = \frac{p_{\text{CO}}}{p_{\text{C}} p_{\text{O}}} \qquad (9\text{-}100)$$

$$K_f(\text{C}_2) = \frac{p_{\text{C}_2}}{(p_{\text{C}})^2} \qquad (9\text{-}101)$$

and

$$K_f(\text{CO}_2) = \frac{p_{\text{CO}_2}}{p_{\text{C}}(p_{\text{O}})^2} \qquad (9\text{-}102)$$

hence, substituting Eqs. (9-100) through (9-102) into Eq. (9-99) we obtain

$$\log K_p = 4 \log p_{\text{CO}} - \log p_{\text{C}_2} - 2 \log p_{\text{CO}} \qquad (9\text{-}103)$$

which is identical with Eq. (9-98) as should be expected. The $K_f(\text{A}_i)$ are determined using the partition functions of the participating gas species.

All the elements of the methods used to calculate thermodynamic properties for gases have now been described. Thermodynamic properties of many gases have been calculated using the modern approach described in this chapter. The results of these calculations are usually tabulated as a function of temperature in terms of the following variables:[1]

$K_f(T)$	The equilibrium constant for formation from the elements
$\Delta H^0_{f(298.16)}$	The standard heat of formation from the elements at 1 atm pressure and at 25°C
$\dfrac{H^0 - H^0_0}{T}$	The enthalpy difference function in the standard state[2] at 1 atm pressure
$\dfrac{F^0 - H^0_0}{T}$	The free-energy function in the standard state at 1 atm pressure
C^0_p	The specific heat at constant pressure in the standard state at 1 atm

[1] See, for example, John S. Gordon, Thermodynamics of High Temperature Gas Mixtures, and Application to Combustion Problems, *U.S. Air Force Wright Air Development Center Tech. Rept.* 57-33, Astia Document 110735, January, 1957; and L. V. Feigenbutz, G. L. Steihl, and G. L. Katz, Combustion Charts for High Energy Fuels, I. Thermodynamic Properties of Combustion Products, *Convair (San Diego Div.) Rept.* ZR-600-001, 1958.

[2] The standard state is the state at 1 atm pressure at the temperature under consideration. The superscript 0 indicates that the thermodynamic function is evaluated for the substance in its standard state.

Several useful equations can be constructed using the data tabulated above. For example, combining Eqs. (9-45), (9-85), and (9-91), it can be shown that for the reaction

$$\sum_{i=1}^{m} a_i A_i \rightarrow \sum_{i=1}^{n} b_i B_i \tag{9-80}$$

we can write an equation for the equilibrium constant in terms of the partition functions of the participating species which is

$$\log K_p = \frac{-\Delta F^0}{RT} = \frac{-\Delta E_0^0}{RT} + \sum_{i=1}^{n} b_i \log [pQ(B_i)] - \sum_{i=1}^{m} a_i \log [pQ(A_i)] \tag{9-104}$$

since, from Eq. (9-55), $Q^0 = pQ$ and where we have used

$$\Delta E_0^0 = \sum_{i=1}^{n} b_i E_0^0(B_i) - \sum_{i=1}^{m} a_i E_0^0(A_i) = \Delta H_0^0 \tag{9-105}$$

since $H_0^0 = E_0^0$ at zero absolute temperature. For the special case of the chemical reaction resulting in the formation of a compound from the elements, Eq. (9-80) becomes

$$\sum_{i=1}^{m} A_i \rightarrow B_1 \tag{9-106}$$

and Eq. (9-104) becomes

$$\log K_f = \frac{-\Delta H_{f(0)}^0}{RT} + \log [pQ(B_1)] - \sum_{i}^{m} \log [pQ(A_i)] \tag{9-107}$$

or

$$K_f = \frac{Q(B_1)}{p^{n-1} \prod_{i=1}^{m} Q(A_i)} \exp \frac{-\Delta H_{f(0)}^0}{RT} \tag{9-108}$$

where, in general, for any value of temperature T, including $T = 0$,

$$\Delta H_{f(T)}^0 = \Delta H_{f(298.16)}^0 - \Delta(H_{298.16}^0 - H_T^0) \tag{9-109}$$

and $\quad \Delta(H_{298.16}^0 - H_T^0) = H_{298.16}^0(B_1) - H_T^0(B_1) - \sum_{i=1}^{m} [H_{298.16}^0(A_i) - H_T^0(A_i)]$

$$\tag{9-110}$$

For the case of the standard temperature of $T = 298.16°$K,

$$\Delta H_{f(298.16)}^0 = H_{298.16}^0(B_1) - \sum_{i=1}^{m} H_{298.16}^0(A_i) \tag{9-111}$$

and this value is usually tabulated.

In using the above equations it is important to remember the following rules:

1. The thermodynamic variables given by capital letters H, E, and S are molar values, that is, the value obtained for 1 mole of the

gas under consideration. They are related to specific values of the thermodynamic variables given by lower-case letters by the following equations:

$$H = Mh$$

$$E = Me$$

and
$$S = Ms$$

where h, e, and s are enthalpy, energy, and entropy per unit mass and M is the molecular weight of the substance under consideration.

2. The numerical subscript refers to the temperature at which the thermodynamic variable is evaluated.

3. The superscript 0 refers to the fact that the variable is given for the substance in the standard state at 1 atm for the temperature under consideration.

When the above equations are used along with the tabulated thermodynamic functions, one is usually applying the results of statistical mechanical calculations using spectrographically determined constants according to refinements of the equations developed in Secs. 9-4, 9-5, and 9-7. The refinements take into account anharmonicity of the vibrational modes and interactions among modes in a straightforward manner, so that the reasoning presented in Secs. 9-4 through 9-7 is not invalidated.

9-9. A Sample Calculation—High-temperature Air. We shall choose a relatively simple mixture of gases to illustrate the application of the equations of Secs. 9-5 through 9-8. In particular, we shall choose a mixture close to air. Air is largely a mixture of O_2, O, N_2, N, O^+, N^+, and e^-, species such as the rare gases being neglected, since they are present in such minute quantities that their presence does not affect calculations for the equilibrium properties of air at high temperatures. Calculations will show that N is largely absent when O_2 is dissociating and O_2 is largely absent when N_2 is dissociating owing to the significant differences in dissociation energy for O_2 and N_2 (5.08 ev for O_2 versus 9.756 ev for N_2). Furthermore, O_2 and N_2 are largely dissociated by the time ionization becomes significant owing to the high ionization potential for O and N (13.62 ev for O and 14.55 ev for N). Thus there will never be more than five species present in significant concentrations. Figure 9-2 taken from the work of Hansen[1] shows what species will be present in appreciable amounts for a range of values of air pressure and air temperature. Because of this separation into regimes, the calculation of equilibrium properties for dry air is simplified.

[1] C. Frederick Hansen, *NACA TN* 4150, 1958.

Let us consider the region where oxygen is dissociating but nitrogen has not yet begun to dissociate. The species present are O_2, O, and N_2, where O is formed through the chemical reaction $O_2 \rightarrow 2O$. Let us use our equations developed in previous sections to calculate the equilibrium concentration and thermodynamic properties of dry air in this region.

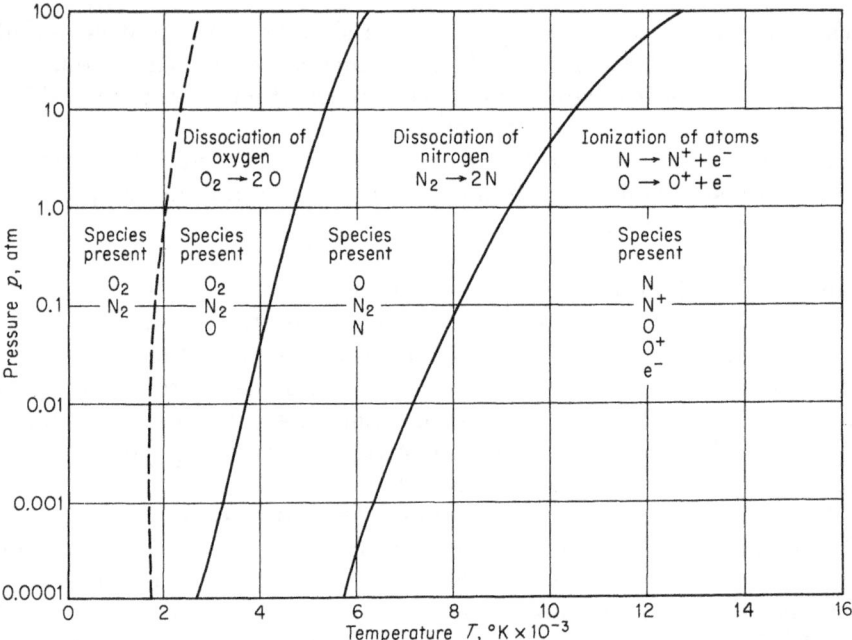

FIG. 9-2. Domains of pressure and temperature for the major chemical reactions in air. (*After C. Frederick Hansen, NACA TN 4150, 1958.*)

If x_i is the mole fraction of species i in our hot gas mixture, then

$$p_i = x_i p \tag{9-112}$$

where, when n_i denotes the number of particles of species i,

$$x_i = \frac{n_i}{\Sigma n_i} \tag{9-113}$$

If we choose to define

$$\sigma = \frac{n_O}{2(\Sigma n_i)_b} \tag{9-114}$$

where the subscript b refers to conditions before dissociation, and if we assume that our air mixture before dissociation took place was such that

$$(x_{N_2})_b = \frac{(n_{N_2})_b}{(\Sigma n_i)_b} = 0.80 \qquad (9\text{-}115)$$

and

$$(x_{O_2})_b = \frac{(n_{O_2})_b}{(\Sigma n_i)_b} = 0.20 \qquad (9\text{-}116)$$

then Eqs. (9-112) through (9-116) can be combined to give

$$p_O = \frac{n_O}{\Sigma n_i} p = \frac{n_O p}{(\Sigma n_i)_b + \frac{1}{2}n_O} = \frac{2\sigma}{1 + \sigma} p \qquad (9\text{-}117)$$

Similarly it will be found that

$$p_{O_2} = \frac{0.2 - \sigma}{1 + \sigma} p \qquad (9\text{-}118)$$

and

$$p_{N_2} = \frac{0.8}{1 + \sigma} p \qquad (9\text{-}119)$$

Now we have the chemical reaction

$$O_2 \to 2O$$

for which

$$K_p = \frac{(p_O)^2}{p_{O_2}} = \frac{1}{K_f} \qquad (9\text{-}120)$$

Thus, using Eqs. (9-117) and (9-118) in Eq. (9-120), we obtain

$$K_p = \frac{1}{K_f} = \frac{4\sigma^2 p}{(1 + \sigma)(0.2 - \sigma)}$$

Solving for σ, we obtain

$$\sigma = \frac{-0.4 + \{0.16 + 2[1 + (4p/K_p)]\}^{1/2}}{1 + (4p/K_p)} \qquad (9\text{-}121)$$

Given p and T, Eq. (9-121) yields σ. Given σ and p, Eqs. (9-117) through (9-119) yield values of p_O, p_{O_2}, and p_{N_2}. We shall now present the equation for K_p in terms of the partition functions. From Eqs. (9-107) and (9-120) we obtain

$$\log K_p = -\log K_f = \frac{\Delta H^0_{f(0)}}{RT} + \log [pQ(O_2)] - 2 \log [pQ(O)] \quad (9\text{-}122)$$

where we obtain $\Delta H^0_{f(0)}$ from spectroscopic data. Its value can be estimated from tabulated thermodynamic data using Eqs. (9-109) and (9-110) as follows:

$$\Delta H^0_{f(0)} = \Delta H^0_{f(298.16)} - [H^0_{298.16}(O_2) - H^0_0(O_2)]$$
$$+ 2[H^0_{298.16}(O) - H^0_0(O)] \quad (9\text{-}123)$$

Thermodynamic tables[1] give

$$H^0_{298.16}(O_2) - H^0_0(O_2) = 2087.6 \text{ cal/mole} \qquad (9\text{-}124)$$

$$H^0_{298.16}(O) - H^0_0(O) = 1617.6 \text{ cal/mole} \qquad (9\text{-}125)$$

and

$$\Delta H^0_{f(298.16)} = -117{,}160 \text{ cal/mole} \qquad (9\text{-}126)$$

hence, by Eq. (9-109),

$$\Delta H^0_{f(0)} = -117{,}630 \text{ cal/mole} \qquad (9\text{-}127)$$

and is exactly the heat of dissociation per molecule which could have been obtained directly from spectroscopic data. Since $R = 1.98717$ cal/mole-°K, Eq. (9-122) becomes, using Eq. (9-127),

$$\log K_p = \frac{-59{,}300}{T} + \log [pQ(O_2)] - 2 \log [pQ(O)] \qquad (9\text{-}128)$$

Equations (9-29), (9-26), (9-55), (9-65), and (9-69) present values for the partition functions for all modes of interest to us here. Using Eqs. (9-29), (9-55), and (9-26) for O with $Q_v = Q_j = 1$ for such a monatomic molecule, we obtain

$$\log [pQ(O)] = 0.5 + \frac{5}{2} \log T + \log \left[\sum_i g_s \exp \frac{-\epsilon(s) + \epsilon_0(s)}{kT} \right] \qquad (9\text{-}129)$$

where p is expressed in atmospheres and T is expressed in degrees Kelvin. Furthermore, using Eqs. (9-29), (9-26), (9-55), (9-65), and (9-69) we obtain for the diatomic molecule O_2

$$\log [pQ(O_2)] = 0.11 + \frac{7}{2} \log T - \log \left(1 - \exp \frac{h\nu'}{kT} \right)$$
$$+ \log \left[\sum g_s \exp \frac{-\epsilon(s) + \epsilon_0(s)}{kT} \right] \qquad (9\text{-}130)$$

with p in units of atmospheres and T in degrees Kelvin as before. Equations (9-129) and (9-130) when combined with Eq. (9-128) give an equation for the desired equilibrium constant K_p once the spectroscopically determined energy-level values are found. Table 9-2 gives these values.

In this manner we calculate the partial pressures of the air species present at different values of temperature and pressure. Let us now proceed to calculate the thermodynamic variables of state for air in the temperature region where oxygen is dissociating. For a perfect gas

$$p_i = \frac{n_i kT}{V} \qquad (9\text{-}131)$$

[1] See, for example, Feigenbutz, Steihl, and Katz, *op. cit.*

TABLE 9-2. PARTITION FUNCTION CONSTANTS FOR O AND O_2†

Species	E_0, in kcal/mole	$\dfrac{hv'}{k}$, in °K	Electronic term s	Degeneracy g_s	Electronic energy‡ $\dfrac{\epsilon(s) - \epsilon_0(s)}{k}$, in °K
O	58.8		0	5	0
			1	3	228
			2	1	326
			3	5	22,800
			4	1	48,600
O_2	0	2270	0	3	0
			1	2	11,390
			2	1	18,990

† After C. Frederick Hansen, *NACA TN* 4150, which, in turn, was taken from Gerhard Herzberg, "Molecular Spectra and Molecular Structure," vol. I, Spectra of Diatomic Molecules, D. Van Nostrand Company, Inc., Princeton, N.J., 1950; and C. E. Moore, Atomic Energy Levels, *Natl. Bur. Standards Circ.* 467, 1948.

‡ Electronic terms with reduced energy above 50,000°K were not included because of the resulting small contribution to the partition function at temperatures of interest to us here.

and from Dalton's law

$$p = \sum p_i = \sum n_i \frac{kT}{V} \tag{9-132}$$

or, since $N_0 k = R$ and $N_0 \bar{m} = \bar{M}$, where $\bar{m}$ and $\bar{M}$ are mean particle masses and mean molecular weight, respectively, and N_0 is Avogadro's number, then, from Eq. (9-132),

$$p = \frac{\bar{m} \sum n_i}{V} \frac{N_0 kT}{N_0 \bar{m}} = \frac{\bar{M}_b}{\bar{M}} \frac{\rho RT}{\bar{M}_b} \tag{9-133}$$

Now

$$\bar{M}_b = N_0 \bar{m}_b = N_0 (\sum x_i m_i)_b = \frac{N_0 (\sum n_i m_i)_b}{(\sum n_i)_b} \tag{9-134}$$

also

$$\bar{M} = \frac{N_0 \sum n_i m_i}{\sum n_i} \tag{9-135}$$

but mass is conserved through any chemical reaction; hence

$$\sum n_i m_i = (\sum n_i m_i)_b \tag{9-136}$$

Hence Eqs. (9-134) to (9-136) can be combined to give

$$\frac{\bar{M}_b}{\bar{M}} = \frac{\sum n_i}{(\sum n_i)_b} = \frac{(\sum n_i)_b + \frac{1}{2} n_O}{(\sum n_i)_b} = 1 + \sigma \tag{9-137}$$

whence, from Eqs. (9-133) and (9-137),

$$\frac{p\bar{M}_b}{\rho RT} = 1 + \sigma \tag{9-138}$$

where $1 + \sigma$ is the total number of moles per original mole of un-dissociated air. The values of $1 + \sigma$ as calculated by Hansen[1] for the entire region shown in Fig. 9-2 in a manner analogous to that just

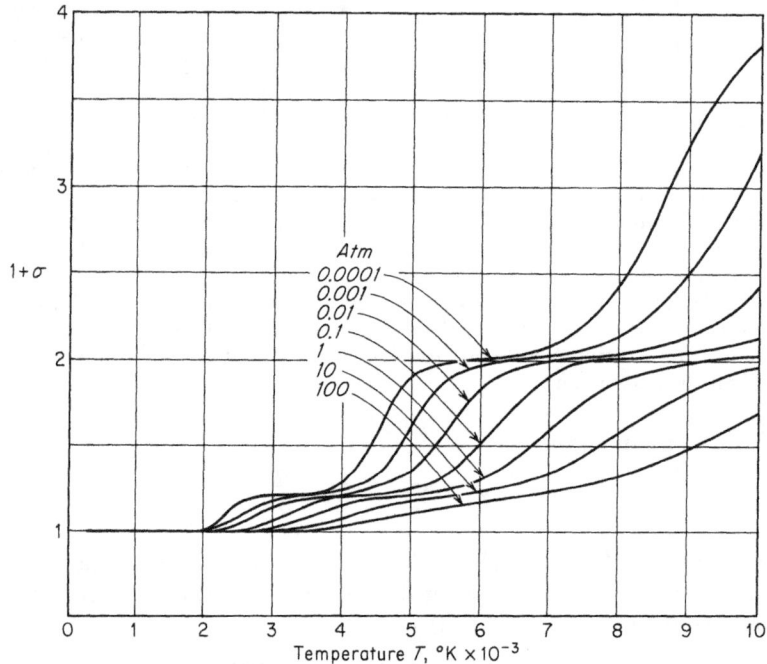

FIG. 9-3. Compressibility of air as a function of temperature. (*After C. Frederick Hansen, NACA TN 4150, 1958.*)

described for the region where oxygen is dissociating are plotted in Fig. 9-3. Note the different plateaus on the curves of $1 + \sigma$ versus temperature which occur as oxygen becomes completely dissociated, then nitrogen becomes completely dissociated, and finally single ionization of nitrogen and oxygen atoms occurs.

The internal energy and enthalpy are calculated from the following equations:

$$E = \sum x_i E_i = \sum \frac{p_i}{p} E_i \tag{9-139}$$

$$H = E + pV \tag{9-140}$$

[1] Hansen, *op. cit.*

and where $E_i - E_{i(0)}$ is given by Eq. (9-35)

$$E_i - E_{i(0)} = RT^2 \left\{ \frac{d}{dt} [\log(Q)] \right\}_V$$

$$(9\text{-}141)$$

$E_{i(0)}$ for O_2 and O are given in Table 9-2. The partial pressures p_i are calculated according to Eqs. (9-117) through (9-119) once the value of σ is determined using Eq. (9-121). Hansen chose to plot the dimensionless quantity $(1 + \sigma)E/RT$, which is easily obtained using Eqs. (9-139) and (9-141) in the form

$$\frac{(1+\sigma)E}{RT} = \sum_i \frac{p_i}{p} \frac{(1+\sigma)E_i}{RT} \quad (9\text{-}142)$$

and $\dfrac{(1+\sigma)E_i}{RT} = \dfrac{(1+\sigma)E_{i(0)}}{RT}$

$$+ (1+\sigma)T \left\{ \frac{d}{dT} [\log(Q)] \right\}_V \quad (9\text{-}143)$$

Once $(1 + \sigma)E/RT$ is obtained in this manner, Eq. (9-140) gives

$$\frac{(1+\sigma)H}{RT} = \frac{(1+\sigma)E}{RT} + (1+\sigma)$$

$$(9\text{-}144)$$

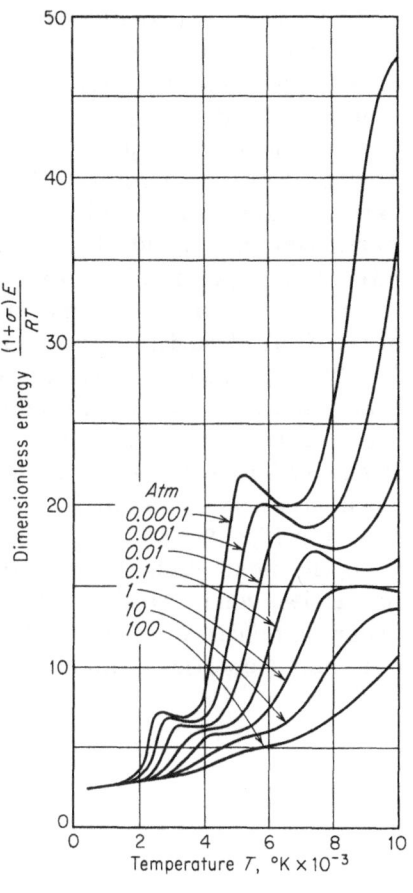

FIG. 9-4. Energy of air as a function of temperature. (*After C. Frederick Hansen, NACA TN* 4150, 1958.)

Figure 9-4 presents the results of Hansen's calculations for the dimensionless energy of dissociating air.

This completes our example calculations, which were intended to illustrate the use of our derived thermodynamic equations, the partition functions, and the basic spectroscopic data. More complex gas mixtures, including polyatomic molecules, are more cumbersome to deal with but, in principle, are treated in a manner analogous to the method of calculation used in this section.

9-10. Conclusions. This chapter presents equations applicable to calculating the thermodynamic properties of equilibrium gas mixtures and explains in a limited way how these equations devolve from quantum theory, wave mechanics, statistical mechanics, and certain physical measurements. The treatment is not exhaustive, and the interested reader is referred to the cited references when a more complete treatment is desired.[1] The author's intention was to present some useful equations in this chapter along with an explanation of the underlying bases for the equations.

It seems worthwhile to restate the limitations of the equations presented in this chapter. First of all, most of the equations for the partition functions apply only to atomic or diatomic molecules. Furthermore, it should be remembered that it was assumed that the interparticle potentials were relatively short ranged compared with the mean free paths. This latter assumption rules out application of the derived partition functions to highly ionized gases and liquids. Lastly, the diatomic molecule model used was idealized in the sense that interactions of vibrational motion with rotational motion and the effects of anharmonicity in the vibrational motion were neglected. It should be mentioned that these latter effects have been taken into account without undue complexity and do not affect the results shown in Figs. 9-3 and 9-4 more than an uncertainty of ±15 per cent.

In the next chapter we shall deal with the subject of transport properties, where we shall find that the interparticle potentials are of prime significance in contrast to their role in determining thermodynamic properties of an equilibrium gas mixture.

---

[1] See, for example, Joseph Edward Mayer and Maria Goeppert Mayer, "Statistical Mechanics," John Wiley & Sons, Inc., New York, 1940.

10

Transport Coefficients of Dilute Gas Mixtures

10-1. Introduction. Before the equations of continuity, momentum, and energy can be applied, it is necessary to define appropriate expressions for the transport coefficients which appear in the mass, momentum, and energy flux terms of those equations. The objective of this chapter is to present the equations for the coefficients of viscosity, diffusion, and thermal conductivity for dilute gas mixtures and show how these transport coefficients are affected by various conditions present in typical gas mixtures. For example, the variation of the transport coefficients of a dissociating gas mixture with composition or temperature and the transport-coefficient variation with the concentration of components of a binary mixture of light gas (such as H_2) with a heavier gas (such as CO) are discussed and treated in this chapter. Curves and tables of transport-coefficient parameters will be presented to illustrate the details and to provide information needed in order to apply skin-friction and heat-transfer equations derived in earlier chapters of this book.

As was described in Chap. 2, the transport coefficients dealt with in this chapter evolve indirectly from Enskog's solution to Boltzmann's equation for the singlet-velocity distribution function.[1] The use of the first two terms in Enskog's series for the singlet-velocity distribution function[2] f_i in the general equations of change for a

[1] See, for example, Sydney Chapman and T. G. Cowling, "The Mathematical Theory of Non-uniform Gases," 2d ed., pp. 107–198, Cambridge University Press, New York, 1952.

[2] By "singlet-velocity distribution function" we mean the classical velocity distribution function for a single particle. For a gas in equilibrium this becomes the Maxwell-Boltzmann distribution function given by Eqs. (9-11), (9-16), and (9-22). Presumably there are doublet distribution functions for pairs of particles, triplet distribution functions, etc. For dilute gas mixtures the behavior of the gas flow field is largely determined by single particles and their movements.

reacting gas mixture results in the equation of continuity for each chemical species, the equation of continuity for the gas as a whole, the Navier-Stokes equations, and the energy equation for a gas mixture, which contain mass, momentum, and energy flux vectors expressible in terms of the transport coefficients and first derivatives of flow-field variables. These transport coefficients, strictly applicable for monatomic gases, are given to a first approximation by the equations

$$\mu_i = \frac{5}{16} \frac{(\pi m_i kT)^{1/2}}{\pi \sigma_i^2 \Omega^{(2,2)*}}$$

$$k_i = \frac{15}{4} R_i \mu_i$$

and
$$D_{12} = \frac{3}{16} \frac{2\pi kT(m_1 + m_2)^{1/2}}{m_1 m_2} \frac{1}{n\pi \sigma_{12}^2 \Omega^{(1,1)*}}$$

where m_i = mass of particle of species i
k = Boltzmann's constant
T = temperature, °K
n = number of particles per unit volume
σ = collision diameter, angstroms

and the functions $\Omega^{(1,1)*}$ and $\Omega^{(2,2)*}$ which appear are complicated integrals which take into account the dynamics of collisions between particles. They are defined so that for rigid elastic spheres of diameter σ they are unity. Their numerical value for other molecular models will be found to vary with T and certain "force constants" which are empirical parameters of the molecular model used. The $\Omega^{(l,s)*}$ integrals will be described later along with their evaluation for three molecular models greatly used in practice. The above equations are first approximations to more complete equations and are known to be accurate to within 3 per cent or less of the full equations which they approximate when compared with existing calculations using the full equations.

10-2. Transport-theory Assumptions. Because two terms only in Enskog's series for the distribution function were used to obtain the gas-dynamic equations and because Boltzmann's equation for the singlet-velocity distribution function only was solved for, a statement of the conditions under which the resulting gas-dynamics transport equations might be expected to be valid is given here. When these statements are examined, one begins to appreciate fully the tenuous basis upon which is built the present-day science of gas dynamics, to understand what a triumph it is to find that a science built upon such restrictive assumptions results in reasonable agreement with experiment under a wide range of conditions, and to realize the immense task being undertaken in extending the theory into regions of applicability not now realistically treated by the theory in its present form.

The assumptions involved in deriving the transport theory for nonuniform, dilute gases as applied in this book include the following:

1. The gradients of the physical quantities, the variables of state and the flow velocity, are "small." That is, they are small enough that the third term and successively higher-order terms of the Enskog series for f_i are negligible relative to those terms retained. Flow systems with large gradients, such as the flow of a gas in the immediate region of a strong shock wave, are thereby, strictly speaking, beyond the range of applicability of the theory.

2. Only binary collisions are important in determining the velocity distribution function f_i. That is, three-body and higher-order collisions are of negligible frequency relative to the frequency of binary collisions in determining the change of the velocity distribution function from one position to another in the flow field. Implied in or related to this assumption are the following statements:

a. The theory is not applicable to liquids where multibody collisions come into play in determining the velocity distribution function.

b. The effective range of intermolecular or interparticle forces is small compared with the mean free path. Thus the treatment of plasmas where long-range Coulomb forces are important is beyond the applicability of the Chapman-Enskog theory as used here.

c. Mass, momentum, and energy are transported entirely by single particles. The transport of mass, momentum, and energy by doublets, triplets, and other multiplets is insignificant.

3. Only monatomic gases are considered. The strict formulation of the theory deals only with gases having no internal degrees of freedom for which the particle-particle interaction potential is spherically symmetric. Since the transfer of mass and momentum is almost unaffected by the presence or absence of internal degrees of freedom, the theory is fairly successful in dealing with mass and momentum flux in the flow of polyatomic gases. The transfer of energy does involve the presence or absence of internal degrees of freedom, and hence the energy flux for polyatomic gases is imperfectly described by the theory in its present form, as will be dwelt upon in that which follows.

4. In the derivation of the transport coefficients, the particle-particle collisions are dealt with, using the methods of classical mechanics. That is, the theory is inadequate in treating dilute gas systems where quantum mechanical effects become important in

dealing with particle-particle interactions. When the de Broglie wavelength $\lambda = h/mv$ (h is Planck's constant, m is the particle mass, and v the particle velocity) is of the order of magnitude of the particle dimensions or greater (about 1 angstrom or greater), quantum-mechanical effects begin to become important. At low temperatures for light particles (helium at temperatures the order of $100°K$ or less), these effects become significant. For the hypersonic aerodynamicist, such effects are negligible for most, if not all, problems of interest.

5. The dimensions of the flow-field boundaries and any objects therein are large compared with the mean free path, so that the great majority of the particles collide far more frequently with one another than with the particles making up the flow-field boundaries or the objects therein. Only under such conditions does the concept of an equilibrium condition or an approach to equilibrium within the gas determined by gas-particle collisions have a meaning. When the gas particles are more likely to collide with the container or immersed object particles than with one another, the distribution of gas-particle velocities is chiefly determined by the mechanics of the gas-particle–object-particle collisions rather than by gas-particle–gas-particle collisions.

The triumph of the Chapman-Enskog theory can now be better appreciated. Most gases of engineering interest are polyatomic. Furthermore, in boundary-layer flow, large gradients of temperature, density, composition, and velocity can exist in narrow regions of the layer. Still, the Navier-Stokes equations and the remaining equations can be used to describe many situations with remarkable success over a wide range of gas pressures and temperatures for the flow of mixtures of polyatomic gases. We must not expect too much of the theory, however. There is no a priori reason to expect that the theory can be applied to ionized gas flows or the flow of liquids with any degree of success, nor is uncritical application to the low-density or nearly free molecule flow of gases warranted. We proceed to treat the description of transport properties for dilute gases for the wide ranges of conditions for which the theory does appear to be applicable.

10-3. The Transport Coefficients. In the equations of continuity, momentum, and energy, there appear the transport coefficients μ, D_{12}, and k. Equations for these coefficients have been presented as they have been derived, to a first approximation, by the Chapman-Enskog[1] theory. The equations will now be expressed

---

[1] See, for example, Joseph O. Hirschfelder, Charles F. Curtiss, and R. Byron Bird, "Molecular Theory of Gases and Liquids," chap. 8, pp. 514–610, John Wiley & Sons, Inc., New York, 1954.

in practical units in terms of collision integrals $\Omega^{(l,s)*}$, molecular weights M_i, the temperature, the pressure, and molecular potential-energy parameters σ and ϵ yet to be described.   A discussion of the meaning and origins of these parameters in the transport-coefficient equations will follow the presentation of the equations.

Coefficient of Viscosity.   For a pure gas of species i,

$$\mu_i = 266.93 \times 10^{-7} \frac{(M_i T)^{1/2}}{\sigma^2 \Omega^{(2,2)*}} \tag{10-1}$$

where μ_i = viscosity of species i, g/cm-sec
 M_i = molecular weight of species i
 σ = collision diameter, angstroms
 T = temperature, °K

and
$$\Omega^{(2,2)*} = \frac{\Omega^{(2,2)}}{[\Omega^{(2,2)}]_{\text{rigid sphere}}} \tag{10-2}$$

and the $\Omega^{(l,s)}$ integrals will be described later.

For a mixture of gases containing ν components, Wilke[1] gives

$$\mu = \sum_{i=1}^{\nu} \mu_i \left(1 + \sum_{\substack{k=1 \\ k \neq i}}^{\nu} G_{ik} \frac{x_k}{x_i}\right)^{-1} \tag{10-3}$$

where μ_i is given by Eq. (10-1) and x_i is the mole fraction of species i; that is,

$$x_i = \frac{C_i}{M_i} \left(\sum_i \frac{C_i}{M_i}\right)^{-1} \tag{10-4}$$

where $C_i = \rho_i/\rho$ is the mass fraction of species i and M_i is the molecular weight of species i

and
$$G_{ik} = \frac{[1 + (\mu_i/\mu_k)^{1/2}(M_k/M_i)^{1/4}]^2}{(2)^{3/2}[1 + (M_i/M_k)]^{1/2}} \tag{10-5}$$

For a binary mixture Eq. (10-3) becomes

$$\mu_{12} = \frac{\mu_1}{1 + G_{12}(x_2/x_1)} + \frac{\mu_2}{1 + G_{21}(x_1/x_2)} \tag{10-6}$$

where, from Eq. (10-4),

$$G_{12} = \frac{[1 + (\mu_1/\mu_2)^{1/2}(M_2/M_1)^{1/4}]^2}{(2)^{3/2}[1 + (M_1/M_2)]^{1/2}} \tag{10-7a}$$

and
$$G_{21} = \frac{[1 + (\mu_2/\mu_1)^{1/2}(M_1/M_2)^{1/4}]^2}{(2)^{3/2}[1 + (M_2/M_1)]^{1/2}} \tag{10-7b}$$

[1] C. R. Wilke, *J. Chem. Phys.*, vol. 18, pp. 517–522, 1950.

and μ_1, μ_2 are given by Eq. (10-1) for a pure gas. Hirschfelder, Curtiss, and Bird[1] show that Eq. (10-3) is a logical approximation to the full, more complex expression for viscosity of a mixture given by the Chapman-Enskog theory. Wilke[2] reduced the complete equation to Eq. (10-3) and showed how Eq. (10-3) agreed with a wide variety of viscosity measurements near room temperature for binary and tertiary mixtures of light and heavy molecules. Amdur and Mason[3] show that for a binary mixture of helium and argon Eq. (10-3) gives excellent agreement with the Chapman-Enskog equation for viscosity of a binary mixture for a temperature range from 1000 to 15,000°K.

Binary Diffusion Coefficient. For most problems of interest to us, we shall be concerned with gas mixtures which are essentially binary in the sense that we shall be able to divide the gas particles into two classes, light particles and heavy particles. For example, for a dissociating air mixture there will be air molecules, O_2 and N_2 principally, and air atoms, O and N. Since the flux of momentum and energy is made up only partially by energy transported by mass diffusion fluxes, any error involved in dealing with the diffusion flux of a multicomponent mixture as though it was an effective binary mixture will be of much smaller magnitude than the total momentum and energy flux. The expression for the binary diffusion coefficient in practical units is

$$D_{12} = D_{21} = 262.8 \times 10^{-5} \frac{[T^3(M_1 + M_2)/2M_1M_2]^{1/2}}{p\sigma_{12}^2 \Omega^{(1,1)*}} \qquad (10\text{-}8)$$

where $D_{21} = D_{12}$ = diffusion coefficient, cm^2/sec

$\sigma_{12} = \frac{1}{2}(\sigma_1 + \sigma_2)$, angstroms (see Sec. 10-9 for further discussion on this equation)

p = pressure, atm

T = temperature, °K

and all other terms have been previously defined. The $\Omega^{(l,s)*}$ integral will be dealt with later.

Thermal Conductivity for Monatomic Gases. As was suggested earlier, the results for thermal conductivity given by the Chapman-Enskog theory are, strictly speaking, valid only for molecules lacking internal-energy modes. Approximations can be made to adjust the results given by the theory for thermal conductivity to account for internal-energy transfer which occurs during binary collisions involving one or two polyatomic molecules. These adjustments will be

[1] Hirschfelder, Curtiss, and Bird, *op. cit.*, p. 533.

[2] Wilke, *op. cit.*

[3] I. Amdur and E. A. Mason, *Phys. Fluids*, vol. 1, no. 5, pp. 370–383, 1958.

given later. For a monatomic gas of species i the Chapman-Enskog theory yields

$$k_i = \frac{15}{4} \frac{R}{M_i} \mu_i = \frac{15}{4} R_i \mu_i \qquad (10\text{-}9)$$

where k_i = thermal conductivity, cal/cm-sec-°K
  μ_i = viscosity given by Eq. (10-1)
  R = universal gas constant
  M_i = molecular weight of species i
  R_i = gas constant for species i
For a mixture of monatomic gases Mason and Saxena[1] give

$$k = \sum_{i=1}^{v} k_i \left(1 + 1.065 \sum_{\substack{k=1 \\ k \neq i}}^{v} G_{i,k} \frac{x_k}{x_i} \right)^{-1} \qquad (10\text{-}10)$$

where k_i is given by Eq. (10-9), G_{ik} by Eq. (10-5), and x_i and x_k by Eq. (10-4). Amdur and Mason[2] show that Eq. (10-10) approximates to within 8 per cent or less the results given by the complete, more complex equations of the Chapman-Enskog theory for a mixture of helium and argon over a temperature range from 1000 to 15,000°K. Mason and Saxena show that Eq. (10-10) gives excellent results when compared with several near-room-temperature measurements of thermal conductivity of binary and tertiary mixtures of light and heavy noble gases.

Thermal Conductivity for Polyatomic Gases. It was stated earlier that the Chapman-Enskog theory treats the gas particles involved in binary collisions leading to mass, momentum, and energy currents in the macroscopic flow field as particles totally lacking internal-energy modes. This assumption is unrealistic in treating real gas molecules. Even the noble gases helium, argon, and krypton possess electronic states which can be excited upon sufficiently energetic collisions with other particles. At low temperatures, however, the number of particles possessing translational energy sufficiently great to excite higher electronic energy states upon collision will be negligible. Since the noble gases are monatomic and spherically symmetric, they lack rotational and vibrational modes, and thus, at temperatures of concern to us here in this book, the Chapman-Enskog theory should yield good results when applied to the noble gases. Comparison of theory with experiment shows this presumption to be a valid one.

The components of air at room temperature include an abundance of diatomic molecules, chiefly N_2 and O_2. Such molecules possess

---

[1] E. A. Mason and S. C. Saxena, *Phys. Fluids*, vol. 1, no. 5, pp. 361–369, 1958.
[2] Amdur and Mason, *op. cit.*

the two rotational and two vibrational internal degrees of freedom in addition to the three translational degrees of freedom assumed by the theory to be possessed by all molecules. The existence of these internal degrees of freedom can lead to nonadiabatic or inelastic collisions in the sense that the total *kinetic* energy of the colliding particles is no longer preserved throughout a collision. Mechanisms exist for the transfer of translational energy into internal energy of rotation and vibration. The problem can be accentuated when dealing with the products of combustion, since polyatomic molecules can be present which possess an even greater number of internal-energy degrees of freedom at all temperatures.

The theory for heat conduction in polyatomic gases has not yet been worked out to the degree to which the theory for monatomic gases has been outlined. However, certain assumptions and approximations can be made which enable one to apply the Chapman-Enskog theory, valid strictly for monatomic gases at low temperatures, to gas mixtures of polyatomic molecules.

We begin by noting that, providing the transfer of translational energy to internal energy or the transfer between internal energy modes occurring upon binary collisions has *no effect* upon the equations describing the change of the singlet-velocity distribution function f_i, the inelastic collisions will have no effect upon mass and momentum transfer. The assumption that such inelastic collisions do not affect the singlet-velocity distribution is a reasonable assumption for many polyatomic molecules at temperatures of interest to us. It can be shown that the velocity distribution function f_i determined for particles possessing no internal energy will represent the velocity distribution function f_i for particles which do possess internal energy for two cases: (1) the case where the transition probability of a change in internal energy upon collision is much less than the probability of no change in internal energy upon collision and (2) the case where the internal energy adjusts to equilibrium upon the order of one collision only.

Calculations by Herzfeld and Schwartz[1] show that the probabilities of transfer of energy between translational modes and vibrational modes during collisions involving diatomic or triatomic molecular pairs are small and of the order of 10^{-5} and smaller at room temperature. Their calculations for N_2 and O_2 have been verified within a factor of 10 by Blackman,[2] who determined vibrational relaxation times using interferometer photographs of the fringe shifts in the

[1] K. F. Herzfeld and R. N. Schwartz, *J. Chem. Phys.*, vol. 22, no. 5, pp. 767–773, 1954.

[2] Vernon Blackman, *J. Fluid. Mech*, vol. 1, part I, pp. 62–85, 1956.

relaxation regions behind shock waves which have passed through N_2 and O_2 in a shock tube. Thus we can assume that case 1 above represents the situation for transfer of energy between the translational and vibrational modes of diatomic and triatomic gases.

Parker, Adams, and Stavseth[1] have reported the results of ultrasonic dispersion measurements in N_2 and O_2 at room temperature which indicate that no more than three collisions are required to bring about equilibrium within the rotational modes of N_2 and O_2. Similar conclusions for CO, CO_2, N_2O, CH_4, Cl_2, NH_3, and HCl are reported by Hornig.[2] It can be concluded that case 2 above applies for the transfer of energy between the translational- and rotational-energy modes.

Accepting the assumption that the velocity distribution function is unaffected by energy transfer between the translational- and internal-energy modes for molecules of interest to us, we can write

$$\dot{q} = -k \text{ grad } T + \sum_i^v \rho_i \bar{\mathbf{V}}_i h_i \qquad (10\text{-}11)$$

where the index i refers to particles in internal quantum state i and $\bar{\mathbf{V}}_i$ is the diffusion velocity vector determined using the singlet-velocity distribution function in the manner indicated in Sec. 2-2. See Eq. (2-12), for example. To proceed, we assume a gas mixture of particles all having the same molecular weight which, however, exist in v internal states of vibration. Assume for clarity that gradients exist in one direction only (such as is approximately true in boundary-layer flow, for example); then Eq. (10-11) is written

$$\dot{q} = -k \frac{dT}{dy} + \sum_{i=1}^v \rho_i \bar{V}_{iy} h_i \qquad (10\text{-}12)$$

where $\bar{V}_{iy}$ = diffusion velocity of species in internal quantum state i
h_i = enthalpy of species in internal quantum state i
k = ordinary coefficient of thermal conductivity

Now, in the absence of pressure gradients and assuming (realistically) that thermal diffusion is of second order, we can write

$$\rho_i \bar{V}_{iy} = -\rho D_i \frac{dC_i}{dy} \qquad (10\text{-}13)$$

[1] J. S. Parker, C. E. Adams, and R. H. Stavseth, *J. Accoust. Soc. Am.*, vol. 25, p. 263, 1953.

[2] Donald F. Hornig, *J. Phys. Chem.*, vol. 61, pp. 856–860, 1957.

where, since we assume species of same molecular weight and essentially the same interaction potentials, D_i is the self-diffusion coefficient. Furthermore,

$$h = \sum_i^v C_i' h_i \tag{10-14}$$

where h is the enthalpy per unit mass and

$$h_i = \int_0^T C_{p_i}\, dT + h_i^0 \tag{10-15}$$

Hence, if we assume that the concentrations are in chemical equilibrium at each point in the flow field,

$$\frac{dh}{dT} = C_p = \sum_{i=1}^v C_i C_{p_i} + \sum_{i=1}^v h_i \frac{dC_i}{dT}$$

or

$$C_p - C_{p_f} = \sum_{i=1}^v h_i \frac{dC_i}{dT} \tag{10-16}$$

where C_{p_f} is the "frozen" specific heat given by

$$C_{p_f} = \sum_{i=1}^v C_i C_{p_i} \tag{10-17}$$

Combining Eqs. (10-12), (10-13), and (10-16) gives

$$\dot{q} = -k' \frac{dT}{dy} \tag{10-18}$$

where

$$k' = k + \rho D_i (C_p - C_{p_f})$$

or

$$\frac{k'}{k} = Eu = 1 - L' + \frac{2}{5} L' \frac{C_p}{R} \tag{10-19}$$

where

$$L' = \frac{\rho D_i C_{p_f}}{k}$$

$$R_i = \frac{R}{M_i}$$

where M_i is the same for all quantum states and

$$C_{p_f} = \tfrac{5}{2} R_i$$

L' can be evaluated using the Chapman-Enskog theory for monatomic molecules., Hirschfelder[1] finds that, for a wide range of

[1] Joseph O. Hirschfelder, *J. Chem. Phys.*, vol. 26, no. 2, pp. 282–285, 1957.

temperatures and assumed intermolecular potential function parameters σ and ϵ, $L' \simeq 0.885$. Using this value, the correction factor Eu becomes

$$Eu = \frac{k'}{k} = 0.115 + 0.354\frac{C_p}{R_i} = 0.469 + 0.354\frac{C_v}{R_i} \qquad (10\text{-}20)$$

since

$$C_p = C_v + R_i \qquad (10\text{-}21a)$$

Eucken[1] first presented a similar expression which can be expressed as

$$Eu = \frac{k'}{k} = \frac{1}{3} + \frac{4}{15}\frac{C_p}{R_i} = \frac{3}{5} + \frac{4}{15}\frac{C_v}{R_i} \qquad (10\text{-}22)$$

which results from using

$$\mu = \frac{4}{15}\frac{k}{R_i} \qquad (10\text{-}9)$$

and

$$C_v = \frac{R_i}{\gamma - 1} \qquad \gamma = \frac{C_p}{C_v} \qquad (10\text{-}21b)$$

in Eucken's original expression

$$k' = \frac{C_v\mu}{4}(9\gamma - 5) \qquad (10\text{-}23)$$

Equation (10-20) is believed to be more accurate than Eq. (10-22), which was derived from Eucken's original equation, because the choice of the transport number L' is more realistic than Eucken's choice.

Equation (10-20) gives us a relation between the thermal conductivity of a pure polyatomic gas and the thermal conductivity for that gas calculated assuming that the polyatomic gas has no internal-energy modes. We are now in a position to present the coefficient of thermal conductivity for a mixture of polyatomic gases.

Thermal Conductivity for Polyatomic Gas Mixtures. It was found by Mason and Saxena[2] that the Chapman-Enskog expression for thermal conductivity of a mixture of monatomic gases can be modified making use of the Eucken factor of Eq. (10-20) or (10-22) to yield an expression for the thermal conductivity of polyatomic gases. Their equation is

$$k = \sum_{i=1}^{v} k_i'\left(1 + 1.065\sum_{\substack{k=1 \\ k \neq i}}^{v} G_{ik}'\frac{x_k}{x_i}\right)^{-1} \qquad (10\text{-}24)$$

where

$$k_i' = k_i Eu \qquad (10\text{-}25)$$

$$Eu = 0.115 + 0.354\frac{C_{p i}}{R_i} \qquad (10\text{-}20)$$

[1] A. Eucken, *Physik. Z.*, vol. 14, p. 324, 1913.
[2] Mason and Saxena, *op. cit.*

and
$$G'_{ik} = \frac{[1 + (k_i/k_k)^{1/2}(M_i/M_k)^{1/4}]^2}{(2)^{3/2}[1 + (M_i/M_k)]^{1/2}} \tag{10-26}$$

and k_i are the "frozen" thermal conductivities given by Eqs. (10-9) and (10-1). Mason and Saxena show that Eq. (10-24) gives excellent results when compared with experimentally determined thermal-conductivity coefficients for a wide variety of binary mixtures of polyatomic gases at temperatures up to 688°K. Amdur and Mason[1] show that Eq. (10-24) agrees well with results calculated using the full Chapman-Enskog theory equation for a mixture of helium in argon over a temperature range of 1000 to 15,000°K. We are thus encouraged that Eq. (10-24) may be used in lieu of the more complex Chapman-Enskog equation for thermal-conductivity coefficient given elsewhere.[2]

We have not yet completed our discussion of the transport properties. Before the expressions presented here can be applied, the intermolecular potential function parameters σ and ϵ and the collision integrals $\Omega^{(l,s)*}$ must be known. The purpose of the next section is to present and describe these quantities.

10-4. The Collision Integrals. Before the transport coefficients can be calculated using the equations presented in the previous section, certain integrals, derived explicitly in the Chapman-Enskog theory, must be evaluated making use of the particular intermolecular potentials chosen. We shall present the appropriate integrals here and discuss their evaluation for a number of different intermolecular potentials in the section following this one.

Fundamental to the Chapman-Enskog theory is the assumption that mass, momentum, and energy fluxes result from binary collisions among the constituent gas particles. The laws of classical mechanics are brought into play in order to describe the trajectories of the particles during the binary collisions. To complete the classical equations, the form of the intermolecular potential must be specified. We define a potential function $\phi(r)$ where

$$\phi(r) = \int_r^\infty F(r)\, dr$$

hence
$$F(r) = -\frac{d\phi}{dr} \tag{10-27}$$

gives the force of interaction on a particle separated a distance r from another particle. In that which follows, we assume that the

[1] Amdur and Mason, *op. cit.*
[2] Hirschfelder, Curtiss, and Bird, *op. cit.*, pp. 501, 537–538.

potential $\phi(r)$, and hence the force $F(r)$, is spherically symmetric. The case of polar molecules possessing a nonspherical force field has been treated elsewhere but is not yet in so useful a state as that for nonpolar molecules. Once the interparticle potential function has been chosen, the laws of classical mechanics are sufficient to describe the trajectories of the two particles while they are within the field of influence of one upon the other.

The binary encounter can be described as an equivalent one-body problem of determining the motion of a particle with reduced mass $\bar{m}$ and initial velocity g within a central force field having spherical symmetry. The equations for this equivalent one-body problem can be easily derived from the equations for the two-body problem by simply shifting the origin from the center of mass of the two colliding particles with mass m_i and m_j to the position of the particle with the mass m_j (or m_i). The equations of motion can be derived from the laws for conservation of angular momentum and energy and are

$$\bar{m}bg = \bar{m}r^2\dot{\theta} \tag{10-28}$$

and

$$\tfrac{1}{2}\bar{m}g^2 = \tfrac{1}{2}\bar{m}(\dot{r}^2 + r^2\dot{\theta}^2) + \phi \tag{10-29}$$

Equations (10-28) and (10-29) can be combined to give

$$\tfrac{1}{2}\bar{m}g^2 = \tfrac{1}{2}\bar{m}\dot{r}^2 + \tfrac{1}{2}\bar{m}g^2\left(\frac{b}{r}\right)^2 + \phi \tag{10-30}$$

where

$$\bar{m} = \frac{m_i m_j}{m_i + m_j} \tag{10-31}$$

$b =$ distance of closest approach of particle i to particle j in the absence of a potential

$g =$ velocity of particle i relative to particle j in absence of intermolecular force field (or at $r = \infty$)

and the coordinates of the motion are as depicted in Fig. 10-1. Also shown in Fig. 10-1 is a representative interparticle potential $\phi(r)$. It is seen that for r large there exists an attractive force upon particle i $[-d\phi/dr = F(r) < 0]$ and, when r is less than r_e, there is a repulsive force upon the particle i $[-d\phi/dr = F(r) > 0]$. We shall discuss possible functional forms for $\phi(r)$ later.

Given b, g, and $\phi(r)$, the trajectory of the particle i relative to particle j is completely determined by Eq. (10-30). We see immediately that the value of $r_{\min}$, the distance of closest approach, results when $\dot{r} = 0$ in Eq. (10-30). Hence

$$\phi(r_{\min}) = \frac{1}{2}\,\bar{m}g^2\left[1 - \left(\frac{b}{r_{\min}}\right)^2\right] \tag{10-32}$$

and $r_{\min}$ is easily determined from Eq. (10-32) once $\phi(r)$ is specified.

Furthermore, as Fig. 10-1 shows, the angle of deflection χ will be determined once $\theta_{\min}$ is determined, where $\theta_{\min}$ is the value of θ

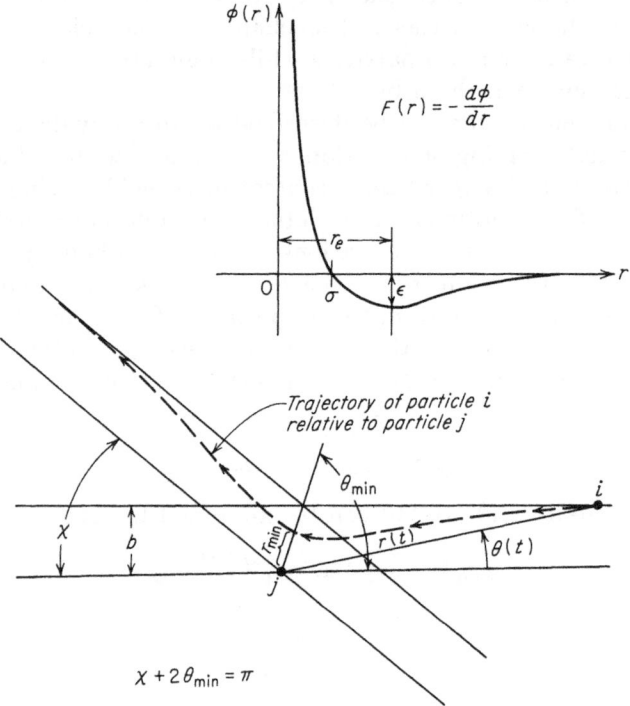

FIG. 10-1. The motion of particle i relative to particle j, with the interparticle force field as depicted.

when $r = r_{\min}$. $\theta_{\min}$ is determined by combining Eqs. (10-28) and (10-30) as follows: From Eq. (10-28)

$$\dot{\theta} = \frac{d\theta}{dt} = \frac{bg}{r^2} \tag{10-28}$$

From Eq. (10-30)

$$\dot{r} = \frac{dr}{dt} = \pm g\left[1 - \left(\frac{b}{r}\right)^2 - \frac{2\phi}{\bar{m}g^2}\right]^{1/2} \tag{10-30}$$

whence

$$\frac{dr}{d\theta} = -\frac{r^2}{b}\left[1 - \left(\frac{b}{r}\right)^2 - \frac{2\phi}{\bar{m}g^2}\right]^{1/2} \tag{10-33}$$

where the minus sign is chosen, since r decreases as θ increases on the incoming part of the trajectory ($r \geq r_{\min}$) (see Fig. 10-1). Thus, from Eq. (10-33),

$$\theta_{min} = \int_0^{\theta_{min}} d\theta = \int_\infty^{r_{min}} \frac{d\theta}{dr} dr$$

or
$$\theta_{min} = -b \int_\infty^{r_{min}} \frac{dr}{r^2[1 - (b/r)^2 - (2\phi/\bar{m}g^2)]^{1/2}}$$

Since, by geometry, $\chi + 2\theta_{min} = \pi$, the angle of deflection is given by

$$\chi(b,g) = \pi - 2b \int_{r_{min}}^\infty \frac{dr}{r^2[1 - (b/r)^2 - (2\phi/\bar{m}g^2)]^{1/2}} \tag{10-34}$$

and r_{min} is determined by solving Eq. (10-32) once $\phi(r)$ is specified. Equation (10-34) can be solved for the angle of deflection χ once initial relative velocity g, the impact parameter b, the reduced mass $\bar{m}$, and the interparticle potential $\phi(r)$ are given. The Chapman-Enskog theory gives the transport properties in terms of these deflection-angle integrals. The pertinent Chapman-Enskog relations are

$$\Omega^{(l,s)} = \left(\frac{kT}{2\pi\bar{m}}\right)^{1/2} \int_0^\infty e^{-\gamma^2} \gamma^{2s+3} Q^{(l)}(g) \, d\gamma \tag{10-35}$$

where l and s appear as exponents in Eqs. (10-37) and (10-35), respectively,

$$\gamma = \frac{\bar{m}g^2}{2kT} \tag{10-36}$$

and
$$Q^{(l)}(g) = 2\pi \int_0^\infty (1 - \cos^l \chi) b \, db \tag{10-37}$$

and $\chi(g,b)$ is given by Eq. (10-34). The transport coefficients in terms of the collision integrals have already been given in Sec. 10-2. The integral $Q^{(l)}(g)$ is referred to as the cross section, and the integral $\Omega^{(l,s)}$ is referred to as the collision integral. It is seen that the cross section involves integration over the range of impact parameters from $b = 0$ to $b = \infty$ for each initial relative velocity g and that the collision integral involves integration over each initial relative velocity from $g = 0$ to $g = \infty$. It will be found that numerical integrations are resorted to in many cases of interest.

The derivation of Eqs. (10-35) and (10-37) is not within the objectives of this book. A description of the derivations of these equations using the Chapman-Enskog theory will be found in Chap. 7 of the book by Hirschfelder, Curtiss, and Bird,[1] to which the interested

[1] Joseph O. Hirschfelder, Charles F. Curtiss, and R. Byron Bird, "Molecular Theory of Gases and Liquids," pp. 441–493, John Wiley & Sons, Inc., New York, 1954.

reader is referred. Our objective in this chapter is to present those portions of the kinetic theory for nonuniform gases which are necessary to understand and calculate the transport coefficients, to understand their limitations, and to understand how the transport properties might be affected by conditions present in hypersonic and/or reacting gas flows.

10-5. The Physics of Intermolecular Forces. Several intermolecular-force contributions to $\phi(r)$ are possible, depending upon the structure of the colliding particles. For electronically neutral

TABLE 10-1. SOME VARIATIONS OF POSSIBLE CONTRIBUTIONS TO THE INTERPARTICLE FORCES WITH INTERPARTICLE DISTANCE r. LONG-RANGE FORCES ONLY†

Force type	Description	Dependency upon r as
Electrostatic (Coulomb)	Charge-charge (ion-ion)	$1/r$
Electrostatic	Charge-dipole	$1/r^2$
Electrostatic	Charge-quadrupole	$1/r^3$
Electrostatic	Dipole-dipole	$1/r^3$
Electrostatic	Dipole-quadrupole	$1/r^4$
Induction	Charge-induced-dipole	$1/r^4$
Induction	Dipole-induced-dipole	$1/r^6$
Dispersion	Induced-dipole–induced-dipole	$1/r^7$
Dispersion	Induced-dipole–induced-quadrupole	$1/r^8$

† See J. C. Slater, "Introduction to Chemical Physics," pp. 352–376, McGraw-Hill Book Company, Inc., New York, 1939, for an excellent discussion of interparticle forces.

molecules it is known that the interparticle potential usually takes the form shown in Fig. 10-1. That is, at long range the forces are attractive and at short range the forces are repulsive. The long-range attractive forces usually decrease in magnitude as r^{-7} with increasing r, and the short-range repulsive forces usually decrease in magnitude as r^{-n} with r, where n is much larger than 7. The long-range attractive force is attributed to forces set up when induced dipoles are created as the electron cloud of each particle is nonsymmetrically dispersed relative to the nucleus of the particle. Most of the components of the long-range forces are electrostatic in origin and can be explained using the laws of electrostatics and quantum mechanics. The short-range forces are most complicated and can be described only with the use of quantum theory. Some notion of the possible contributions to long-range components of the interparticle potentials can be obtained from examining Table 10-1 where several possible interparticle forces are listed.

At room temperature, most molecules are electrically neutral in gas mixtures of concern to us here. At higher temperatures, ionization will take place, but we shall concern ourselves with temperatures below the threshold where appreciable ionization is present. The magnitude of the induction forces, while present, are lesser in magnitude than the principal dispersion force, the induced-dipole–induced-dipole force varying as r^{-7}, and this latter force is the principal long-range attractive force of interest to us here.

At high temperatures of concern to us in this book, dissociation of molecules will occur, resulting in a significant concentration of radicals.[1] At long ranges the radical-radical interparticle force is largely composed of the dispersion force components, which vary as r^{-7} as we have seen. As the two radicals approach each other, restrictions will be placed upon them by quantum-mechanical rules which will determine whether the resulting collision complex is in one of the attractive or repulsive states. According to the spin valence theory, the interaction of two radicals the unpaired electrons of which upon collision pair off in orbitals of the collision complex (in such a way that the electron pairs have antiparallel spins) will form a complex in an attractive state. In fact, two such particles having zero relative velocity at large distance ($r \rightarrow \infty$) will form a stable complex, since the process described is that of forming a chemical bond. However, if during collision some of the previously unpaired electrons pair off in orbitals of the collision complex (with antiparallel spins) and some do not, there is the possibility that the collision complex will not be in an attractive state.

As an example of the above, consider the nitrogen atom which in its normal or ground electronic state will have three unpaired electrons all of which occupy different orbitals and have parallel spins. Upon collision with another such nitrogen atom, there are four spin states of the combined complex possible. These four states are described in the language of quantum mechanics as the $^1\Sigma$, $^3\Sigma$, $^5\Sigma$, and $^7\Sigma$ states. The superscript is $2S + 1$, where S is the spin quantum number of the combined complex and Σ denotes a state of the combined complex having zero total angular momentum. The $^1\Sigma$ state is the ground state of the nitrogen molecule having spin 0. As seen on the schematic of Fig. 10-2, its potential has a deep well appropriate to many bound vibrational states, and thus has been well determined spectroscopically. The triplet, or $^3\Sigma$, state may or may not be stable and as yet has not been determined spectroscopically or otherwise. The $^5\Sigma$ and $^7\Sigma$ states are strongly repulsive and have not been determined

---

[1] A radical is defined here as an atom or molecule (not necessarily, but usually, electrically neutral) possessing an unpaired orbital electron.

by any means.[1]   The superscript $2S + 1$ also denotes the statistical weight or the degeneracy of the Σ state.   That is, statistically, if 16 such nitrogen atoms approach one another, it is most probable that only 1 of the 16 resulting collision complexes will be described by the lowest-lying or ground-state potential curve shown in Fig. 10-2, whereas 7 of the resulting collision complexes will be represented by the presently quantitatively unknown curve for the $^7\Sigma$ state shown in

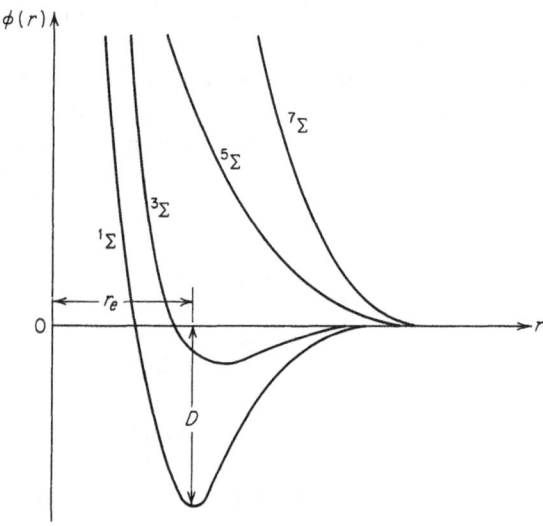

Fig. 10-2. Schematic of interaction potentials for two colliding ground-state N atoms.

Fig. 10-2.   Thus is described our dilemma.   The least frequently occurring collision among the four types possible is well described, and the most frequently occurring are unknown.   Of course, what is ideally desired is some average curve which represents all such collisions.   It seems reasonable to suggest that measurements of radical-radical atomic beam scattering might yield such information.

Normal oxygen radicals have two unpaired electrons, and thus their collision complex potentials are represented by the $^1\Sigma$, $^3\Sigma$, and $^5\Sigma$ states of the O_2 molecules.   The singlet, $^1\Sigma$, and triplet, $^3\Sigma$, states are well determined spectroscopically, while the other is not, since it represents an unstable or repulsive state, and a problem presents itself in describing O-O interactions.   Similar observations

[1] Recent evidence indicates that the $^5\Sigma$ state is ·weakly bound.   Progress toward quantitatively determining potential curves for the $^3\Sigma$, $^5\Sigma$, and $^7\Sigma$ states of nitrogen is reported by Joseph T. Vanderslice, Edward A. Mason, and Ellis R. Lippincott, *J. Chem. Phys.*, vol. 30, no. 1, pp. 129–136, 1959.

can be made for radical-radical interactions of other gas-atom pairs where the multiplicity of unknown interaction potentials will be determined by the number of unpaired electrons possessed by the colliding radicals (position of the elements in the periodic table for atom-atom collisions) and the combination rules of quantum mechanics.

We take note here that the depth of the ground-state potential well for radical-radical collisions is the dissocation energy D and is much greater than the depth of the potential well for chemically neutral molecules and atoms. For example, for the N-N interactions forming a complex in the $^1\Sigma$ state, $\epsilon/k = 113, 200°K$ ($\epsilon = D$), whereas for N_2-N_2 interactions, $\epsilon/k = 79.8°K$, where D and ϵ are the respective depths of the potential wells (see Figs. 10-1 and 10-2). Obviously, at high temperatures, $500°K < T < 10,000°K$, such as we shall be concerned with later, there will be differing influences of the attactive part of the interaction potential depending upon whether radical-radical or chemically neutral-neutral particle interactions are involved. For the former, the attractive forces may be important; for the latter, the potential well will have little bearing upon particle trajectories during the collision process.

Collisions involving atoms and/or molecules in excited electronic states pose problems similar to those described above for interacting radicals. In general, it can be stated that molecules in excited electronic states possess large cross sections because of the larger orbits of the displaced electrons compared with those of their ground-state relatives. Fortunately, for the diatomic molecules including the components of air, there are not an appreciable number of electronically excited molecules or atoms present at temperatures of interest to us here and their effect will be negligible.

10-6. Various Intermolecular Potentials. In principle, the interparticle potentials can be determined using the methods of quantum mechanics, although the task is immense for all but the simplest atoms and molecules. It is much simpler and more practical to represent the interparticle potential with a simple functional form which preserves the essential physical details and also allows the integrals in Eqs. (10-34), (10-35), and (10-37) to be evaluated in a reasonably straightforward manner. Three such potentials will be described here and are illustrated in Fig. 10-3.

Rigid-sphere Potential. This potential ignores completely the long-range forces and gives a crude representation to the strong, short-range repulsive forces by forbidding both colliding particles to occupy the same volume. Mathematically,

$$\phi(r) = \infty \quad \text{for } r \leq \sigma$$

and

$$\phi(r) = 0 \quad \text{for } r > \sigma \tag{10-38}$$

When Eq. (10-38) is used in Eq. (10-34), where $r_{\min} = \sigma$ if $b < \sigma$ and $r_{\min} = b$ if $b > \sigma$, there results

$$\chi(g,b) = 2 \cos^{-1} \frac{b}{\sigma} \qquad b \leq \sigma \qquad (10\text{-}39)$$

and

$$\chi(g,b) = 0 \qquad b \geq \sigma \qquad (10\text{-}40)$$

Substitution of Eqs. (10-39) and (10-40) into Eq. (10-37) results in

$$Q^{(l)} = \left[1 - \frac{1}{2} \frac{1 + (-1)^{(l)}}{1 + l} \right] \pi \sigma^2 \qquad (10\text{-}41)$$

and when Eq. (10-41) is substituted into Eq. (10-35), there results

$$\Omega^{(l,s)} = \left(\frac{kT}{2\pi \bar{m}} \right)^{1/2} \frac{(s + 1)!}{2} Q^{(l)} \qquad (10\text{-}42)$$

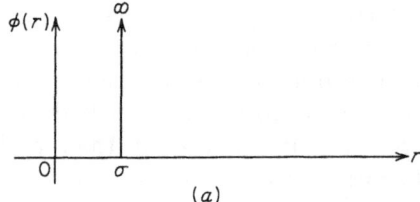

(a)

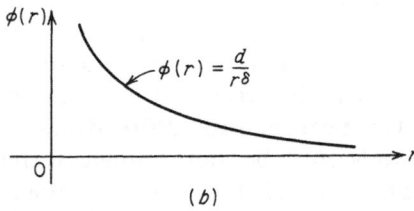

(b)

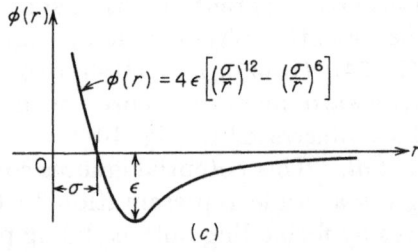

(c)

FIG. 10-3. Three useful intermolecular potential functions used in this book: (a) Rigid elastic spheres; (b) point centers of repulsion; (c) Lennard-Jones potential.

Equation (10-41) gives for $l = 1$

$$Q^{(1)} = \pi\sigma^2 \tag{10-43}$$

and for $l = 2$

$$Q^{(2)} = \tfrac{2}{3}\pi\sigma^2 \tag{10-44}$$

Thus Eq. (10-42) combined with Eqs. (10-43) and (10-44) gives the desired collision integrals $\Omega^{(1,1)}$ and $\Omega^{(2,2)}$; viz.,

$$[\Omega^{(1,1)}]_{r.s.} = \left(\frac{kT}{2\pi\bar{m}}\right)^{1/2}\pi\sigma^2 \tag{10-45}$$

and

$$[\Omega^{(2,2)}]_{r.s.} = \left(\frac{kT}{2\pi\bar{m}}\right)^{1/2}2\pi\sigma^2 \tag{10-46}$$

where the subscripts $r.s.$ refer to rigid-sphere values of the collision integrals.

Therefore, when the rigid-sphere potential is used,

$$\Omega^{(1,1)*} = \frac{\Omega^{(1,1)}}{[\Omega^{(1,1)}]_{r.s.}} = 1$$

and

$$\Omega^{(2,2)*} = \frac{\Omega^{(2,2)}}{[\Omega^{(2,2)}]_{r.s.}} = 1$$

and Eqs. (10-1), (10-8), and (10-9) show that

$$\mu_i \propto \frac{(m_i T)^{1/2}}{\sigma_i^2}$$

$$D_{12} \propto \frac{T^{3/2}}{\bar{m}^{1/2}p\sigma_{12}^2}$$

and

$$k_i \propto \frac{(m_i T)^{1/2}}{\sigma_i^2}$$

σ_i is a free parameter determined by matching measured transport coefficients with values calculated using Eqs. (10-1), (10-8), and (10-9).

High-temperature transport properties for the noble gases will approach the above behavior with temperature and pressure, since as temperature increases, the attractive portion of the interparticle potential becomes less and less influential in determining particle trajectories during the collision process and the particles behave more and more as rigid elastic spheres.

Point Centers of Repulsion. Kinetic theory tells us that for a gas composed of particles with mass $\bar{m}$

$$\tfrac{1}{2}\bar{m}v_s^2 = \tfrac{3}{2}kT \tag{10-47}$$

where v_s is the root-mean-square speed and the expression on the left-hand side of Eq. (10-47) is the mean kinetic energy of all kinetic energies possessed by the molecules in equilibrium at temperature T. The left-hand side is, then, for our two-body collision

$$\tfrac{1}{2}\bar{m}v_s^2 \simeq \tfrac{1}{2}\bar{m}(g_{\text{mean}})^2$$

hence, on the average, the kinetic energy possessed by one colliding particle in a coordinate system fixed with the other is

$$\tfrac{1}{2}\bar{m}(g_{\text{mean}})^2 \simeq \tfrac{3}{2}kT$$

Now most interparticle potentials for chemically and electrically neutral particles take the form of that shown on Fig. 10-3 for the Lennard-Jones potential. Furthermore, the depth of the energy well, ϵ in Fig. 10-3, is of the order of $100k$ for chemically neutral gases of the type we are concerned with here. Thus, as $T \to 10,000°K$ we find that the kinetic energy of a colliding particle far exceeds the potential energy in the vicinity of the potential well. That is,

$$\frac{\bar{m}(g_{\text{mean}})^2}{2k} \sim \frac{3}{2}T \gg \frac{\epsilon}{k} \simeq 100°K$$

and the potential well will have very little bearing upon the interparticle trajectories. In fact it will be the repulsive region of the potential curve in the neighborhood of $|\phi| = |kT|$ which will determine the collision dynamics at high temperatures. This suggests that a simple potential function representing point centers of repulsion might very well be appropriate for dealing with gas mixtures for which $T \gg \epsilon/k$. Such a potential is

$$\phi(r) = \frac{d}{r^\delta} \tag{10-48}$$

When Eq. (10-48) is used with (10-34), (10-35), and (10-37), there results

$$\Omega^{(l,s)} = \left(\frac{2\pi kT}{\bar{m}}\right)^{1/2} \frac{A^{(l)}}{2}\left(\frac{\delta d}{kT}\right)^{2/\delta} \Gamma\left(s + 2 - \frac{2}{\delta}\right) \tag{10-49}$$

where Γ is the gamma function[1] with argument $s + 2 - 2/\delta$. The number $A^{(l)}$ results from the numerical integration necessary when Eq. (10-48) is used with Eqs. (10-34), (10-35), and (10-37). Table 10-2 gives values of $A^{(1)}$ and $A^{(2)}$ for various values of δ.

---

[1] Tabulated, for example, in B. O. Peirce, "A Short Table of Integrals," pp. 136–137, Ginn & Company, Boston, 1957.

Making use of Eqs. (10-42) and (10-49) we obtain, for point centers of repulsion,

$$\Omega^{(l,s)*} = \frac{2\pi A^{(l)}(\delta d/kT)^{2/\delta}\Gamma(s + 2 - 2/\delta)}{(s + 1)! \, \pi\sigma^2\{1 - \frac{1}{2}[1 + (-1)^l]/(1 + l)\}} \tag{10-50}$$

These values are then used with Eqs. (10-1), (10-8), and (10-9) to obtain transport properties when the potential for point centers of repulsion is used.

TABLE 10-2. VALUES OF $A^{(l)}$ FOR VARIOUS VALUES OF δ IN
THE POTENTIAL LAW $\phi(r) = dr^{-\delta}$, EQ. (10-48)†

δ	$A^{(1)}(\delta)$	$A^{(2)}(\delta)$
4	0.298	0.308
6	0.306	0.283
8	0.321	0.279
10	0.333	0.278
12	0.346	0.279
14	0.356	0.280
∞	0.500	0.333

† Values given in Sidney Chapman and T. G. Cowling, "The Mathematical Theory of Non-uniform Gases," p. 172, Cambridge University Press, New York, 1958, where $A^{(l)}(\delta) = 2^{-2/\delta}A_l(\nu + 1)$.

δ and d are free parameters determined by matching measured transport coefficients with calculated coefficients, obtained using Eq. (10-50) with Eqs. (10-1), (10-8), and (10-9).

The Lennard-Jones Potential. For gas mixtures near room temperature the Lennard-Jones potential shown in Fig. 10-3 is quite appropriate. This potential is

$$\phi(r) = 4\epsilon\left[\left(\frac{\sigma}{r}\right)^{12} - \left(\frac{\sigma}{r}\right)^{6}\right] \tag{10-51}$$

where ϵ is the depth of the potential well (see Fig. 10-3) and σ is the value of r for which $\phi = 0$.

When Eq. (10-51) is used with Eqs. (10-34), (10-35), and (10-37), numerical integration is required. The integrals $\Omega^{(l,s)*}$ have been calculated for this potential function by numerous workers and are tabulated as a function of reduced temperature $T^* = kT/\epsilon$ in Table I-M of the book by Hirschfelder, Curtiss, and Bird.[1] Plots of $\Omega^{(1,1)*}$

[1] Joseph O. Hirschfelder, Charles F. Curtiss, and R. Byron Bird, "Molecular Theory of Gases and Liquids," pp. 1126–1127, John Wiley & Sons, Inc., New York, 1954.

and $\Omega^{(2,2)}*$ versus $T*$ are shown in Fig. 10-4, which can be used in conjunction with Eqs. (10-1), (10-8), and (10-9) to calculate transport properties of various gases once the free parameters ϵ and σ are given.

Before transport properties for pure gases and mixtures of gases can be calculated using the methods of this book, the empirically determined force constants σ and ϵ/k must be known. These constants have been determined for a number of gases by numerous

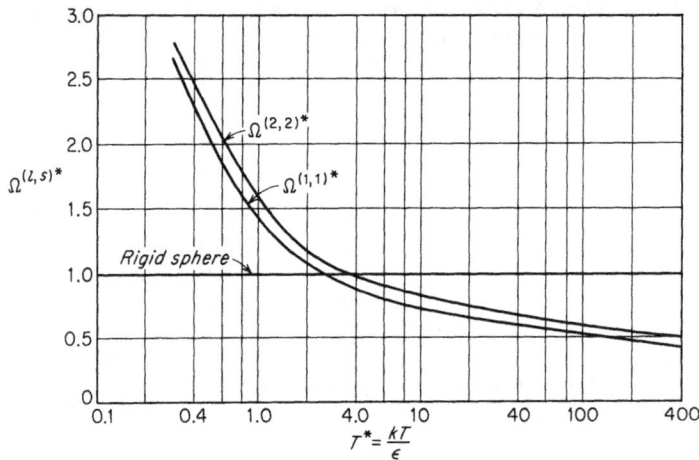

FIG. 10-4. The reduced collision integrals $\Omega^{(1,1)}*$ and $\Omega^{(2,2)}*$ plotted versus reduced temperature $T*$ for the Lennard-Jones interparticle potential function. (*After Joseph O. Hirschfelder, Charles F. Curtiss, and R. Byron Bird, "Molecular Theory of Gases and Liquids," chap. 8, pp. 514–610, John Wiley & Sons, Inc., New York, 1954.*)

investigations over the years. To attempt to tabulate them all would be an undertaking beyond the intentions of this book. However, such constants for several gases of interest to engineering application are presented on Table 10-3. Table I-A of the book by Hirschfelder, Curtiss, and Bird[1] presents force constants for several additional gases for the Lennard-Jones potential model.

In the next section we shall discuss the application of the transport-property equations to calculating the transport properties of a particular mixture, that of high-temperature air.

10-7. Transport Properties for Dissociated Air. The transport properties for high-temperature air can be calculated using the equations presented in Secs. 10-2 and 10-6. One would begin by

---

[1] *Ibid.*, pp. 1110–1112.

TABLE 10-3. EMPIRICAL FORCE CONSTANTS FOR USE WITH EQ. (10-50) FOR POINT REPULSION AND EQ. (10-51) FOR LENNARD-JONES POTENTIAL

Point repulsion (Fig. 10-3)

Interaction pair	δ	d, in ev $\times$ (angstroms)$^\delta$	Reference
He—He	5.94	4.71	1
N_2—N_2	7.27	595	1
He—A	7.14	63	3
A—A	8.33	848	1

Lennard-Jones (6:12)(Fig. 10-3)

Interaction pair	ϵ/k, in °K	σ, in angstroms	Reference
H_2—H_2	38	2.915	4
O_2—O_2	128	3.398	2
N_2—N_2	79.8	3.749	4
O—O($^3\Sigma$ state)	59,000	$\sim$1.07	5
N—N($^1\Sigma$ state)	113,200	$\sim$0.97	5
NO—NO	119	3.47	6
CO—CO	110	3.59	6
CO_2—CO_2	190	3.996	6
NO—NO	119	3.470	6
N_2O—N_2O	220	3.879	6
C_2H_6—C_2H_6	230	4.418	7
C_3H_8—C_3H_8	254	5.061	7
C_2N_2—C_2N_2	339	4.38	7

References:
[1] I. Amdur and E. A. Mason, *Phys. Fluids*, vol. 1, no. 5, pp. 370–383, 1958.

[2] C. T. G. Raw and C. P. Ellis, *J. Chem. Phys.*, vol. 28, p. 127, 1958.

[3] I. Amdur, E. A. Mason, and A. C. Harknen, *J. Chem. Phys.*, vol. 22, p. 1071, 1954.

[4] M. Trautz, A. Melster, and R. Zink, *Ann. Physik*, vol. 5, no. 7, pp. 409–452, 1930.

[5] G. Herzberg, "Spectra of Diatomic Molecules," D. Van Nostrand Company, Inc., Princeton, N.J., 1957.

[6] H. L. Johnston and K. E. McCloskey, *J. Chem. Phys.*, vol. 44, p. 1038, 1940.

[7] Landholt-Börnstein, "Physikalisch-Chemische Tabellen," Springer-Verlag, Berlin.

specifying the equilibrium gas composition for the mixture for which the transport properties are desired. This equilibrium composition can be determined by applying the methods of statistical thermodynamics described in Chap. 9. This has already been done by many authors, resulting in information on the composition and thermodynamic properties of air at temperatures ranging from room

temperature to 24,000°K at various pressures. Figure 10-5 presents
a plot of the mole-fraction composition of air versus temperature for
a temperature range from 0 to 15,000°K at a density of 10^{-2} normal
atmospheric density. Figure 10-5 was prepared using the equilibrium air calculations of Gilmore[1] which were presented graphically
by Moeckel and Weston.[2] Figure 10-5 shows that below about
10,000°K there is an insufficient concentration of electrons (e⁻) and

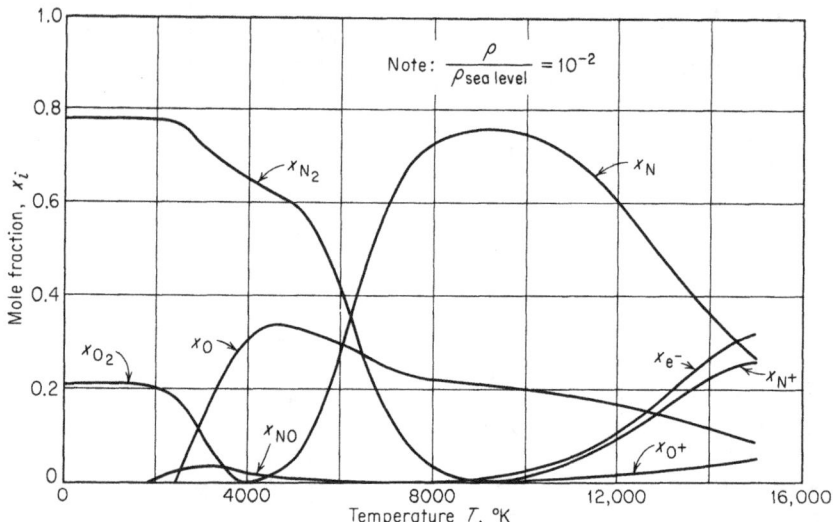

FIG. 10-5. Equilibrium mole fractions of the components of air versus temperature for 10^{-2} atmospheric density. (*After F. R. Gilmore, Rand. Corp. Rept.*
1543, Santa Monica, Calif., 1955.)

ions (O⁺ and N⁺) to affect transport-property calculations at this
density. Above 10,000°K at this density, the methods of this chapter are inadequate, since an appreciable concentration of ions and
electrons with their long-range Coulomb forces are present. Note,
however, that both the oxygen and nitrogen molecules are almost
completely dissociated at 10,000°K. Hence, we might inquire as to the
possible effects of dissociation of air molecules upon the transport properties of the air mixture at temperatures below ionization temperature
but high enough that dissociation is occurring. We shall treat the
ideal dissociating gas model in an approximate manner first and then
deal with dissociating air in a more exact manner.

The Ideal Dissociating Gas. Consider, for example, a simple

[1] F. R. Gilmore, *Rand Corp. Rept.* 1543, Santa Monica, Calif., 1955.
[2] W. E. Moeckel and Kenneth C. Weston, *NACA TN* 4265, 1958.

dissociating gas mixture of diatomic molecules.   Such a gas might
be, for example, $N_2 \rightleftharpoons 2N$ or $O_2 \rightleftharpoons 2O$.   We choose to deal with

$$O_2 \rightleftharpoons 2O$$

Let us examine the variation of the transport parameters P, L,
and S with temperature at a constant density for a range of temperatures within which a simple ideal gas varies from a condition of no
dissociation to that of a completely dissociated gas.   For oxygen, for
example, this temperature range would vary from 2000 to 8000°K
depending upon the pressure.   Here

$$P = \frac{C_{p_f}\mu}{k} = \text{Prandtl number} \tag{10-52}$$

$$S = \frac{\mu}{\rho D_{12}} = \text{Schmidt number} \tag{10-53}$$

$$L = \frac{P}{S} = \frac{\rho C_{p_f} D_{12}}{k} = \text{Lewis number} \tag{10-54}$$

Equations (10-1), (10-8), and (10-9) can be used to show that to a
first approximation these transport parameters are constant with
increasing temperature when no dissociation occurs.   Let us see how
dissociation will affect them.

Consider first the viscosity.   From Eq. (10-1)

$$\mu_i \propto \frac{(M_i T)^{1/2}}{\sigma^2 \Omega^{(2,2)*}}$$

Choosing a representative temperature of $T = 5000°K$ and making
use of the data given in Table 10-3 and the values of $\Omega^{(2,2)*}$ plotted in
Fig. 10-4 versus $T^* = kT/\epsilon$, we obtain the following ratio:

$$\frac{\mu_{O_2}}{\mu_O} = \left(\frac{M_{O_2}}{M_O}\right)^{1/2}\left(\frac{\sigma_O}{\sigma_{O_2}}\right)^2 \frac{\Omega_O^{(2,2)*}}{\Omega_{O_2}^{(2,2)*}} = 0.95$$

and it can be shown that this ratio is close to 1.0 for other temperatures as well.   Hence $\mu_{O_2} \simeq \mu_O$.

Now to a first approximation appropriate to the approximate
nature of the analysis of this section,

$$\mu = \sum_i \mu_i x_i \tag{10-55}$$

where, from Eq. (10-4),

$$x_i = \frac{C_i}{M_i}\left(\sum_i \frac{C_i}{M_i}\right)^{-1}$$

whence, if we let

$$1 - \alpha = C_1 = \text{mass fraction of molecules}$$
$$\alpha = C_2 = \text{mass fraction of atoms}$$

it follows that

$$x_1 = \frac{1 - \alpha}{1 + \alpha} \tag{10-56}$$

$$x_2 = \frac{2\alpha}{1 + \alpha} \tag{10-57}$$

Hence, from Eqs. (10-55) to (10-57)

$$\mu = \mu_{O_2} \frac{1 - \alpha}{1 + \alpha} + \mu_O \frac{2\alpha}{1 + \alpha}$$

or, since $\mu_{O_2} = \mu_O$,

$$\mu = \mu_{O_2} = \mu_O \tag{10-58}$$

Furthermore, to a first approximation,

$$k = \sum x_i k_i = k_{O_2} \frac{1 - \alpha}{1 + \alpha} + k_O \frac{2\alpha}{1 + \alpha} \tag{10-59}$$

But, from Eq. (10-9),

$$k_O = \frac{15}{4} R_O \mu_O \tag{10-60}$$

and, from Eqs. (10-9), (10-20), and (10-25),

$$k_{O_2} = \frac{15}{4} R_{O_2} \mu_{O_2} \left(0.115 + 0.354 \frac{C_{pO2}}{R_{O_2}} \right) \tag{10-61}$$

now

$$R_O = \frac{R}{M_O} \tag{10-62}$$

$$R_{O_2} = \frac{R}{M_{O_2}} = \frac{R}{2M_O} = \frac{1}{2} R_O \tag{10-63}$$

Furthermore,

$$C_{pO_2} = \frac{d(h_{O_2})}{dT} \tag{10-64}$$

where

$$h_{O_2} = e_{O_2} + R_{O_2} T \tag{10-65}$$

and

$$e_{O_2} = (e_{O_2})_l + (e_{O_2})_j + (e_{O_2})_v \tag{10-66}$$

where $l, j,$ and v refer to the translational, rotational, and vibrational modes, respectively, or

$$e_{O_2} = \tfrac{3}{2} R_{O_2} T + R_{O_2} T + \tfrac{1}{2} R_{O_2} T \tag{10-67}$$

where it is arbitrarily assumed the vibrational mode is excited to half its full classical value.   Combining Eqs. (10-64) to (10-67), we find that

$$\frac{C_{p_{O_2}}}{R_{O_2}} = 4 \tag{10-68}$$

Combining Eqs. (10-58) to (10-63) and (10-68) gives

$$k = k_{O_2} \frac{1 + 1.62\alpha}{1 + \alpha} = 0.765 k_0 \frac{1 + 1.62\alpha}{1 + \alpha} \tag{10-69}$$

where

$$k_{O_2} = 1.53 \times \frac{15}{4} R_{O_2}\mu_{O_2} = 0.765 k_0 \tag{10-70}$$

Equations (10-58) and (10-59) show that, while the viscosity of the mixture is given reasonably well by the viscosity one might calculate at any temperature using Eq. (10-1) with force constants for O, the thermal conductivity will be less than that calculated using Eq. (10-9) for O by a factor varying from 0.765 when $\alpha = 0$ to 1.00 when $\alpha = 1$ because of the Eucken factor given by Eq. (10-20) which applies to the molecules but not to the atoms and because $R_O = 2R_{O_2}$.

Consider the Prandtl number P:

$$P = \frac{C_{p_f}\mu}{k}$$

Now, from Eq. (10-17),

$$C_{p_f} = \sum_i C_i C_{p_i} = (1 - \alpha)C_{p_{O_2}} + \alpha C_{p_O} \tag{10-71}$$

where, since the atoms possess only translational degrees of freedom,

$$C_{p_O} = \frac{d}{dT}(e_0 + R_0 T) = \frac{d}{dT}\left(\frac{3}{2} R_0 T + R_0 T\right) = \frac{5}{2} R_0$$

or, using Eq. (10-63),

$$C_{p_O} = 5R_{O_2} \tag{10-72}$$

Combining Eqs. (10-52), (10-58), and (10-68) to (10-72), we arrive at

$$P = 0.70 \frac{(1 + 0.25\alpha)(1 + \alpha)}{1 + 1.62\alpha} \tag{10-73}$$

For α varying from 0 to 1, P varies from 0.70 to 0.667 as shown in Fig. 10-6.

Consider the Schmidt number S.   From Eq. (10-8) it can be seen that pD_{12} is independent of composition.   Also,

$$p = \sum_i p_i = p_O + p_{O_2}$$

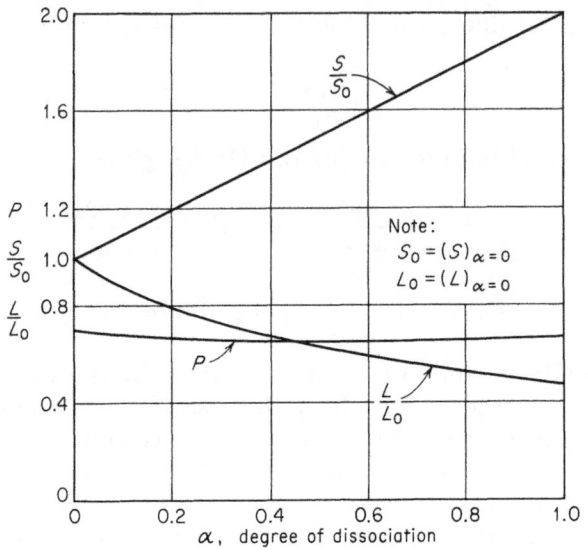

FIG. 10-6. Variation of Prandtl number P, Schmidt number S, and Lewis number L with degree of dissociation for a dissociating diatomic gas.

or, since

$$p_0 = \rho_0 R_0 T = \rho_0 2 R_{O_2} T = 2\alpha\rho R_{O_2} T$$

and

$$p_{O_2} = \rho_{O_2} R_{O_2} T = (1 - \alpha)\rho R_{O_2} T \tag{10-74}$$

then

$$p = \rho(1 + \alpha)R_{O_2} T \tag{10-75}$$

Thus, by Eqs. (10-53), (10-58), and (10-75),

$$S = \frac{\mu}{\rho D_{12}} = \frac{\mu_{O_2}}{p_{O_2} D_{12}} (1 + \alpha)R_{O_2} T$$

but

$$p_{O_2} D_{12} = (p_{O_2} D_{12})_{\alpha=0} = (\rho_{O_2} R_{O_2} T D_{12})_{\alpha=0}$$

because $p D_{12}$ is independent of composition.   Hence

$$S = S_0(1 + \alpha) \tag{10-76}$$

where

$$S_0 = \left(\frac{\mu_{O_2}}{\rho_{O_2} D_{12}}\right)_{\alpha=0} \tag{10-77}$$

Furthermore, from Eqs. (10-54), (10-69), (10-70), and (10-76),

$$L = L_0 \frac{1 + 0.25\alpha}{1 + 1.62\alpha} \tag{10-78}$$

where, by Eq. (10-73),

$$L_0 = \frac{P_0}{S_0} = 0.70 \frac{\rho_{O_2} D_{12}}{\mu_{O_2}}$$

$L/L_0 = S/S_0$ and P are all plotted versus α on Fig. 10-6. We find that the Lewis number should decrease as dissociation increases, the Schmidt number should increase with dissociation increasing, and the Prandtl number should vary very little. Let us see whether detailed calculations bear out the results of this analysis of the problem.

Calculations for Dissociated Air. The results of calculations of transport coefficients for high-temperature air components and mixtures will be no better than approximations made during the calculations. In calculations for air at temperatures high enough that appreciable dissociation occurs, the approximations made are mostly concerned with the interparticle potential function. The potential function parameters for chemically inert molecules are known at room temperature. Perhaps their uncritical use at high temperatures is a minor risk compared with the state of affairs for radical-radical interactions (O-O, N-N). As Fig. 10-2 and the discussion in Sec. 10-5 have suggested, the interparticle potentials for radical-radical interactions in the various spin states other than the ground state are unknown. Under these circumstances calculations of transport properties for high-temperature air must be carefully qualified as to their probable accuracy.

Several calculations have been made at the time of writing this book. Scala and Baulknight[1] reported calculations made using the Lennard-Jones potential for molecules and the rigid-sphere model for the atoms. Low-temperature values of ϵ/k and σ were used with the Lennard-Jones model, and the temperature dependence of viscosity for the rigid-sphere model is subject to question notwithstanding the fact that empirical rigid-sphere diameters σ are unknown for radicals. Stupochenko, Dotsenko, Stakhanov, and Samuilov[2] present calculations of the viscosity and thermal conductivity of dissociating air mixtures for temperatures between 2000 and 8000°K and pressures from 10^{-3} to 10^3 atm using the rigid-sphere model for atom-atom and atom-molecule interactions. An interesting contribution of this work is a suggestion of an interpolation rule to determine effective rigid-sphere diameters σ for intermediate elements for which measurements of viscosity are not available making use of measurements for H, Na, and the noble gases which are available. However, as stated before, the temperature

[1] Sinclaire M. Scala and Charles W. Baulknight, *ARS J.*, vol. 29, no. 1, pp. 39–45, 1959.

[2] E. P. Stupochenko, B. V. Dotsenko, I. P. Stakhanov, and E. V. Samuilov, *Fiz. Gazodinamika, USSR Acad. Sci.*, pp. 39–58; translation given in *ARS J.*, vol. 30, no. 4, pp. 394–402, 1960.

dependence of the viscosity law given by the rigid-sphere model is open to question, even though these authors attempted to correct for variation in diameter of a rigid-sphere molecule with temperature.

Bauer and Zlotnick[1] calculated transport properties for equilibrium dissociated air using the point center of repulsion interparticle potential given by Eq. (10-48) for temperatures between 3000 and 8000°K and densities between 10^{-2} and 10 times sea-level density at 273°K. Bauer and Zlotnick used values of the potential function parameter d which seemed reasonable based upon scattering cross-section measurements made using chemically inert and electronically unexcited molecules. The chief merit of their approach appears to be that their chosen potential model, the point center of repulsion model, accentuates the proper portion of the interparticle potential function at high temperatures for molecule-molecule interactions, and furthermore, the scattering measurements should give reasonable values of the parameter d for the molecules O_2 and N_2. However, as Sec. 10-3 made clear, this model may not be the best model for radical-radical collisions (O-O and N-N) which in their ground state, at least, form a collision complex the potential of which possesses a deep potential well. However, Bauer and Zlotnick acknowledge the uncertainties involved and at least estimated the probable errors involved in their calculations.

Bauer and Zlotnick proceeded as follows. They argued that the principal components present at temperatures between 3000 and 8000°K are O_2, O, N_2, N, and NO. Figure 10-5 certainly supports this assumption. Furthermore, as Fig. 10-5 shows, by the time N is present in appreciable concentration, O_2 has disappeared, so that there are only three or four components present in any given temperature range: O_2, O, and N_2 while O_2 is dissociating and O, N_2, N, and NO while N_2 is dissociating. Bauer and Zlotnick assumed that the interaction potential parameters for O-O, O-N, and N-N are all the same, being given by d_A and δ_A, and make a similar assumption for the molecules involved, that is, for O_2-O_2, N_2-N_2, and O_2-N_2 interactions. When diffusion-coefficient calculations were made, they used the empirically confirmed combination rules given below. That is, for the interaction potential model

$$\phi(r) = \frac{d}{r^\delta} \tag{10-48}$$

$$d_{A-M} = (d_A d_M)^{1/2} \tag{10-79}$$

and

$$\delta_{A-M} = \tfrac{1}{2}(\delta_A + \delta_M) \tag{10-80}$$

[1] Ernst Bauer and Martin Zlotnick, *ARS J.*, vol. 29, no. 10, pp. 721–776, 1959.

Because they were uncertain as to the correct variation of ϕ with r, they left δ as a free parameter and used $\delta = 6$ and 8. The values of d used are given in Table 10-4 along with the empirically determined value for N_2-N_2 interactions given in Table 10-3, which is included for comparison.

Bauer and Zlotnick carried out their calculations assuming that the gas was a binary mixture of atoms with atomic weight 15 and molecules of molecular weight 30 with specific heats the weighted

TABLE 10-4. VALUES OF d USED BY BAUER AND ZLOTNICK IN INTERPARTICLE POTENTIAL GIVEN BY EQ. (10-48) $\phi(r) = dr^{-\delta}$

δ	d_A, in ev $\times$ (angstroms)$^\delta$	d_M, in ev $\times$ (angstroms)$^\delta$	$d_{N_2-N_2}$ in ev $\times$ (angstroms)$^\delta$
6	24.8	158	
8	190	1215	
7.27	...	...	595

averages between oxygen and nitrogen values. The "Eucken" correction factor given by Eq. (10-20) was used in calculating thermal conductivity for the molecules.

The calculation procedure is straightforward. Choosing a δ, and thus d_A or d_M, one calculates the collision integrals $\Omega^{(1,1)*}$ and $\Omega^{(2,2)*}$ using Eq. (10-50) with values of $A^{(1)}$ and $A^{(2)}$ determined by Table 10-2. The pure species μ_i and k_i are determined using Eqs. (10-1) and (10-9). Equation (10-25) is used to determine μ_i for the molecules. One then calculates the μ and k for the mixture by using Eqs. (10-3) and (10-10) and the equilibrium concentrations x_i which are prepared beforehand using the methods described in Chap. 9. See, for example, Fig. 10-5. D_{12} is calculated using Eq. (10-8) and mean molecular weights for the atoms and molecules of 15 and 30, respectively.

It is clear from Eqs. (10-1), (10-8), and (10-50) that the point center of repulsion model gives, for constant density and variable pressure,

$$\mu \text{ and } D_{12} \propto T^{1/2 + 2/\delta}$$

Thus, for the present calculation,

$$\mu \text{ and } D_{12} \propto T^{0.833} \quad \text{if } \delta = 6$$

and

$$\mu \text{ and } D_{12} \propto T^{0.75} \quad \text{if } \delta = 8$$

The results of the calculations for μ, k, S, P, and L for $\rho/\rho_{SL} = 10^{-2}$ are given in Figs. 10-7 through 10-11. Upon examining these results along with the previous approximate results summarized in Fig. 10-6, we conclude that the trends exhibited by S, P, and L in Figs. 10-9

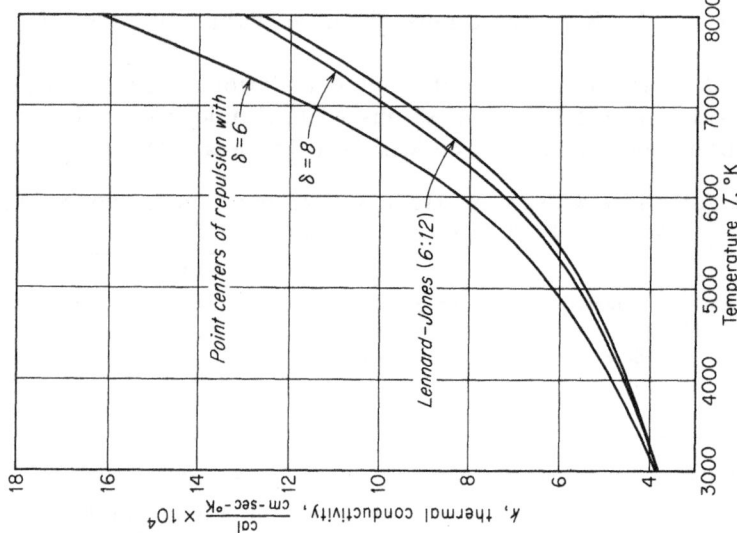

Fig. 10-8. Thermal conductivity for equilibrium air for $\rho/\rho_{SL} = 0.01$. (After Ernst Bauer and Martin Zlotnick, ARS J., vol. 29, no. 10, pp. 721–776, 1959.)

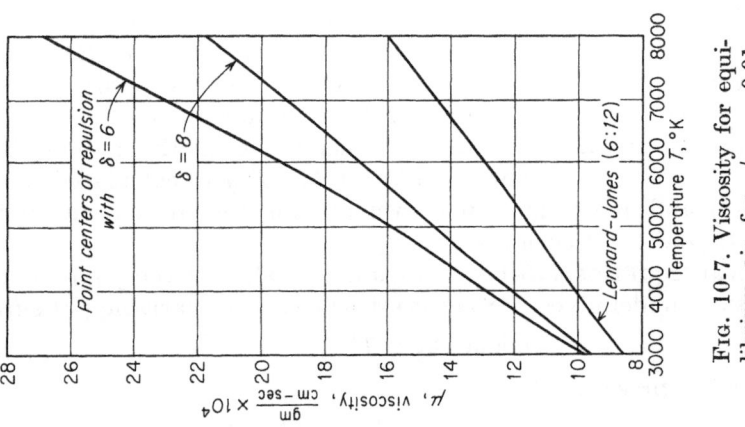

Fig. 10-7. Viscosity for equilibrium air for $\rho/\rho_{SL} = 0.01$. (After Ernst Bauer and Martin Zlotnick, ARS J., vol. 29, no. 10, pp. 721–776, 1959.)

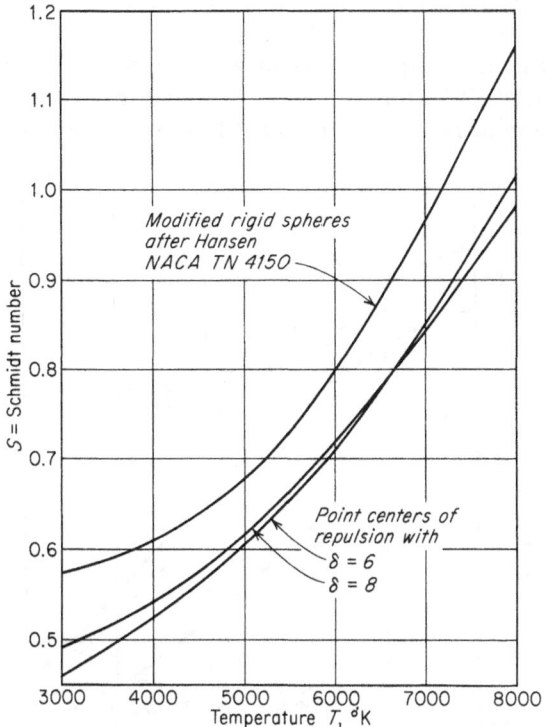

FIG. 10-9. Schmidt number for equilibrium air versus temperature for $\rho/\rho_{SL} = 0.01$. (*After Ernst Bauer and Martin Zlotnick, ARS J., vol. 29, no. 10, pp. 721–776, 1959.*)

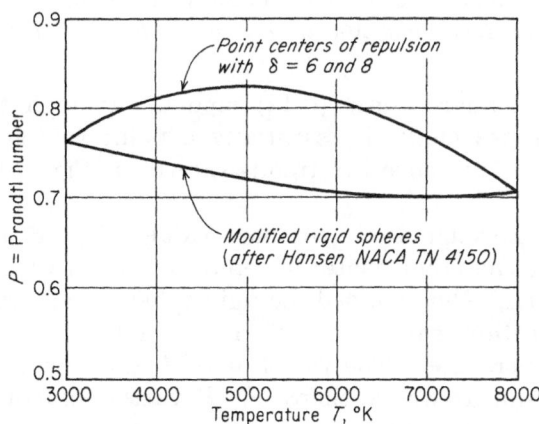

FIG. 10-10. Prandtl number for equilibrium air versus temperature for $\rho/\rho_{SL} = 0.01$. (*After Ernst Bauer and Martin Zlotnick, ARS J., vol. 29, no. 10, pp. 721–776, 1959.*)

through 10-11 agree well with the anticipations given in Fig. 10-6. Furthermore, it is apparent from inspection of Figs. 10-7 through 10-11 that the results are sensitive to both the interparticle potential model used and the empirical parameters used with a given potential model.   Figures 10-7 and 10-8 show how μ and k calculated assuming the point center of repulsion model differ from the values calculated

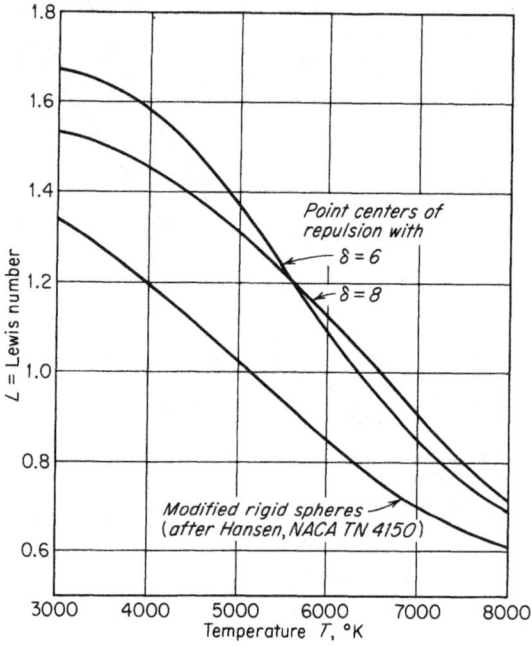

FIG. 10-11. Lewis number for equilibrium air versus temperature for $\rho/\rho_{SL} = 0.01$.   (*After Ernst Bauer and Martin Zlotnick, ARS J., vol. 29, no. 10, pp. 721–776, 1959.*)

assuming the Lennard-Jones (6:12) model.   At high temperatures the Lennard-Jones model is essentially a point center of repulsion model with $\delta = 12$; hence the trends shown on Figs. 10-7 and 10-8 are consistent.

   Figures 10-9 through 10-11 compare values of S, P, and L calculated assuming the point center of repulsion model with calculations made by Hansen,[1] who assumed the rigid-sphere model with the rigid-sphere cross-section parameter, σ^2 in Eq. (10-4), decreasing with increasing temperature.   The variation of L and S with temperature is consistent, and as was anticipated, P varies comparatively little with increasing temperature.

[1] C. F. Hansen, *NACA TN* 4150, 1958.

Bauer and Zlotnick estimate that their calculations are accurate to no better than ± 25 per cent at the high temperatures. These inaccuracies are chiefly due to

1. Possible inaccuracy in representing the radical-radical interactions

2. Probable inaccuracies in representing molecules and atoms in electronically excited states at the higher temperatures

3. Use of the empirical mixing rules given by Eqs. (10-79) and (10-80)

This section has illustrated how the transport properties for high-temperature air can be calculated using the equations and data presented in Secs. 10-2 and 10-3. It can be concluded that more experimental data are needed on particle-particle interactions at high temperature before the present calculations can be improved upon. Until these data are forthcoming, the transport properties of high-temperature air will remain one of the principal sources of uncertainty in applying the equations for skin friction and heat transfer to reacting gas mixtures.

10-8. Transport Properties for Gas Mixtures. Many problems of engineering interest are concerned with situations involving the viscous flow of multicomponent gas mixtures other than air. One example is the calculation of the effect upon skin friction and heat transfer to the porous wall of injecting a light gas such as helium through the porous wall and into the boundary layer. Another example is the calculation of heat transfer from a hypersonic boundary layer made up of high-temperature air and the products of combustion. Both problems have been described in earlier chapters of this book.

The purpose of this section is to show how calculations of transport properties for various mixtures compare with experimentally determined transport properties. Because of the paucity of data we shall be using binary mixtures of components which are unlikely to be used in aerodynamic analyses. However, the mixtures chosen for comparison will be mixtures of light and heavy gases and of gases made up of polyatomic molecules, which when compared with the theoretically determined values should provide a severe test of the theory.

Consider the binary mixture of H_2 and CO. Both components are diatomic, and the molecular weights differ by a factor of 14. The viscosity for this binary mixture is calculated using Eq. (10-6), and the viscosity of its components calculated according to Eq. (10-1) with the force constants given for H_2 and CO in Table 10-3. A comparison of the viscosity of the mixture as a function of mole fraction of H_2 measured at a temperature of 273°K with values calculated according to Eq. (10-6) is given in Fig. 10-12. The experimental values were reported by van Itterbeek, van Paemel, and van Lierde.[1] It appears that Eq. (10-6) gives satisfactory agreement between theory and experiment. Wilke[2] shows similar agreement between Eq. (10-6) and experiment for a number of binary mixtures of polyatomic molecules.

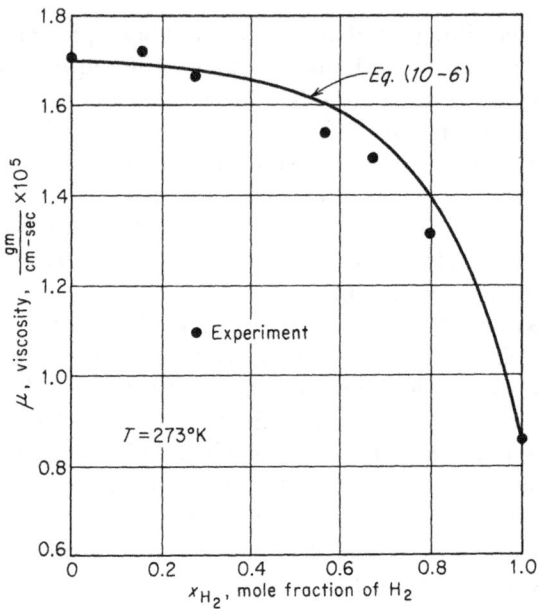

FIG. 10-12. Viscosity of a binary mixture of H_2 and CO versus mole fraction of H_2. (*Experimental data after A. van Itterbeek, O. van Paemel, and J. van Lierde, Physica, vol. 13, p. 88, 1947.*)

Equation (10-24) was used to calculate the thermal conductivity for the H_2-CO mixture. The component k_i can be calculated using Eqs. (10-29), (10-20), and (10-25), where $C_{p_i}/R_i = \frac{7}{2}$ for both CO and H_2 in Eq. (10-20). However, since it is the validity of Eq. (10-24) which we wish to examine, the experimentally determined values of k'_{CO} and k'_{H_2} were used. (The calculated values of μ_{CO}, k'_{CO}, μ_{H_2}, and k'_{H_2} differ very little from the experimental values.) The comparison between the values of k given for the mixture by

[1] A. van Itterbeek, O. van Paemel, and J. van Lierde, *Physica*, vol. 13, p. 88, 1947.

[2] C. R. Wilke, *J. Chem. Phys.*, vol. 18, pp. 517–522, 1950.

Eq. (10-24) with the measurements of Ibbs and Hirst[1] is shown in Fig. 10-13. Again satisfactory agreement between theory and experiment results.

Some data on measurements of viscosity and thermal conductivity of ternary mixtures are available. It is found that the accuracy of Eqs. (10-6) for μ and (10-25) for k is as good or better than is shown by Figs. 10-12 and 10-13 for a binary mixture. There is a lack of

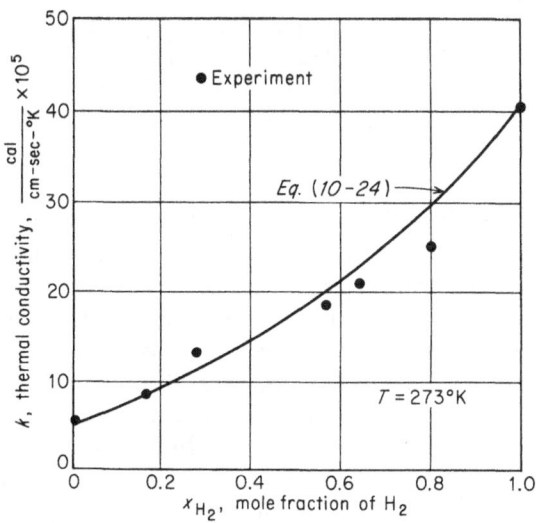

FIG. 10-13. Thermal conductivity of a binary mixture of H_2 and CO versus mole fraction of H_2. (*Experimental data after T. L. Ibbs and A. A. Hirst, Proc. Roy. Soc. (London), ser. A, vol. 123, p. 134, 1929.*)

measurements at high temperatures. However, there is no reason to believe that Eqs. (10-6) and (10-25) will be any less accurate at high temperatures providing the component values of μ_i and k_i are accurately known.

10-9. Combination Rules. In the calculation of the coefficient of diffusion for a binary mixture using Eq. (10-8), an interparticle potential for unlike particles is needed. In discussing the calculation of diffusion coefficient for a dissociating air mixture in Sec. 10-7, it was related that use was made of the point center of repulsion potential model given by Eq. (10-48) modified for dissimilar particles according to the combination rules

$$\phi_{12}(r) = \frac{d_{12}}{r^{\delta_{12}}}$$

[1] T. L. Ibbs and A. A. Hirst, *Proc. Roy. Soc. (London)*, ser. A, vol. 123, p. 134, 1929.

where
$$d_{12} = (d_1 d_2)^{1/2}$$

and
$$\delta_{12} = \tfrac{1}{2}(\delta_1 + \delta_2)$$

Such a combination rule has little justification other than the fact that it gives results for the binary diffusion coefficient D_{12} which agree reasonably well with experiment.

When the Lennard-Jones potential model is used for the interaction of dissimilar particles, we have

$$\phi_{12}(r) = 4\epsilon_{12}\left[\left(\frac{\sigma_{12}}{r}\right)^{12} - \left(\frac{\sigma_{12}}{r}\right)^6\right] \tag{10-81}$$

where
$$\epsilon_{12} = (\epsilon_1 \epsilon_2)^{1/2} \tag{10-82}$$

and
$$\sigma_{12} = \tfrac{1}{2}(\sigma_1 + \sigma_2) \tag{10-83}$$

and again the chief justification for using these expressions has been that their use in Eq. (10-8) to obtain the binary diffusion coefficient gives results in reasonable agreement with experiment. Weissman, Saxena, and Mason,[1] for example, have shown that, when Eqs. (10-81) through (10-83) are used in Eq. (10-8) to calculate D_{12} for He-CO_2 and H_2-CO_2 mixtures, values of D_{12} result in good agreement with measurements over a range of temperatures from 200 to 500°K. Because of these results it may be concluded that the combination rules given by Eqs. (10-82) and (10-83) are reasonable for binary mixtures of electrically and chemically neutral molecules.

The use of Eqs. (10-81) through (10-83) to describe the interparticle potentials for molecule-radical or radical-radical interactions is not so well substantiated, however. The combination rules can be checked on the ground-state potentials using the spectroscopically determined data. Herzberg[2] tabulates D and r_e for numerous diatomic molecules. If we assume that the Lennard-Jones potential given by Eq. (10-81) is appropriate for describing the ground-state potential, then it can be shown that (see Fig. 10-2)

$$\epsilon = D$$

and, by finding r for $d\phi/dr = 0$ in Eq. (10-81),

$$r_e = 1.122\sigma$$

Hence we can test the combination rules by comparing D_{12} with

[1] S. Weissman, S. C. Saxena, and E. A. Mason, *Phys. Fluids*, vol. 3, no. 4, pp. 510–518, 1960.
[2] Gerhard Herzberg, "Molecular Spectra and Molecular Structure—I. Spectra of Diatomic Molecules," Table 39, pp. 501–580, D. Van Nostrand Company, Inc., Princeton, N.J., 1950.

$(D_1 D_2)^{1/2}$ and $(r_e)_{12}$ with $\frac{1}{2}[(r_e)_1 + (r_e)_2]$. This is done in Table 10-5 for a number of diatomic molecules for which D and r_e are reasonably well known.

As can be seen, Eq. (10-83) gives more accurate values for σ_{12} than does Eq. (10-82) for ϵ_{12}. In general, considering the wide variety of cases with which the combination rules have been tested, it can be concluded that Eqs. (10-81) through (10-83) are reasonably

TABLE 10-5. A Comparison of the Spectroscopically Obtained Values of D and r_e with Those Given by Combination Rule Eqs. (10-82) and (10-83)

Species	D, in ev Experiment	D, in ev Eq. (10-82)	r_e, in angstroms Experiment	r_e, in angstroms Eq. (10-83)
O—O	5.080	...	1.207	
N—N	9.756	...	1.094	
H—H	4.476	...	0.741	
I—I	1.541	...	2.666	
C—C	3.60	...	1.311	
Cl—Cl	2.475	...	1.988	
O—N	6.487	7.02	1.150	1.150
O—H	4.35	4.77	0.970	0.974
H—I	3.056	2.63	1.604	1.703
C—H	3.47	3.98	1.119	1.026
H—Cl	4.431	3.33	1.274	1.364
I—Cl	2.152	1.952	2.320	2.327

accurate for radical-radical interactions of the type covered in Table 10-5 as well as for chemically neutral molecule-molecule interactions of the type examined by Weissman, Saxena, and Mason.

There remains the question of the application of the combination rules to radical-molecule interactions. It can be stated that, because of the absence of large numbers of N_2O, NO_2, O_3, and N_3 molecules under most conditions of engineering significance, O_2-O, N_2-O, N-O_2, and O_2-N interactions do not form a collision complex with a potential having a relatively deep well ϵ. Rather, this suggests that such interactions are strongly repulsive. However, the use of combination rule, Eq. (10-82), would suggest that $\epsilon_{O\text{-}O_2}$ is large simply because ϵ_O is large. Furthermore, Eq. (10-83) suggests that $\sigma_{O\text{-}O_2}$ is small simply because σ_O is small. Both such results are contrary to the observation that O-O_2 interactions appear to be strongly repulsive, since O_3 is not formed in a significant amount in a dissociating O_2 mixture.

Lees[1] has used a modification of the combination rules for the case

[1] Lester Lees, *Jet Propulsion*, vol. 26, no. 4, pp. 267–268, 1956.

of radical-diatomic molecule interactions of this type. Lees used the assumption that the potential minimum for diatomic molecule-atom interactions was probably close to that for molecule-molecule interactions of the same molecular species, which, of course, has a relatively shallow potential well. This is in agreement with the observation that molecules such as O_3, N_3, NO_2, and N_2O are infrequently observed at temperatures of interest to us here. Furthermore, the atom in the atom-molecule interaction approaches the molecule to within a distance nearly equal to the distance of approach of the nearest atom of a molecule in the related molecule-molecule interaction. That is, for example, for the N_2-N interaction, according to this assumption,

$$\sigma_{N_2\text{-}N} = \sigma_{N_2} - \frac{3}{4}\frac{r_e}{2}$$

where r_e is the equilibrium separation of N atoms in N_2. Also, according to this assumption, as discussed above,

$$\epsilon_{N_2\text{-}N} = \epsilon_{N_2}$$

and these rules might be reasonably applied to all diatomic molecule-monatomic radical interactions. The factor $\frac{3}{4}$ applied to $r_e/2$ was found to be reasonable based upon the analysis of the H_2-H interaction by Margenau.[1]

It can be concluded that the combination rules given by Eqs. (10-81) through (10-83) are reasonably accurate for chemically and electrically neutral molecule-molecule interactions. Furthermore, it appears that the combination rules are reasonably accurate for ground-state monatomic radical-radical interactions of the type listed in Table 10-5. The accuracy of the rule is unknown for diatomic molecule-radical interactions, but a reasonable modification has been suggested which agrees with the only sound theoretical analysis now available, that for the interaction of H_2 with H.

10-10. Conclusions. This chapter has been concerned with presenting equations to use in the calculation of transport properties for monatomic and polyatomic gas mixtures. The transport-property equations are those given in the first approximation by the Chapman-Enskog theory. The limitations placed upon this theory are thoroughly discussed in Sec. 10-2. It was found that the calculation of the most prominent transport properties is a straightforward process using the equations of this chapter and, furthermore, the results obtained using these equations compare favorably with experiment whenever experimental results are available.

---

[1] H. Margenau, *Phys. Rev.*, vol. 66, nos. 11 and 12, pp. 303–306, 1944.

The principal uncertainty in applying these equations under conditions of interest to the hypersonic aerodynamicist rests in the empirically determined force constants for the interparticle potential chosen and, indeed, in the form of the interparticle potential itself. It was seen that the interparticle potentials for radical-radical and molecule-radical collisions are uncertainly known, particularly at the high temperatures of interest to us in this book. Because of this, the results available for transport properties of air at temperatures high enough that dissociation is appreciable but low enough that ionization is negligible have an uncertain accuracy, perhaps ±25 per cent.

Of course, the equations of this chapter are not to be employed when ionization is appreciable, nor are they appropriate for application to liquids. There remains, however, an appreciable number of gas mixtures of engineering interest for which the equations of this chapter are entirely adequate.

Symbols and Nomenclature

In this appendix there are defined letter and Greek symbols, subscripts, and superscripts used in this book. Only those symbols, subscripts, and superscripts used repeatedly are defined here, since many symbols used are defined locally and used but once. The symbols and nomenclature are arranged alphabetically in the following sequence: letter symbols, Greek symbols, subscripts, and superscripts. Equation numbers are referred to whenever the symbol represents a mathematical expression or is more clearly defined by a mathematical relation.

Letter Symbols

a	Velocity of sound; also often a constant of integration
a_i, b_i	Stoichiometric coefficients
A	Interaction theory solution parameter, see Eq. (6-28)
A_1	See Eq. (8-25b)
A_i, B_i	Chemical symbols for reactants and products
B	Interaction theory solution parameter, see Eq. (6-29)
B_1	$\dfrac{2(\rho v)_w}{\rho_e u_e C_{f_0}}$, mass-transfer parameter
B_2	$\dfrac{(\rho v)_w}{\rho_e u_e C_{H_0}}$, mass-transfer parameter
B_3	$\dfrac{(\rho v)_w}{\rho_e u_e C_H}$, mass-transfer parameter
B_4	$\dfrac{G(\infty;S)}{G(\infty;P)} B_3$, mass-transfer parameter
B_5	$\dfrac{2(\rho v)_w}{\rho_e u_e C_f}$, mass-transfer parameter

316

B_6 $\dfrac{(\rho v)_w}{\rho_e u_e C_{H_d}}$, mass-transfer parameter

c Velocity of light $= 2.998 \times 10^{10}$ cm/sec; also constant of integration

C $\dfrac{\rho \mu}{\rho_e \mu_e}$

C_F $s^{-1}\displaystyle\int_0^s C_f\, ds =$ mean skin-friction coefficient

C_f $\dfrac{2\tau_w}{\rho_e u_e^2} =$ local skin-friction coefficient

$\bar{C}_{D_f}$ Average friction coefficient, see Eq. (1-7)

C_{f_0} Local skin-friction coefficient with zero mass transfer in Chaps. 3 and 8 and skin-friction coefficient with no dissociation in Chap. 7

C_H $\dfrac{-\dot{q}_w}{\rho_e u_e (I_r - h_w)} =$ heat-transfer coefficient

$\bar{C}_H$ Average heat-transfer coefficient, see Eq. (1-2)

C_{H_d} $\dfrac{-(\dot{q}_w)_d}{\rho_e u_e \left\{ \sum_i h_i^0 [(C_i)_e - (C_i)_w] \right\}}$, diffusion heat-transfer coefficient

C_{H_0} Heat-transfer coefficient with zero mass transfer in Chaps. 3 and 8 and heat-transfer coefficient with zero dissociation in Chap. 7

$\bar{C}_i$ $\sum_i r_{i,k} C_k =$ element mass fraction, see Eq. (2-108)

C_i $\dfrac{\rho_i}{\rho} =$ species mass fraction

C_p Specific heat at constant pressure

$\bar{C}_p$ $M_i C_p =$ molar heat capacity

C_v Specific heat at constant volume

$\bar{C}_v$ $M_i C_v =$ molar heat capacity

C_1 Reaction-rate parameter, see Eq. (4-81)

C_2 Interaction-theory constant, see Eq. (6-33)

C_3 Constant of integration, see Eq. (8-10)

d Parameter in point center of repulsion potential, see Eq. (10-48)

$\left(\dfrac{du}{ds}\right)_0$ Stagnation-point velocity gradient

D Dissociation energy per molecule

D Drag force

D Interaction-theory constant, see Eq. (6-38)

D_i^T Coefficient of thermal diffusion of species i

D_{ij} Binary diffusion coefficient

$D_T(s,y)$ Turbulent eddy diffusion coefficient, see Eq. (7-23)

D_{12} Binary diffusion coefficient

E Energy of N_0 particles or molar internal energy, see Eq. (7-69)

E $\dfrac{u_e^2}{h_e}$ = Eckert number

E_u Correction factor to thermal conductivity for internal degrees of freedom, see Eq. (10-20)

f $\dfrac{(\rho v)_w}{\dot{m}_L}$ = fraction of melt which vaporizes or gasifies

f_i Velocity distribution function for particles of kind i

$f(\eta)$ Reduced stream function, see Eqs. (2-92) and (2-93)

F Free energy $= H - TS$

$F(r)$ $-\dfrac{d\theta}{dr}$ = force of interaction between two colliding particles, see Eq. (10-27)

$F(s)$ Body-shape correction factor, see Eq. (4-112)

$F(\gamma)$ See Eq. (6-62)

F_1 See Eq. (7-56b)

F_2 See Eq. (8-44b)

g Acceleration due to gravity

$g(\eta)$ $\dfrac{I}{I_e}$ = reduced total enthalpy

g_f $\dfrac{I_f}{(I_f)_e}$ = reduced partial enthalpy

g_n Degeneracy or multiplicity of quantum-energy state n

G $\dfrac{Y(s)}{\delta^*(s)}$, see Eq. (6-52)

G See Eqs. (7-41) and (7-42)

$G(\infty;Z)$ See Eq. (4-75)

G_{ik} Transport-property parameter, see Eq. (10-5)

h $\sum\limits_i C_i h_i$ = enthalpy of mixture

h Planck's constant

h_c Chemical enthalpy, see Eqs. (5-50) and (5-77)

h_i $\displaystyle\int_0^T C_{p_i}\, dT + h_i^0$ = species enthalpy

h_i^0 Enthalpy of formation of species i

H Effective heating capacity of melting material, see Eq. (3-40)

H Molar enthalpy $= Mh$

H_{eff}	Effective heat capacity of ablating material, see Eq. (3-62)
H_f	$\dfrac{\delta^*}{\theta}$ form factor, see Eq. (8-124)
i, j, k,	Unit vectors
I	Moment of inertia of rigid rotator, see Eq. (9-56)
I	$h + \dfrac{u^2}{2} = $ total enthalpy
I_f	$I - \sum_i h_i^0 C_i = $ partial enthalpy, see Eq. (4-60)
I_n	Various integrals where $n = 0, 1, 2, 3, 4,$ and 5 in Eqs. (6-21), (9-15), (9-19), (9-54), (7-70b), and (7-87), respectively
j	Rotational quantum number
J_{ij}	Collision integral, see Eq. (2-19)
k	Boltzmann's constant
k	Thermal conductivity
k_i	Thermal conductivity for species i
k_r	Reaction conductivity, see Eqs. (4-36) and (4-37)
k_D	Dissociation-rate constant
k_R	Recombination-rate constant
k_T	See Eq. (2-49)
K	Mixing-length constant, see Eq. (7-68)
K_f	Equilibrium constant for formation of a molecule from the elements, see Eq. (9-93)
K_i	Rate of change of species i per unit volume
K_i	Equilibrium constant when $i = 1, 2, 3, 4, 5,$ or 6
K_p	Equilibrium constant, see Eq. (9-81)
l	Translational quantum number
L	$\dfrac{\rho D_{12} C_{p_f}}{k} = \dfrac{P}{S} = $ Lewis number
L_m	Heat of fusion
L_v	Heat of vaporization
$\bar{m}$	$m_1 m_2 (m_1 + m_2)^{-1} = $ reduced mass
m_i	Mass of species i
$\dot{m}_L$	Liquid melting rate per unit area
M	Mach number
M	Molecular weight, g/mole
M_i	Molecular weight of species i
n_i	Number density of species
n_0	Number of particles possessing minimum energy ϵ_0, see Eq. (9-24)
N	Constant of proportionality, see Eq. (3-41)
$N(s)$	Normalizing parameter, see Eqs. (2-79a) and (2-89)

N_0	Avogadro's number $= 6.02472 \times 10^{23}$ molecules/g-mole
Nu	$\dfrac{-\dot{q}_w s C_{p_w}}{k_w (I_r - h_w)} =$ Nusselt's number
p	Pressure
p_i	Partial pressure of species i
P	$\dfrac{C_{p_f} \mu}{k} =$ Prandtl number
P_{eq}	$\dfrac{C_p \mu}{k + k_r} =$ equilibrium mixture Prandtl number
$\dot{q}$	Heat-transfer rate
Q	Any dependent variable
Q	Heat
Q	Partition function, see Eq. (9-26)
Q'	$Q - \bar{Q}$
$Q^{(l)}(g)$	Collision cross section, see Eq. (10-37)
Q_i	See Eq. (7-10)
$\bar{Q}_i$	See Eq. (7-11)
r	Length of rigid rotator, see Eqs. (9-57) and (9-58)
r	Radial distance between two colliding particles
r	Recovery factor
$\mathbf{r}$	$x\mathbf{i} + y\mathbf{j} + z\mathbf{k} =$ position vector
r_e	Radius between particles at position of minimum potential energy, see Fig. 10-1
$r_{i,k}$	Fraction of mass of species k which is contributed by element i, see Eq. (5-81)
$r_0(s)$	Radius of body of revolution in meridian plane
R	$N_0 k = M_i R_i =$ universal gas constant
$\bar{R}$	Gas constant for mixture $= \displaystyle\sum_i C_i R_i$
R_e	$\dfrac{\rho_e u_e s}{\mu_e} =$ Reynolds number based on local free-stream properties
$\bar{R}_e$	Reduced Reynolds number, see Eq. (8-135)
R_i	Gas constant for species $i = \dfrac{R}{M_i} = \dfrac{k}{m_i}$
R_w	$\dfrac{\rho_w u_e s}{\mu_w} =$ Reynolds number based on wall properties
R_∞	$\dfrac{\rho_\infty u_\infty s}{\mu_\infty} =$ Reynolds number based on free-stream properties
s	Electric quantum number
s	Specific entropy $= \dfrac{S}{M}$

s, y	Orthogonal coordinates
$\bar{s}(s)$	Transformed coordinate, see Eq. (2-91)
S	sM = molar entropy
S	Wetted area
S	$\dfrac{\mu}{\rho D_{12}}$ = Schmidt number
t	Blunted-leading-edge thickness
t	Time
T	Temperature
T'	Reference temperature, see Eqs. (5-32) and (8-43)
u, v, w	Velocity components
v	Vibrational quantum number
$\mathbf{v}$	$v_x\mathbf{i} + v_y\mathbf{j} + v_z\mathbf{k}$ = particle velocity vector
$\mathbf{v}_0$	Mass average velocity, see Eq. (2-9)
$\bar{\mathbf{v}}_i$	Average velocity, see Eq. (2-8)
V	Velocity
V	$\dfrac{p}{RT}$ = molar volume, see Eq. (9-40)
$\mathbf{V}$	$u\mathbf{i} + v\mathbf{j} + w\mathbf{k}$ = mass average velocity vector
$\bar{\mathbf{V}}_i$	Average diffusion velocity of species i, see Eq. (2-12)
$\mathbf{V}_i$	Diffusion velocity vector for particle of kind i
w	Total energy
$\dot{w}_i$	$K_i m_i$ = mass rate of change of species i per unit volume
W	Weight
W	Probability function, see Eq. (9-3)
x, y, z	Coordinate system
z	$\dfrac{u}{u_e}$; also $z = (\Sigma\, h_i^0 C_i)(\Sigma\, h_i^0 C_i)_e^{-1}$
x_i	$\dfrac{p_i}{p}$ = mole fraction of species i
$Y(s)$	Locus of leading-edge shock wave
z_A	$\dfrac{\alpha}{\alpha_e}$ = reduced atom mass fraction
$\bar{z}_i$	$\dfrac{\bar{C}_i}{(C_i)_e}$ = reduced element mass fraction
z_i	$\dfrac{C_i}{(C_i)_e}$ = reduced mass fraction

Greek Symbols

α	Symmetry number, see Eqs. (9-65) and (9-66)

α $\qquad$ $\dfrac{k}{C_p\rho}$ = thermal diffusivity

α $\qquad$ $N_0\left(\dfrac{\beta m}{2\pi}\right)^{3/2}$ = Maxwell-Boltzmann distribution

constant of integration, see Eqs. (9-9a) and (9-16)

α $\qquad$ $\dfrac{\rho_A}{\rho}$ = mass fraction of atoms

α^* $\qquad$ Equilibrium value of α

β $\qquad$ $\dfrac{1}{kT}$ = Maxwell-Boltzmann distribution parameter, see

Eqs. (9-9c) and (9-22)

β $\qquad$ Pressure-gradient parameter, see Eq. (4-105)

γ $\qquad$ Ratio of specific heats

γ $\qquad$ $\dfrac{\bar m g^2}{2kT}$ = reduced kinetic energy, see Eq. (10-36)

$\Gamma(n)$ $\qquad$ Gamma function

δ $\qquad$ Exponent on point center of repulsion potential, see Eq. (10-48)

δ $\qquad$ Momentum-boundary-layer thickness

δ_c $\qquad$ Species-concentration boundary-layer thickness

δ_T $\qquad$ Energy-boundary-layer thickness

δ^* $\qquad$ Displacement thickness, see Eq. (6-1)

$-\Delta H$ $\qquad$ Heat of reaction, see Eq. (3-7)

ΔQ $\qquad$ Heat release, see Eq. (3-8)

ϵ $\qquad$ Depth of potential well in Lennard-Jones potential, see Eq. (10-51)

$\epsilon(s,y)$ $\qquad$ Turbulent eddy viscosity, see Eq. (7-24)

ϵ_i $\qquad$ Energy of particle of kind i, see Eq. (9-2)

$\eta(s,y)$ $\qquad$ Transformed coordinate, see Eq. (2-90)

θ $\qquad$ Momentum thickness, see Eq. (8-126)

$\theta(\eta)$ $\qquad$ $\dfrac{T}{T_e}$ = reduced temperature

$\bar\theta_i$ $\qquad$ Mean value of a dependent variable θ_i in velocity space, see Eq. (2-7)

$\kappa(s,y)$ $\qquad$ Turbulent eddy thermal conductivity, see Eq. (7-25)

λ $\qquad$ Dependent variable, see Eq. (5-24)

λ $\qquad$ Wavelength, see Eq. (9-78)

μ $\qquad$ Viscosity

μ_i $\qquad$ Viscosity of species i, see Eq. (10-1)

ρ_i $\qquad$ $n_i m_i$ = partial density of species i

σ $\qquad$ See Eq. (9-114)

σ_i $\qquad$ Transport-property collision diameter

τ	Shear stress
ϕ	Potential energy
Φ	Surface-recombination-rate parameter, see Eq. (4-96)
$\phi(0)$	Constant of integration, see Eq. (7-70c)
χ	$(\Sigma\, p_i M_i)^{-1}$
$\bar{\chi}$	$\dfrac{C^{1/2} M_\infty^3}{(R_\infty)^{1/2}}$
$\chi(b,g)$	Deflection angle, see Eq. (10-34) and Fig. 10-1
Ψ	Wave function
Ψ^*	Complex conjugate of wave function
$\psi(s,y)$	Stream function, see Eq. (2-75)
ψ_1	$\dfrac{C_f}{C_{f_0}}$
ψ_2	$\dfrac{C_H}{C_{H_0}}$
Ω	Vorticity parameter, see Eq. (6-74)
$\Omega^{(l,s)}$	Collision integrals, see Eq. (10-35)
$\Omega^{(l,s)*}$	$\dfrac{\Omega^{(l,s)}}{[\Omega^{(l,s)}]_{r.s.}}$ = reduced collision integral
ω	Cone semivertex angle
ω	Rotational frequency of rigid rotator, see Eq. (9-56)
$\boldsymbol{\omega}$	Vorticity, see Eq. (6-67)

Subscripts

A	Atom
e	Evaluated at edge of boundary layer
eq	Equilibrium value
f	Friction
f	Frozen or sometimes partial value
i	Species i
j	Rotational quantum state
l	Translational quantum state; sometimes value at lower energy level
L	Liquid
m	Melting
M	Molecule
0	Initial value at $t = 0$, as in T_0; value with zero blowing on C_f and C_H in Chaps. 3 and 8; value with no dissociation on C_f in Chap. 7; and value at absolute zero when applied to thermodynamic functions
p	Pressure

r	Evaluated at recovery temperature
$r.s.$	Rigid-sphere value
s	Electric quantum state
u	Value at upper energy level
v	Vibrational quantum state
w	At surface $\eta = 0$ or $y = 0$
∞	Free stream

Superscripts

o	Evaluated at standard state (STP) when applied to thermodynamic properties
$*$	Equilibrium value
$*$	Dimensionless value

Name Index

Adams, C. E., 281
Adams, M. C., 49, 61, 62, 189, 203
Amdur, I., 278, 279, 284, 297

Bartle, E. R., 59, 217
Bauer, E., 194, 304
Baulknight, C. W., 87, 88, 194, 303
Bertram, M. H., 156, 161
Bethe, H., 49
Bird, R. B., 12, 14, 17, 276, 278, 284, 287, 295
Blackman, V., 280
Blasius, H., 30
Blyholder, G., 128
Bohr, N., 258
Bradfield, W. S., 117
Brokaw, R. S., 76
Bromberg, R., 47
Butler, J. N., 76

Carslaw, H. S., 63, 65
Chapman, D. R., 202, 273
Chapman, S., 12, 18
Cheng, H. K., 163
Chung, P. M., 94
Clarke, J. F., 187
Clauser, F. H., 241
Cohen, C. B., 47, 83, 96, 151
Cohen, N. B., 138
Coles, D., 202

Cowling, T. G., 12, 18, 273
Crocco, L., 33, 169
Curtiss, C. F., 12, 14, 17, 276, 278, 284, 287, 295

Davis, R. S., 213
Denison, M. R., 236
Detra, R. W., 97
Diaconis, N. S., 90
Dooley, D. A., 104
Dore, F. J., 195, 212, 237
Dorrance, W. H., 7, 134, 146, 189, 190, 195, 212, 237
Dotsenko, B. V., 303

Eckert, E. R. G., 138, 139, 220
Enskog, D., 17–19, 273–276
Eucken, A., 283
Eyring, H., 128

Fay, J. A., 86, 87
Feigenbutz, L. V., 120, 123, 263, 268
Feldman, S., 163
Ferri, A., 161
Forstall, W., Jr., 188

Gazley, C., Jr., 174, 230
Gilmore, F. R., 298

Subject Index

Engineering

FUNDAMENTALS OF ASTRODYNAMICS, Roger R. Bate, Donald D. Mueller, and Jerry E. White. Teaching text developed by U.S. Air Force Academy develops the basic two-body and n-body equations of motion; orbit determination; classical orbital elements, coordinate transformations; differential correction; more. 1971 edition. 455pp. 5 3/8 x 8 1/2. 0-486-60061-0

INTRODUCTION TO CONTINUUM MECHANICS FOR ENGINEERS: Revised Edition, Ray M. Bowen. This self-contained text introduces classical continuum models within a modern framework. Its numerous exercises illustrate the governing principles, linearizations, and other approximations that constitute classical continuum models. 2007 edition. 320pp. 6 1/8 x 9 1/4. 0-486-47460-7

ENGINEERING MECHANICS FOR STRUCTURES, Louis L. Bucciarelli. This text explores the mechanics of solids and statics as well as the strength of materials and elasticity theory. Its many design exercises encourage creative initiative and systems thinking. 2009 edition. 320pp. 6 1/8 x 9 1/4. 0-486-46855-0

FEEDBACK CONTROL THEORY, John C. Doyle, Bruce A. Francis and Allen R. Tannenbaum. This excellent introduction to feedback control system design offers a theoretical approach that captures the essential issues and can be applied to a wide range of practical problems. 1992 edition. 224pp. 6 1/2 x 9 1/4. 0-486-46933-6

THE FORCES OF MATTER, Michael Faraday. These lectures by a famous inventor offer an easy-to-understand introduction to the interactions of the universe's physical forces. Six essays explore gravitation, cohesion, chemical affinity, heat, magnetism, and electricity. 1993 edition. 96pp. 5 3/8 x 8 1/2. 0-486-47482-8

DYNAMICS, Lawrence E. Goodman and William H. Warner. Beginning engineering text introduces calculus of vectors, particle motion, dynamics of particle systems and plane rigid bodies, technical applications in plane motions, and more. Exercises and answers in every chapter. 619pp. 5 3/8 x 8 1/2. 0-486-42006-X

ADAPTIVE FILTERING PREDICTION AND CONTROL, Graham C. Goodwin and Kwai Sang Sin. This unified survey focuses on linear discrete-time systems and explores natural extensions to nonlinear systems. It emphasizes discrete-time systems, summarizing theoretical and practical aspects of a large class of adaptive algorithms. 1984 edition. 560pp. 6 1/2 x 9 1/4. 0-486-46932-8

INDUCTANCE CALCULATIONS, Frederick W. Grover. This authoritative reference enables the design of virtually every type of inductor. It features a single simple formula for each type of inductor, together with tables containing essential numerical factors. 1946 edition. 304pp. 5 3/8 x 8 1/2. 0-486-47440-2

THERMODYNAMICS: Foundations and Applications, Elias P. Gyftopoulos and Gian Paolo Beretta. Designed by two MIT professors, this authoritative text discusses basic concepts and applications in detail, emphasizing generality, definitions, and logical consistency. More than 300 solved problems cover realistic energy systems and processes. 800pp. 6 1/8 x 9 1/4. 0-486-43932-1

THE FINITE ELEMENT METHOD: Linear Static and Dynamic Finite Element Analysis, Thomas J. R. Hughes. Text for students without in-depth mathematical training, this text includes a comprehensive presentation and analysis of algorithms of time-dependent phenomena plus beam, plate, and shell theories. Solution guide available upon request. 672pp. 6 1/2 x 9 1/4. 0-486-41181-8

HELICOPTER THEORY, Wayne Johnson. Monumental engineering text covers vertical flight, forward flight, performance, mathematics of rotating systems, rotary wing dynamics and aerodynamics, aeroelasticity, stability and control, stall, noise, and more. 189 illustrations. 1980 edition. 1089pp. 5 5/8 x 8 1/4.   0-486-68230-7

MATHEMATICAL HANDBOOK FOR SCIENTISTS AND ENGINEERS: Definitions, Theorems, and Formulas for Reference and Review, Granino A. Korn and Theresa M. Korn. Convenient access to information from every area of mathematics: Fourier transforms, Z transforms, linear and nonlinear programming, calculus of variations, random-process theory, special functions, combinatorial analysis, game theory, much more. 1152pp. 5 3/8 x 8 1/2.   0-486-41147-8

A HEAT TRANSFER TEXTBOOK: Fourth Edition, John H. Lienhard V and John H. Lienhard IV. This introduction to heat and mass transfer for engineering students features worked examples and end-of-chapter exercises. Worked examples and end-of-chapter exercises appear throughout the book, along with well-drawn, illuminating figures. 768pp. 7 x 9 1/4.   0-486-47931-5

BASIC ELECTRICITY, U.S. Bureau of Naval Personnel. Originally a training course; best nontechnical coverage. Topics include batteries, circuits, conductors, AC and DC, inductance and capacitance, generators, motors, transformers, amplifiers, etc. Many questions with answers. 349 illustrations. 1969 edition. 448pp. 6 1/2 x 9 1/4.

0-486-20973-3

BASIC ELECTRONICS, U.S. Bureau of Naval Personnel. Clear, well-illustrated introduction to electronic equipment covers numerous essential topics: electron tubes, semiconductors, electronic power supplies, tuned circuits, amplifiers, receivers, ranging and navigation systems, computers, antennas, more. 560 illustrations. 567pp. 6 1/2 x 9 1/4.   0-486-21076-6

BASIC WING AND AIRFOIL THEORY, Alan Pope. This self-contained treatment by a pioneer in the study of wind effects covers flow functions, airfoil construction and pressure distribution, finite and monoplane wings, and many other subjects. 1951 edition. 320pp. 5 3/8 x 8 1/2.   0-486-47188-8

SYNTHETIC FUELS, Ronald F. Probstein and R. Edwin Hicks. This unified presentation examines the methods and processes for converting coal, oil, shale, tar sands, and various forms of biomass into liquid, gaseous, and clean solid fuels. 1982 edition. 512pp. 6 1/8 x 9 1/4.   0-486-44977-7

THEORY OF ELASTIC STABILITY, Stephen P. Timoshenko and James M. Gere. Written by world-renowned authorities on mechanics, this classic ranges from theoretical explanations of 2- and 3-D stress and strain to practical applications such as torsion, bending, and thermal stress. 1961 edition. 560pp. 5 3/8 x 8 1/2.   0-486-47207-8

PRINCIPLES OF DIGITAL COMMUNICATION AND CODING, Andrew J. Viterbi and Jim K. Omura. This classic by two digital communications experts is geared toward students of communications theory and to designers of channels, links, terminals, modems, or networks used to transmit and receive digital messages. 1979 edition. 576pp. 6 1/8 x 9 1/4.   0-486-46901-8

LINEAR SYSTEM THEORY: The State Space Approach, Lotfi A. Zadeh and Charles A. Desoer. Written by two pioneers in the field, this exploration of the state space approach focuses on problems of stability and control, plus connections between this approach and classical techniques. 1963 edition. 656pp. 6 1/8 x 9 1/4.

0-486-46663-9

Browse over 9,000 books at www.doverpublications.com

Mathematics–Bestsellers

HANDBOOK OF MATHEMATICAL FUNCTIONS: with Formulas, Graphs, and Mathematical Tables, Edited by Milton Abramowitz and Irene A. Stegun. A classic resource for working with special functions, standard trig, and exponential logarithmic definitions and extensions, it features 29 sets of tables, some to as high as 20 places. 1046pp. 8 x 10 1/2.   0-486-61272-4

ABSTRACT AND CONCRETE CATEGORIES: The Joy of Cats, Jiri Adamek, Horst Herrlich, and George E. Strecker. This up-to-date introductory treatment employs category theory to explore the theory of structures. Its unique approach stresses concrete categories and presents a systematic view of factorization structures. Numerous examples. 1990 edition, updated 2004. 528pp. 6 1/8 x 9 1/4.   0-486-46934-4

MATHEMATICS: Its Content, Methods and Meaning, A. D. Aleksandrov, A. N. Kolmogorov, and M. A. Lavrent'ev. Major survey offers comprehensive, coherent discussions of analytic geometry, algebra, differential equations, calculus of variations, functions of a complex variable, prime numbers, linear and non-Euclidean geometry, topology, functional analysis, more. 1963 edition. 1120pp. 5 3/8 x 8 1/2.   0-486-40916-3

INTRODUCTION TO VECTORS AND TENSORS: Second Edition--Two Volumes Bound as One, Ray M. Bowen and C.-C. Wang. Convenient single-volume compilation of two texts offers both introduction and in-depth survey. Geared toward engineering and science students rather than mathematicians, it focuses on physics and engineering applications. 1976 edition. 560pp. 6 1/2 x 9 1/4.   0-486-46914-X

AN INTRODUCTION TO ORTHOGONAL POLYNOMIALS, Theodore S. Chihara. Concise introduction covers general elementary theory, including the representation theorem and distribution functions, continued fractions and chain sequences, the recurrence formula, special functions, and some specific systems. 1978 edition. 272pp. 5 3/8 x 8 1/2.
                                                                              0-486-47929-3

ADVANCED MATHEMATICS FOR ENGINEERS AND SCIENTISTS, Paul DuChateau. This primary text and supplemental reference focuses on linear algebra, calculus, and ordinary differential equations. Additional topics include partial differential equations and approximation methods. Includes solved problems. 1992 edition. 400pp. 7 1/2 x 9 1/4.   0-486-47930-7

PARTIAL DIFFERENTIAL EQUATIONS FOR SCIENTISTS AND ENGINEERS, Stanley J. Farlow. Practical text shows how to formulate and solve partial differential equations. Coverage of diffusion-type problems, hyperbolic-type problems, elliptic-type problems, numerical and approximate methods. Solution guide available upon request. 1982 edition. 414pp. 6 1/8 x 9 1/4.   0-486-67620-X

VARIATIONAL PRINCIPLES AND FREE-BOUNDARY PROBLEMS, Avner Friedman. Advanced graduate-level text examines variational methods in partial differential equations and illustrates their applications to free-boundary problems. Features detailed statements of standard theory of elliptic and parabolic operators. 1982 edition. 720pp. 6 1/8 x 9 1/4.   0-486-47853-X

LINEAR ANALYSIS AND REPRESENTATION THEORY, Steven A. Gaal. Unified treatment covers topics from the theory of operators and operator algebras on Hilbert spaces; integration and representation theory for topological groups; and the theory of Lie algebras, Lie groups, and transform groups. 1973 edition. 704pp. 6 1/8 x 9 1/4.
                                                                              0-486-47851-3

Browse over 9,000 books at www.doverpublications.com

A SURVEY OF INDUSTRIAL MATHEMATICS, Charles R. MacCluer. Students learn how to solve problems they'll encounter in their professional lives with this concise single-volume treatment. It employs MATLAB and other strategies to explore typical industrial problems. 2000 edition. 384pp. 5 3/8 x 8 1/2.   0-486-47702-9

NUMBER SYSTEMS AND THE FOUNDATIONS OF ANALYSIS, Elliott Mendelson. Geared toward undergraduate and beginning graduate students, this study explores natural numbers, integers, rational numbers, real numbers, and complex numbers. Numerous exercises and appendixes supplement the text. 1973 edition. 368pp. 5 3/8 x 8 1/2.   0-486-45792-3

A FIRST LOOK AT NUMERICAL FUNCTIONAL ANALYSIS, W. W. Sawyer. Text by renowned educator shows how problems in numerical analysis lead to concepts of functional analysis. Topics include Banach and Hilbert spaces, contraction mappings, convergence, differentiation and integration, and Euclidean space. 1978 edition. 208pp. 5 3/8 x 8 1/2.   0-486-47882-3

FRACTALS, CHAOS, POWER LAWS: Minutes from an Infinite Paradise, Manfred Schroeder. A fascinating exploration of the connections between chaos theory, physics, biology, and mathematics, this book abounds in award-winning computer graphics, optical illusions, and games that clarify memorable insights into self-similarity. 1992 edition. 448pp. 6 1/8 x 9 1/4.   0-486-47204-3

SET THEORY AND THE CONTINUUM PROBLEM, Raymond M. Smullyan and Melvin Fitting. A lucid, elegant, and complete survey of set theory, this three-part treatment explores axiomatic set theory, the consistency of the continuum hypothesis, and forcing and independence results. 1996 edition. 336pp. 6 x 9.   0-486-47484-4

DYNAMICAL SYSTEMS, Shlomo Sternberg. A pioneer in the field of dynamical systems discusses one-dimensional dynamics, differential equations, random walks, iterated function systems, symbolic dynamics, and Markov chains. Supplementary materials include PowerPoint slides and MATLAB exercises. 2010 edition. 272pp. 6 1/8 x 9 1/4.   0-486-47705-3

ORDINARY DIFFERENTIAL EQUATIONS, Morris Tenenbaum and Harry Pollard. Skillfully organized introductory text examines origin of differential equations, then defines basic terms and outlines general solution of a differential equation. Explores integrating factors; dilution and accretion problems; Laplace Transforms; Newton's Interpolation Formulas, more. 818pp. 5 3/8 x 8 1/2.   0-486-64940-7

MATROID THEORY, D. J. A. Welsh. Text by a noted expert describes standard examples and investigation results, using elementary proofs to develop basic matroid properties before advancing to a more sophisticated treatment. Includes numerous exercises. 1976 edition. 448pp. 5 3/8 x 8 1/2.   0-486-47439-9

THE CONCEPT OF A RIEMANN SURFACE, Hermann Weyl. This classic on the general history of functions combines function theory and geometry, forming the basis of the modern approach to analysis, geometry, and topology. 1955 edition. 208pp. 5 3/8 x 8 1/2.   0-486-47004-0

THE LAPLACE TRANSFORM, David Vernon Widder. This volume focuses on the Laplace and Stieltjes transforms, offering a highly theoretical treatment. Topics include fundamental formulas, the moment problem, monotonic functions, and Tauberian theorems. 1941 edition. 416pp. 5 3/8 x 8 1/2.   0-486-47755-X

Browse over 9,000 books at www.doverpublications.com